COMPUTATIONAL ENGINEERING GEOLOGY

Edward Derringh

Department of Applied Mathematics and Sciences
Wentworth Institute of Technology

Prentice Hall
Upper Saddle River, New Jersey 07458

Library of Congress Cataloging-in-Publication Data

Derringh, Edward.
Computational engineering geology/Edward Derringh.
p. cm.
Includes bibliographical references and index.
ISBN 0-13-834236-9
1. Engineering geology--Mathematics. I. Title
TA705.D47 1998
624.1'51--dc21 97-34433
CIP

EXECUTIVE EDITOR: ***Robert McConnin***
PRODUCTION EDITOR: ***Alison Aquino***
MANUFACTURING MANAGER: ***Trudy Pisciotti***
COVER DESIGNER: ***Bruce Kenselaar***
COVER ART: ***Patrice van Acker***

Simon & Schuster/A Viacom Company
Upper Saddle River, New Jersey 07458

Printed in the United States of America
10 9 8 7 6 5 4 3 2 1

ISBN 0-13-834236-9

Prentice-Hall International (UK) Limited, *London*
Prentice-Hall of Australia Pty. Limited, *Sydney*
Prentice-Hall Canada Inc., *Toronto*
Prentice-Hall Hispanoamericana, S.A., *Mexico*
Prentice-Hall of India Private Limited, *New Delhi*
Prentice-Hall of Japan, Inc., *Tokyo*
Simon & Schuster Asia Pte. Ltd., *Singapore*
Editora Prentice-Hall do Brasil, Ltda., *Rio de Janeiro*

Contents

Preface

This volume is intended as a computational manual for an introductory survey course in engineering geology, or a course in physical geology with an engineering slant. By "computational" is meant an emphasis on the solving of problems as the primary method to acquiring a competent grasp of the principles introduced. The book's philosophy is that such an approach, grounded only in first-and second-year undergraduate physics and mathematics, is a realistic prelude to more advanced study in one or more of the disciplines necessarily treated here in an introductory fashion.

The book deals only with the quantitative aspects of the subject. Any descriptive background can be provided by the instructor, either directly in recitation, through additional reading material accessed by a reserved library shelf, or with a companion course textbook. However, there is enough material in this work for a full one-semester course.

By "first-and second-year undergraduate physics and mathematics" is meant at least one semester of an algebra-based physics course at the level of Wilson/Buffa: COLLEGE PHYSICS, 3/E (1997), Prentice Hall; or, Giancoli: PHYSICS: PRINCIPLES WITH APPLICATIONS, 5/E (1998), Prentice Hall. If the course is calculus based, then at the level of Halliday/Resnick/Walker: FUNDAMENTALS OF PHYSICS, 5/E (1997), John Wiley; or Fishbane/Gariorowicz/Thornton: PHYSICS FOR SCIENTISTS AND ENGINEERS, 2/E (1996), Prentice Hall. Sadly, many of today's undergraduates take only one semester of physics. Basic mechanics should have been covered. The mathematics preparation should be, at the minimum, through "pre-calculus" mathematics; to wit, trigonometric functions, logarithms, exponentials, etc. This book uses a little first-semester calculus in a few spots, which can be sidestepped if use of calculus is contraindicated for a particular class.

This monograph originated in my dissatisfaction with the introductory texts in engineering geology that have been available over about the last decade, the period in which I have been teaching a survey course in this subject at Wentworth Institute of Technology to students in the civil engineering, building construction, and architecture programs. My specific complaint was, and still is, the meager amount of computational material contained in these books, as though the complexity of the natural world, with its refusal to fit simple models, actually precludes a serious attempt to apply science and mathematics to illuminate specific problems.

This reluctance to apply science (mainly physics) and mathematics in these books is important to me, not the least because for many (not all) of my students, the engineering geology course is one of the last "quasi-science" courses in their curriculum, and hence my aim is to include as much useful information as possible. Graphic descriptions of landslides, floods, and bursting dams are not the need. The inclusion of practice problems is. And there must be enough of them so that the reader can attain competence in all aspects of the various chapter contents. The problems should not be questions masquerading as problems.

The problems in this book are mainly of the easier variety. Most are numerical, reflecting the engineer's need for numerical answers to specific questions. But some are algebraic. The problems use, I believe, realistic numerical data. I have resisted the current fashion of arranging problems according to the relevant section or attaching a marker to the harder problems. However, they are arranged roughly in the order that the corresponding material appears in the chapter text. There are 357 problems, with answers to the odd-numbered ones in Appendix B. Each chapter also contains several worked-out examples, a total of 67 in all. These have been selected to be neither so simple as to be virtually valueless, nor so hard as to discourage the reader.

Only a basic scientific calculator is needed to work the problems. Neither a graphing nor a programmable calculator is required.

This book uses the pristine SI metric unit system exclusively. By "pristine" is meant the "internationally recognized conventions for standard usage of SI units" as described in Nelson, R.A., PHYSICS TODAY, Vol. 50, No. 8, Part 2, pp. BG13-14. Hence, disfiguring contrivances such as kilogram-force are strictly prohibited. (The SI Unit System is summarized in Appendix A.) I do not give USCS (British system) equivalences: Most students in an introductory engineering geology course subsequently take more specific engineering courses and are exposed, when necessary, to USCS units in those course. I am unwilling to inflict a tedious, dual-track display of units on the readers of this work.

The topics covered are those that most mainstream undergraduate books on engineering geology cover in at least a semi-quantitative fashion, even if no problem sets are included. These are: bulk properties of rock, rock deformation, slopes, soil, "fluvial" processes, groundwater, dams, subsurface exploration, earthquakes, rock age, and radioactivity in earth materials. The depth of treatment is intended for the second-year undergraduate level.

For reviewing a previous edition of this manuscript, I gratefully acknowledge: Prof. John H. Foster, California State University, Fullerton; Prof. John F. Lewis, George Washington University; and Prof. Nels F. Forsman, University of North Dakota. Also, thanks go to the many students who endured an earlier version of this material. It is a pleasure to be indebted to my wife, Shirley, for her patience and encouragement.

Edward Derringh

FOR INSTRUCTORS: A Solution Manual accompanying COMPUTATIONAL ENGINEERING GEOLOGY may be obtained directly from Robert A. McConnin, Executive Editor-Geology, Prentice Hall, Inc., 1633 Broadway, 6th Floor, New York, NY 10019-6785. Requests should be submitted on school letterhead.

Chapter 1

Bulk Properties of Rock

1.1 Bulk Density

Suppose that a sample of rock is under examination. Perhaps the sample was cut from rock in a quarry, or from an outcrop produced in the course of an engineering project. Call the volume of the sample V; i.e., V is the volume enclosed by the outer surface of the sample. The volume can be calculated from the dimensions of the sample if it has a shape for which there is a formula for the volume (e.g., a cylinder). If the sample has an irregular shape, then the volume may be harder to determine. It might be found from the displacement of water in a graduated cylinder into which the sample is immersed. Allowance may have to be made for any water absorbed by the sample.

On the other hand, the mass M of the sample is easy to measure, with a suitable balance, for example.

From the mass M and volume V the *density* ρ of the sample can be calculated. By definition

$$\rho = \frac{M}{V}. \tag{1.1}$$

The density defined by Eq.(1.1) is the average density of the sample: there is no way to tell from just the total mass and volume what is the internal distribution of density in the sample. This average density is also known as the *bulk density* .

In SI base units, mass is in kg (kilogram)and volume in m^3 (cubic meter), so that, by Eq. (1.1), the units of density are kg/m^3. In the cgs set of metric units, the unit of mass is g (gram) and of volume cm^3 (cubic centimeter); the density therefore has units g/cm^3. For future use, note that

$$\begin{aligned}1 \text{ kg/m}^3 &= (1000 \text{ g})/(100 \text{ cm})^3,\\ 1 \text{ kg/m}^3 &= 1 \text{ X } 10^{-3} \text{ g/cm}^3,\\ 1 \text{ g/cm}^3 &= 1 \text{ X } 10^3 \text{ kg/m}^3.\end{aligned}$$

Most of the rocks found near the surface of the Earth have densities in the range from 1.5 g/cm^3 to 3 g/cm^3.

1.2 Specific Gravity

The densities of substances making up rocks, or of a rock itself, are sometimes expressed in terms of the density ρ_w of liquid water. The *specific gravity* G of a substance is defined by

$$G = \frac{\rho}{\rho_w}, \tag{1.2}$$

where ρ is the density of the substance. Since G is the ratio of two densities, the density units cancel, so that G has no units. (This requires that both densities in Eq.(1.2) must be expressed in the same units, either both kg/m^3 or g/cm^3; it would lead to meaningless confusion if one density is in kg/m^3 and the other g/cm^3.)

For the density of water it is common to use $\rho_w = 1.00$ g/cm^3. The density of water varies slightly with temperature (as does the density of other substances), but this variation can often (not always) be ignored. In the present work, the density of water will be taken as 1000 kg/m^3 exactly for the purpose of calculating specific gravity.

****

EXAMPLE 1
A block of rock with edge lengths 85.5 cm, 79.0 cm, 43.8 cm has a mass of 953 kg. Find the specific gravity of the rock.

Some of the data are in SI base units (the mass), the rest are in cgs units (the edge lengths). Calculations should be done either with all base units or all cgs units. In engineering, the base units are more commonly used. (But not always: sometimes SI base units and cgs units are mixed!) Using all SI base units means that the edge lengths must be converted to meters. But this is easy since 1 m = 100 cm. By "block" of rock is implied a rectangular block whose volume is the product of the edge lengths. Therefore,

$$V = (0.855 \text{ m})(0.790 \text{ m})(0.438 \text{ m}),$$
$$V = 0.2958 \text{ m}^3.$$

By Eq.(1.1), the density is

$$\rho = \frac{M}{V},$$
$$\rho = \frac{953 \text{ kg}}{0.2958 \text{ m}^3},$$
$$\rho = 3222 \text{ kg/m}^3.$$

Now use Eq.(1.2) to find the specific gravity G. Since SI base units are being used here, the density of water ρ_w must be entered in these units. Therefore,

$$G = \frac{\rho}{\rho_w},$$

$$G = \frac{3222 \text{ kg/m}^3}{1000 \text{ kg/m}^3},$$

$$G = 3.22.$$

Since the data has 3 significant figures (sig fig), the final result can be given to no more than 3 sig fig, although 1 extra sig fig is carried within the calculation to guard against round-off error. As mentioned previously, the density of water is considered to be exact.

****

1.3 Unit Weight

The weight W of an object is the gravitational force exerted on it by the rest of the planet Earth. Dividing the weight W of an object by its volume V yields the *unit weight* γ of the object or of the material of which the object is made; that is,

$$\gamma = \frac{W}{V}. \tag{1.3}$$

The SI base unit of weight is the newton (N). Since the base unit of volume is m^3, the SI base unit of unit weight is N/m^3. (Note the two uses of the word *unit* in the last sentence.) The cgs unit of weight, the dyne, will not be used in this book; therefore, cgs units of unit weight will not be encountered in the present work.

Looking back at Eq.(1.1), it can be seen that the unit weight is defined very like the bulk density, except that the weight W replaces the mass M. (For this reason, unit weight is also known as *weight density*.) There is a relation between weight and mass. From physics,

$$W = Mg, \tag{1.4}$$

where g is the acceleration due to gravity (often called simply *gravity*). Hence,

$$\gamma = Mg/V,$$

$$\gamma = (M/V)g,$$

$$\gamma = \rho g. \tag{1.5}$$

Equation (1.5) can be used to calculate unit weight from density, and vice versa. The numerical value of g to be used is

$$g = 9.8 \text{ m/s}^2.$$

This value is often adopted as the value of gravity averaged over the surface of the Earth. For the purpose of evaluating the significant figures in any calculation, this value for g shall be considered exact.

Note that Eq.(1.2) for specific gravity can be written in terms of unit weight:

$$G = \rho/\rho_w,$$

$$G = \rho g/\rho_w g,$$

$$G = \gamma/\gamma_w. \quad (1.6)$$

The unit weight of water is, by Eq.(1.5),

$$\gamma_w = \rho_w g,$$

$$\gamma_w = (1000 \text{ kg/m}^3)(9.8 \text{ m/s}^2),$$

$$\gamma_w = 9.8 \text{ kN/m}^3.$$

The SI prefix k stands for 1 X 10^3.

****

EXAMPLE 2
Calculate the density in g/cm^3 of a rock with unit weight 27.6 kN/m^3.

As noted just above, the SI prefix k means a factor of 1000. By Eq.(1.5),

$$\gamma = \rho g,$$
$$27.6 \text{ X } 10^3 \text{ N/m}^3 = \rho(9.8 \text{ m/s}^2),$$
$$\rho = 2820 \text{ kg/m}^3,$$
$$\rho = 2.82 \text{ g/cm}^3.$$

The answer is given to 3 sig fig since the data is given to 3 sig fig (remember that the value $g = 9.8$ m/s^2 is considered to be of infinite precision). Also, the density conversion factor between kg/m^3 and g/cm^3 is used in the last step.

****

1.4 Porosity

Under a microscope, most types of rock are seen to contain small open spaces, called *pores.* These pores can originate in various ways. For example, many sedimentary rocks seem to be assembled from many small, solid particles, called *grains.* These grains are irregularly shaped. However, unlike a jigsaw puzzle in which the irregularly shaped pieces completely interlock leaving no spaces, the grains in rock may not fit together perfectly, leaving the gaps called pores. See Fig.(1.1).

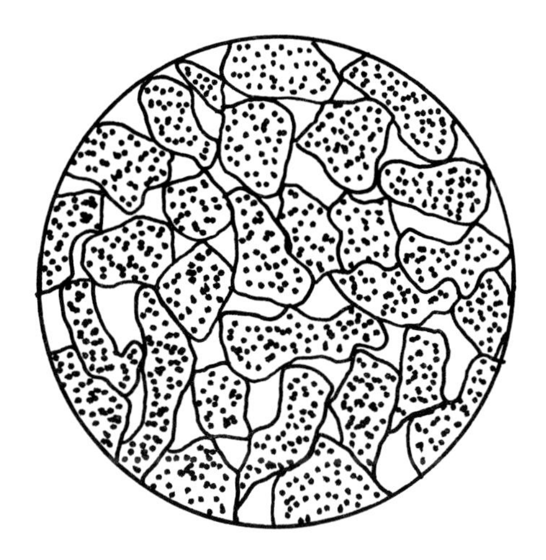

Fig.(1.1) Section of Rock Showing Grains Enclosing Pores.

Pores may simply be cracks in the rock, the result of mechanical or thermal forces exerted on the rock sometime in the past. They could be bubbles frozen into the rock when it solidified.

Whatever the origin of the pores, the truly solid part or parts of the rock are called the *grains* or the *matrix*. The terms grains and matrix often are used interchangeably.

To quantitatively express the degree of porosity (volume of pores vs. volume of rock), the *porosity* n is defined by

$$n = V_{\text{pores}}/V. \tag{1.7}$$

In Eq.(1.7), V_{pores} is the total volume of all the pores present in a rock sample whose volume is V. It is important to recognize that V is the volume of the rock sample as found in nature (in the field) and includes both the pores and the matrix; that is

$$V = V_{\text{pores}} + V_{\text{matrix}}, \tag{1.8}$$

where V_{matrix} is the total volume of all the truly solid portions of the rock. Put another way, as in the definition of bulk density, V is the volume that would be calculated from the external dimensions of the rock sample, with no regard for how much of the volume is occupied by grains or pores.

If the sample of rock being examined for porosity has been broken off from a much larger formation of the rock in the field, then the sample must be large enough to include a great many pores, to ensure that the sample is representative of the rock formation.

Since the total pore volume must be less than the volume of the rock (i.e., $V_{\text{pores}} < V$), it follows that $n < 1$. The situation $V_{\text{pores}} = V$ indicates just empty space ($V_{\text{matrix}} = 0$, by Eq.(1.8), so that $n = 1$ is never realized for a real rock sample. On the other hand, in some very fine-grained rock, the grains can interlock like a jigsaw puzzle, yielding zero porosity. Hence, the values of porosity actually encountered are in the range $0 \leq n < 1$.

Often, porosity is expressed as a percent (%). However, in the equations in this book (such as Eq.(1.7)), it is always assumed that n is expressed as a decimal. Therefore, if n is given as a percent, it must be divided by 100 to obtain the corresponding decimal value before being used in any equation.

The density ρ of a rock defined by Eq.(1.1), and the unit weight γ (weight density) defined by Eq.(1.3), refer to the overall rock sample; i.e., the volume V in both of these equations encloses both grains and pores. These densities are average densities.

The density ρ_{g} of the grains (or matrix) is given by a relation analogous to Eq.(1.1), but applied to the grains:

$$\rho_{\text{g}} = \frac{M_{\text{grains}}}{V_{\text{grains}}}, \tag{1.9}$$

where M_{grains} is the total mass of the grains in the rock sample containing a volume V_{grains} of grains.

If the pores are unoccupied, or filled only with a gas, then $M_{\text{grains}} = M$, where M is the mass of the rock sample. (A gas contributes negligible weight to the sample.) By Eqs.(1.7) and (1.8),

$$\begin{aligned} V_{\text{grains}} &= V - V_{\text{pores}}, \\ V_{\text{grains}} &= V - nV, \\ V_{\text{grains}} &= (1-n)V. \end{aligned} \tag{1.10}$$

Substituting this into Eq.(1.9) and invoking Eq.(1.1) yields

$$\begin{aligned} \rho_{\text{g}} &= \frac{M}{(1-n)V}, \\ \rho_{\text{g}} &= \frac{\rho}{1-n}. \end{aligned} \tag{1.11}$$

This equation for the density ρ_{g} of the grains in terms of the bulk density ρ of the rock and of its porosity n applies only if the pores are either empty or occupied only by a gas. In practice, it is unlikely that the pores will be found to be truly empty (i.e., enclosing a vacuum). The term *empty* implies, therefore, that the pores are filled only with a gas (e.g., air, or methane).

An equation like Eq.(1.11) can be written in terms of unit weights, rather than mass densities. Multiply Eq.(1.11) by gravity g and use Eq.(1.5) to get

$$\gamma_{\text{g}} = \frac{\gamma}{1-n}. \tag{1.12}$$

****

EXAMPLE 3
A 0.885-m^3 block of sandstone has a mass of 1752 kg. When the block is crushed just sufficiently to close all the pores, which are empty, the volume of the rock becomes 0.584 m^3. Find (*a*) the porosity of the sandstone and (*b*) the density of the grains. (Assume that the density of the grains is not changed in the crushing process.)

(*a*) The original block has a volume $V = 0.885\ \text{m}^3$. The volume of the crushed rock must equal the volume of all the grains in the original block, since the crushed rock has zero pore volume. That is, in the original block, $V_{\text{grains}} = 0.584\ \text{m}^3$. Therefore, the volume of the pores in the original block is, by Eq.(1.8),

$$
\begin{aligned}
V &= V_{\text{pores}} + V_{\text{grains}}, \\
0.885\ \text{m}^3 &= V_{\text{pores}} + 0.584\ \text{m}^3, \\
V_{\text{pores}} &= 0.301\ \text{m}^3.
\end{aligned}
$$

Now calculate the porosity by Eq.(1.7):

$$
\begin{aligned}
n &= V_{\text{pores}}/V, \\
n &= (0.301\ \text{m}^3)/(0.885\ \text{m}^3), \\
n &= 0.340\ (34.0\%).
\end{aligned}
$$

(*b*) Since the pores are empty, the mass of the crushed rock is the same as that of the original block, 1752 kg. The volume of the crushed rock is 0.584 m^3. But the crushed rock is entirely grains, and therefore, by Eq.(1.9),

$$
\begin{aligned}
\rho_g &= M_{\text{grains}}/V_{\text{grains}}, \\
\rho_g &= (1752\ \text{kg})/(0.584\ \text{m}^3), \\
\rho_g &= 3000\ \text{kg/m}^3, \\
\rho_g &= 3.00\ \text{g/cm}^3.
\end{aligned}
$$

****

1.5 Dry and Saturated Unit Weights

The pores of *in situ* rock (rock as found in the Earth, undisturbed by human activity) may be filled with gas or liquid. The densities of gases found in rocks are very much less than the densities of the grains or matrix of the rocks. This means that, as already mentioned, it is safe to ignore the contribution of the gas trapped in the pores to the total weight of a rock sample.

A similar statement cannot be made for liquids. The densities of the liquids commonly found in the pores of rocks, although less than the densities of the grains, are not very much less. If the porosity of the rock is large enough, and a significant fraction of the pores contain liquid, then the weight (or mass) of the liquid is likely to be an appreciable part of the total weight (or mass) of the rock.

If all the pores in a rock sample are completely filled with liquid, then the rock is said to be *saturated.* If all the pores are empty, then the rock is said to be *dry.* Saturated rock can be rendered dry by heating the rock in an oven; at sufficiently high temperature the liquid vaporizes and the vapor is driven out of the rock.

An important relation is that between the unit weight γ_{sat} of a saturated rock sample, the unit weight γ_{dry} of the same sample when dry, and the unit weight γ_{L} of the liquid occupying the pores of the saturated sample.

(It may be tempting to write $\gamma_{\text{sat}} = \gamma_{\text{dry}} + \gamma_{\text{L}}$, but this is not correct because of the different volumes involved.)

To obtain the actual relation, note that the weight W_{sat} of the saturated rock sample is just the sum of the dry weight W_{dry} and the weight W_{L} of the liquid in the saturated rock:

$$W_{\text{sat}} = W_{\text{dry}} + W_{\text{L}}.$$

The volume V of the rock sample is the same whether it is dry or saturated (just as the volume of your car's gas tank, metal with one large pore, is the same whether the tank is empty or full). Dividing the preceding equation by V gives

$$\frac{W_{\text{sat}}}{V} = \frac{W_{\text{dry}}}{V} + \frac{W_{\text{L}}}{V},$$

$$\gamma_{\text{sat}} = \gamma_{\text{dry}} + \frac{W_{\text{L}}}{V}, \tag{1.13}$$

by Eq.(1.3). The unit weight of the liquid γ_{L} is

$$\gamma_{\text{L}} = \frac{W_{\text{L}}}{V_{\text{L}}}, \tag{1.14}$$

where V_{L} is the volume of the liquid with weight W_{L}. But, since the liquid fills all the pores,

$$V_{\text{L}} = V_{\text{pores}}. \tag{1.15}$$

Hence, by the definition of porosity n

$$V_{\text{L}} = nV. \tag{1.16}$$

Solve this last equation for V (easy!) and substitute into Eq.(1.13). Then use the definition of γ_{L} given in Eq.(1.14) to obtain

$$\gamma_{\text{sat}} = \gamma_{\text{dry}} + n\frac{W_{\text{L}}}{V_{\text{L}}},$$

$$\gamma_{\text{sat}} = \gamma_{\text{dry}} + n\gamma_{\text{L}}. \tag{1.17}$$

A similar relation holds between the mass densities. By Eq.(1.5), Eq.(1.17) becomes

$$\rho_{sat} g = \rho_{dry} g + n(\rho_L g),$$

$$\rho_{sat} = \rho_{dry} + n\rho_L. \quad (1.18)$$

Example 4 below describes how these relations can be used to determine the porosity of a rock sample by injecting it with mercury Hg. Liquid mercury is much denser than water: $G_{Hg} = 13.6$ compared with $G_w = 1$ for water. This implies that, even if the porosity of the rock sample is quite small, saturating the rock with mercury could change the unit weight significantly, making accurate laboratory measurements of the weights and their differences relatively easy. A disadvantage is that liquid mercury is a hazardous substance, mainly because of its vapor; great care must be exercised with its use.

****

EXAMPLE 4

A test cylinder of rock has a diameter of 12.6 cm and a length of 14.0 cm. When dry its weight is 50.3 N. When saturated with mercury the weight of the sample is 62.8 N. The specific gravity of mercury is 13.6. Find the porosity of the rock.

The volume of the rock sample is

$$V = \pi D^2 L/4,$$

$$V = \pi (0.126 \text{ m})^2 (0.140 \text{ m})/4,$$

$$V = 1.746 \text{ X } 10^{-3} \text{ m}^3.$$

Therefore, the dry and saturated unit weights are

$$\gamma_{dry} = W_{dry}/V,$$

$$\gamma_{dry} = (50.3 \text{ N})/(1.746 \text{ X } 10^{-3} \text{ m}^3),$$

$$\gamma_{dry} = 28.81 \text{ kN/m}^3;$$

$$\gamma_{sat} = W_{sat}/V,$$

$$\gamma_{sat} = (62.8 \text{ N})/(1.746 \text{ X } 10^{-3} \text{ m}^3),$$

$$\gamma_{sat} = 35.97 \text{ kN/m}^3.$$

The unit weight of the liquid mercury γ_L follows from Eq.(1.6):

$$\gamma_L = G_L \gamma_w,$$

$$\gamma_L = (13.6)(9.8 \text{ kN/m}^3),$$

$$\gamma_L = 133.3 \text{ kN/m}^3.$$

Note that the unit weight of the mercury is greater than the unit weight of the rock, whether dry or saturated. Now use Eq.(1.17) to solve for the porosity n. Note that the units of unit weight cancel, so that

$$\gamma_{sat} = \gamma_{dry} + n\gamma_L,$$

$$35.97 \text{ kN/m}^3 = 28.81 \text{ kN/m}^3 + n(133.3 \text{ kN/m}^3),$$

$$n = 0.0537 \ (5.37\%).$$

****

1.6 Subsidence

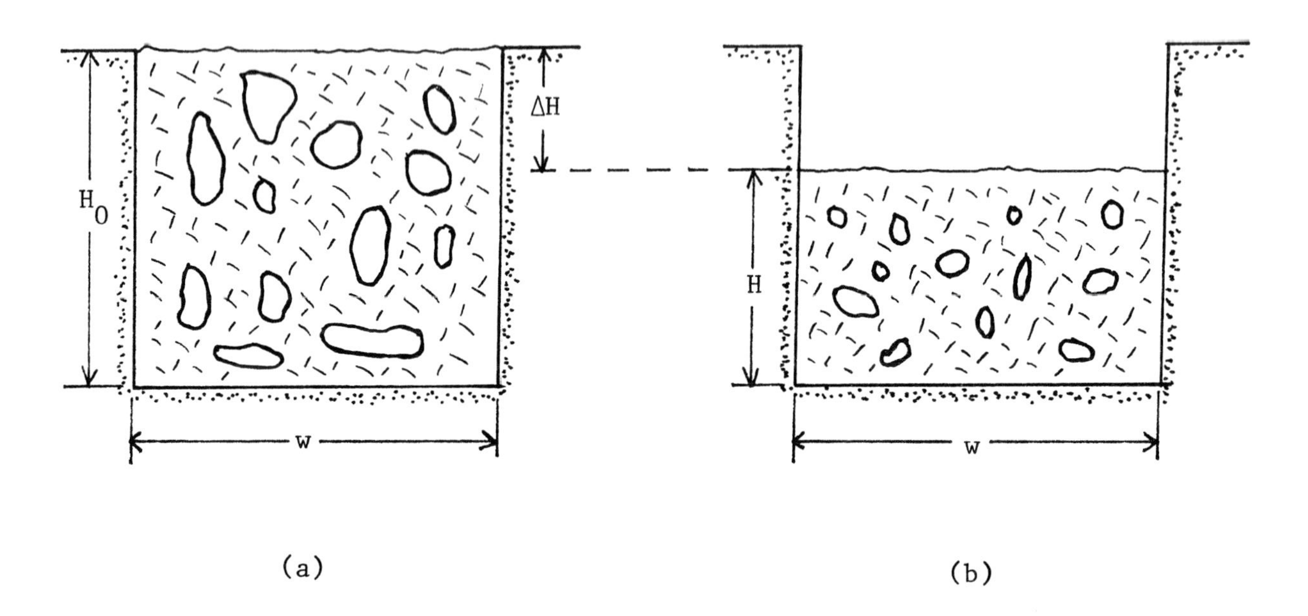

Fig.(1.2) Consolidation and Subsidence

Suppose that loose, unconsolidated, rock (or soil) is dumped into a rectangular trench of length L, width w, and depth H_0, as shown in Fig.(1.2a). Initially the material, of porosity n_0 just fills the trench. However, because of an applied external load (such as the weight of a building constructed thereon), the material is compacted, or consolidated. That is, its pores partially collapse, squeezing out any water that might have occupied them (this requires that the water has a place to go, and that the consolidation is not so rapid that the water cannot get there). Eventually, the consolidation ceases, with the porosity reduced to $n < n_0$. It shall be presumed that the density of the matrix is unaffected; this is reasonable, since the force needed to shrink the pores is much less than that needed to densify the matrix. The material now occupies a region of smaller volume, so that it fills the trench only to a height $H < H_0$. This means that the surface of the material (the ground) is lowered by an amount $\Delta H = H - H_0$. See Fig.(1.2b). A lowering of the ground surface is called *subsidence* . It can occur for reasons other than a change in porosity, and can take place slowly or quickly.

The task now is to calculate the subsidence ΔH due to a reduction in porosity. To do this, write expressions for the volume of the matrix before and after the consolidation. Since, by assumption, the density of the matrix is unchanged, its volume is also (provided there is no loss of matrix). Hence, the two expressions must be equivalent. By the definition of porosity, they are

$$V_{\text{matrix}} = (1 - n_0)(H_0 wL),$$

$$V_{\text{matrix}} = (1 - n)(HwL).$$

Setting these expressions equal gives

$$(1 - n_0)(H_0 wL) = (1 - n)(HwL),$$

$$(1 - n_0)H_0 = (1 - n)H,$$

$$(1 - n_0)H_0 = (1 - n)(H_0 - \Delta H),$$

$$\Delta H = H_0[\frac{n_0 - n}{1 - n}]. \tag{1.19}$$

Equation(1.19) applies only to a trench with a rectangular cross section, since the volume of the material was presumed to be given by the product of the three edge lengths.

****

EXAMPLE 5

A layer of clay with a porosity of 47.0% and saturated with water is deposited into a rectangular trench 260 m long and 17.5 m wide to a depth of 2.72 m. Later, it is found that the clay has settled by 15.0 cm. Find the volume of water squeezed out of the clay.

The subsidence and the original height must be expressed in the same units, so that these units will cancel. Choosing meters, and remembering to express the porosity in decimal form, Eq.(1.19) becomes, after substitution of the data,

$$0.150 = 2.72[\frac{0.470 - n}{1 - n}],$$

$$n = 0.4391.$$

Since the clay was saturated, the volume V_w of water squeezed out equals the loss of pore space in the clay due to the compaction. Therefore,

$$V_w = n_0(H_0 wL) - n(HwL),$$

$$V_w = (n_0 H_0 - nH)wL.$$

Now H = 2.72 m - 0.15 m = 2.57 m, so that

$$V_w = [(0.470)(2.72 \text{ m}) - (0.4391)(2.57 \text{ m})](17.5 \text{ m})(260 \text{ m}),$$
$$V_w = 682 \text{ m}^3.$$

****

1.7 Multimineral Rock

Suppose that the porosity of a particular rock specimen is to be found by measuring the bulk density ρ, the grain density ρ_g and then applying Eq.(1.11). The bulk density is easy to measure. What about the grain density?

If all of the grains in the rock are of the same mineral, and the density of the mineral as it occurs in nature is known (many have been measured in the laboratory), then the grain density simply equals the density of that mineral.

For a rock that contains several minerals, the value of ρ_g to use in equations like Eq.(1.11) is the average of the densities of the individual minerals present. As an example, consider a rock made up of three minerals, the densities of the minerals grains being denoted by ρ_1, ρ_2 and ρ_3. The bulk grain density ρ_g will not, in general, be simply $\frac{1}{3}(\rho_1+\rho_2+\rho_3)$, because the minerals may be present in different amounts. A *weighted* average must be used, the precise nature of which must now be deduced.

Let the total mass of all the grains in the rock sample be M_g and the total volume of all the grains V_g. By Eq.(1.9),

$$M_g = \rho_g V_g.$$

If M_1 be the total mass and V_1 the total volume of mineral 1 in the rock, with similar notation for the other two minerals present (assuming that the rock contains three minerals), then since

$$M_g = M_1 + M_2 + M_3,$$

it follows that

$$\rho_g V_g = \rho_1 V_1 + \rho_2 V_2 + \rho_3 V_3,$$

$$\rho_g = \rho_1(\frac{V_1}{V_g}) + \rho_2(\frac{V_2}{V_g}) + \rho_3(\frac{V_3}{V_g}).$$

Finally, write

$$f_1 = \frac{V_1}{V_g}, \tag{1.20}$$

for the fractional volume abundance of the first mineral, with similar notation for the other minerals), so that the bulk grain density becomes

$$\rho_g = f_1\rho_1 + f_2\rho_2 + f_3\rho_3. \tag{1.21}$$

Equation (1.21) is the relation sought for the bulk grain density; it is a volume-weighted average of the densities of the individual minerals present.

Writing equations like Eq.(1.20) for the three minerals and then adding gives

$$f_1 + f_2 + f_3 = 1. \tag{1.22}$$

The volume abundances can be measured in the laboratory by examining a representative piece of the rock with a microscope powerful enough so that the individual grains can be seen and distinguished from each other. The number of grains of each mineral present must then be counted and their total volume estimated.

Identifying a mineral by name does not necessarily specify its chemical composition. For example, a grain of the mineral olivine can contain some or all of the molecules Fe_2SiO_4, $FeMgSiO_4$, and Mg_2SiO_4. Different olivine grains will include these molecules with varying abundancies.

This leads to the expectation that different olivine grains have different densities. This is indeed the case. However, the variation in densities is fairly small, between about 3.2 g/cm^3 and 3.6 g/cm^3.

This is the situation for many other (but not all) minerals. Although the densities of their grains shows some variation, the variation is small enough that, to a good approximation, the average of these densities can be taken as the density of all grains of the mineral. Table (1.1) lists some minerals which show small variation in density and their average density.

Mineral	Density (g/cm^3)
Gypsum	2.35
Orthoclase	2.55
Chalcedony	2.62
Quartz	2.65
Plagioclase	2.70
Chlorite	2.80
Muscovite	2.85
Anhydrite	2.95
Pyroxene	3.40
Barite	4.45
Pyrite	5.05
Galena	7.54

Table (1.1) Average Density of Some Minerals

****

EXAMPLE 6

A shale consists of 34.1% chlorite and 65.9% pyrite, and has a porosity of 38.8%. Find the bulk density of the shale.

First, use Eq.(1.21), suitably modified for a rock that consists of only two minerals, and Table (1.1) to calculate the bulk grain density:

$$\rho_g = f_{ch}\rho_{ch} + f_{py}\rho_{py},$$
$$\rho_g = (0.341)(2.80\ \mathrm{g/cm^3}) + (0.659)(5.05\ \mathrm{g/cm^3}),$$
$$\rho_g = 4.283\ \mathrm{g/cm^3}.$$

From Eq.(1.11) the rock's bulk density ρ is

$$\rho = \rho_g(1 - n),$$
$$\rho = (4.283\ \mathrm{g/cm^3})(1 - 0.388),$$
$$\rho = 2.62\ \mathrm{g/cm^3}.$$

****

1.8 Triangular Composition Diagrams

The mineral composition of a rock is specified by giving the abundances of all the minerals that it contains. In many rocks, it is sufficient to give the abundances of just three minerals, either because the rock contains only three minerals, or that if it contains more, the abundances of the others are so small that they can be ignored.

For example, many igneous rocks contain mainly the minerals quartz, alkali feldspar, and plagioclase. The rocks are classified into different named groups in part on the range of abundances of these three minerals.

For these three-mineral rocks, it is fashionable to display the abundances of the minerals in the associated rock groups on a *triangular composition diagram*. In this way, the compositional relationships between the various associated rock groups can be visualized without examining tables of their mineral abundances.

To see how the compositional display can be constructed, it is only necessary to become convinced of this, otherwise rather obscure, property of equilateral triangles: From any point inside the triangle, draw perpendiculars to the three median lines; the sum of the three distances from the midpoint of each side to the foot of the perpendicular drawn to it equals the length of a median line.

Figure (1.3) shows an equilateral triangle with one vertex at the origin of an x,y coordinate system. Each side of the triangle has the same length L and each vertex angle is 60°. The x,y coordinates of the vertices and of the midpoints of the three sides are given. An arbitrarily selected interior point is shown, labelled with its coordinates x_p, y_p. The dashed lines r_1, r_2, r_3 mark the perpendiculars from the interior point to the three median lines, and also represent the lengths of the perpendiculars. The distances from the foot of each perpendicular to the midpoint of the side of the triangle that the associated median line is drawn to are D_1, D_2, D_3.

Now the length of each median line is $L\cos 30° = L\sqrt{3}/2$. Hence, the property referred to above, in equation form, is

$$D_1 + D_2 + D_3 = L\sqrt{3}/2. \tag{1.23}$$

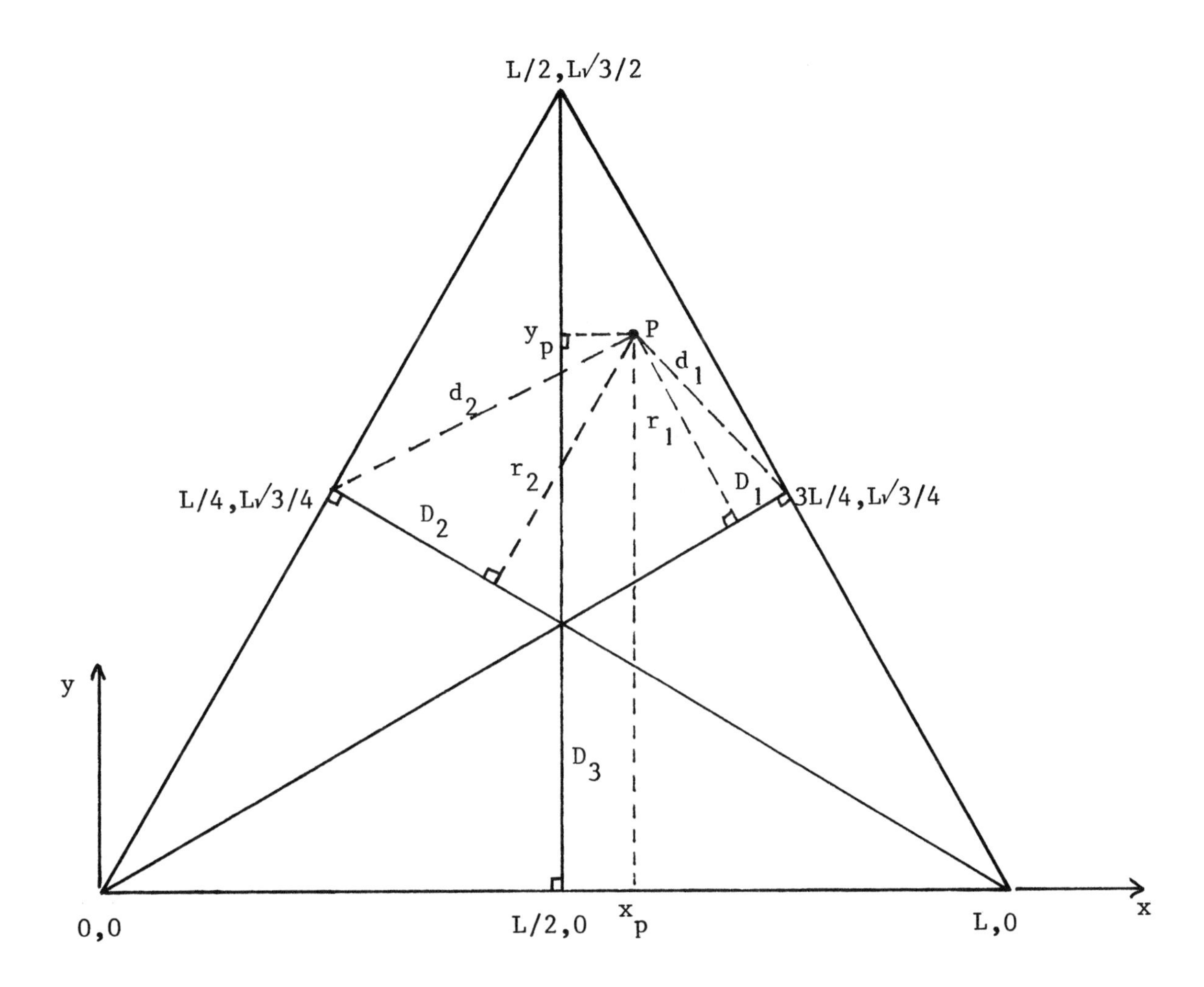

Fig.(1.3) An Equilateral Triangle

To establish this, first recall from algebra and analytic geometry that the perpendicular distance r from a point on the x, y plane with coordinates x_p, y_p to the line whose equation is $y = mx + b$ is given by

$$r = \pm \frac{y_p - (mx_p + b)}{\sqrt{m^2 + 1}}, \tag{1.24}$$

where the choice indicated by $\pm$ is made such that r is positive.

Determine D_1 first. To do this, draw the line marked, and of length, d_1 from the interior point to the midpoint of the side intersected by the median line to which r_1 is drawn; see

Fig.(1.3). Using the coordinates of thc midpoint of this side, and the (should be) well-known formula for the distance between two points on the x, y plane, it follows that

$$d_1^2 = (\frac{3}{4}L - x_p)^2 + (\frac{\sqrt{3}}{4}L - y_p)^2. \tag{1.25}$$

The equation of the median line on which the distance D_1 is measured is

$$y = (\frac{\sqrt{3}}{3})x,$$

since the slope of the line is $\tan 30° = \sqrt{3}/3$. Therefore, by Eq.(1.24),

$$r_1 = \pm(\frac{\sqrt{3}}{2}y_p - \frac{1}{2}x_p). \tag{1.26}$$

Invoking the Pythagorean theorem and then using Eqs.(1.25) and (1.26), it follows that

$$D_1^2 = d_1^2 - r_1^2,$$

$$D_1^2 = (\frac{3}{4}L - x_p)^2 + (\frac{\sqrt{3}}{4}L - y_p)^2 - (\frac{\sqrt{3}}{2}y_p - \frac{1}{2}x_p)^2,$$

$$D_1 = \frac{\sqrt{3}}{2}L - \frac{\sqrt{3}}{2}x_p - \frac{1}{2}y_p. \tag{1.27}$$

The distance D_2 is found in a similar manner. First, note that

$$d_2^2 = (\frac{1}{4}L - x_p)^2 - (\frac{\sqrt{3}}{4}L - y_p)^2. \tag{1.28}$$

The equation of the median line on which D_2 is measured is

$$y = -\frac{\sqrt{3}}{3}x + \frac{\sqrt{3}}{3}L,$$

so that, again by Eq.(1.24),

$$r_2 = \frac{\sqrt{3}}{2}y_p + \frac{1}{2}x_p - \frac{1}{2}L. \tag{1.29}$$

Equations (1.28) and (1.29) now yield

$$D_2^2 = d_2^2 - r_2^2,$$

$$D_2^2 = (\frac{1}{4}L - x_p)^2 + (\frac{\sqrt{3}}{4}L - y_p)^2 - (\frac{\sqrt{3}}{2}y_p + \frac{1}{2}x_p - \frac{1}{2}L)^2,$$

$$D_2 = \frac{\sqrt{3}}{2}x_p - \frac{1}{2}y_p. \tag{1.30}$$

The last distance is easy (really!); look at Fig.(1.3) to see that

$$D_3 = y_p. \tag{1.31}$$

Now add Eqs.(1.27), (1.30), and (1.31) to get $D_1 + D_2 + D_3 = \frac{\sqrt{3}}{2}L$. That is, the distances add up to the length of a median line.

Now label each vertex of the triangle with one of the three minerals in the rocks. Since the intercepted distances add up to the length of a median line, each median line can be marked from zero at the midpoint of the side to which it is drawn evenly to 100 at the vertex. The distances D_1, D_2, D_3 can then be interpreted on these scales as the percent abundance of the mineral shown at the vertex. Equation (1.23), the validity of which has just been proven, ensures that the three abundances add up to 100%, as they must.

Figure (1.4) shows a resulting composition diagram for rocks consisting only of three minerals, called simply A, B, C. Each vertex is labelled with one of the minerals. A rock whose plotted composition point lies at a vertex is composed of 100% of that mineral. A rock whose composition point is located at the intersection of the three medians has a composition that is $33\frac{1}{3}$% each of the three minerals. The point P shown on Fig.(1.4) indicates a rock with composition 10% mineral A, 30% mineral B and 60% mineral C.

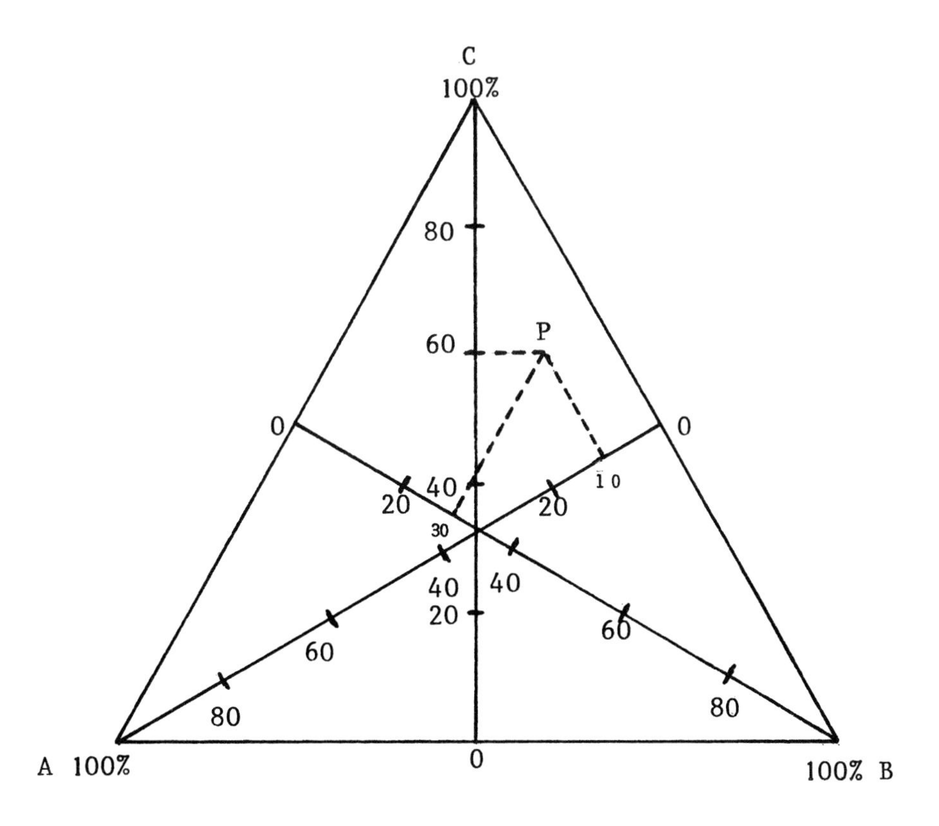

Fig.(1.4) Triangular Composition Diagram

****

EXAMPLE 7

Draw a triangular composition diagram for rocks consisting of the three minerals called R, S, and T. On the diagram, locate rocks with compositions (*a*) 50% R, 50% S; (*b*) 70%R, 10% S; (*c*) 60% T.

Draw an equilateral triangle; label the vertices R, S, T in any order. Draw the three median lines and divide into convenient intervals from 0% at the intersection with the side to 100% at the vertex. See Fig.(1.5). (*a*) A composition 50% R, 50% S means that the abundance of T is 100% - 50% - 50% = 0. Hence, the composition point falls at the 0% mark opposite the T vertex, on the side of the triangle that connects the R and S vertices. (*b*) Through the medians from the R and S vertices, draw perpendiculars through the marks at the proper percentages. The composition point sought must be at the intersection of these perpendiculars. That point will automatically indicate a T abundance of 100% - 70% - 10% = 20%. (*c*) All rocks with composition 60% T must lie on the perpendicular through the T median at the 60% mark. Since neither the R nor S composition is given, a specific point on this perpendicular cannot be identified.

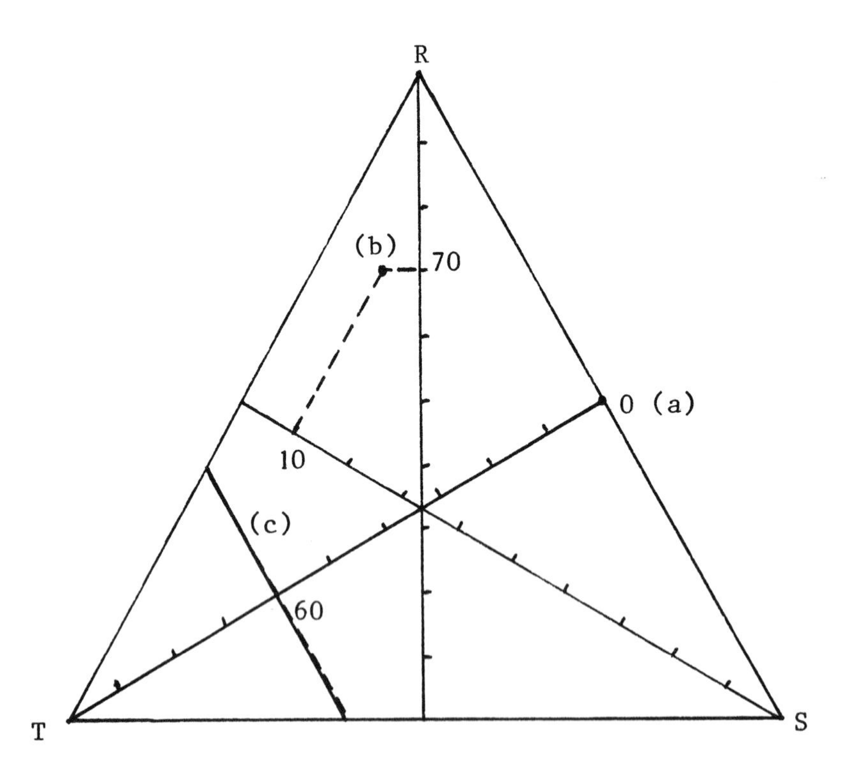

Fig.(1.5) Example 7

****

1.9 Problems

1. A cylindrical sample of rock has a length of 37.7 cm and a diameter of 7.50 cm. The mass of the sample is 4747 g. Find the unit weight of the rock, in kN/m^3.

2. Find the unit weight of rock with a specific gravity of 3.08.

3. Calculate the volume of a rock slab with unit weight 29.5 kN/m^3 and mass 4570 kg.

4. A block of granite has edge lengths 1.24 m, 0.820 m, 0.933 m. It weighs 24.7 kN. (*a*) Find the unit weight of the granite. (*b*) Find its specific gravity.

5. A block of rock has edge lengths 45.0 cm, 37.2 cm, 12.8 cm. Its porosity is 38.4%. Find the total volume of the pores in the block.

6. The specific gravity of a rock is 2.94. The porosity of the rock is 0.344. Calculate the specific gravity of the grains.

7. A 12.74 m^3 block of rock has a porosity of 26.40%. What is the volume of this rock after the block is crushed just sufficiently to close all the pores?

8. A block of rock has edge lengths 1.22 m, 2.40 m, 1.81 m. When dry its mass is 14.7 Mg; when saturated with water its mass is 16.6 Mg. Find the porosity of the rock. (The SI prefix M stands for 1 X 10^6.)

9. A slab of rock has a volume of 5.56 m^3 and a porosity of 0.417. It is saturated with oil of density 0.620 g/cm^3. Find the weight of the oil in the slab.

10. A rock saturated with oil has a unit weight of 29.3 kN/m^3. When dry the rock has a unit weight of 26.4 kN/m^3. The porosity of the rock is 0.370. Determine the density of the oil.

11. Calculate the porosity of a 92.0 cm^3 sample of rock containing 1270 spherical pores, each with a diameter of 3.82 mm.

12. A sample of rock has a very irregular shape for which there is no formula for the volume. When placed into a graduated cylinder containing water filled to the 65.0 mL mark, the water level rises to the 107.5 mL mark. Immediately upon removal from the water, the sample is found to have a mass of 118.3 g. The sample is then dried in an oven, thereby

driving out any water that seeped into the rock while it was submerged. When reweighed, the mass of the rock sample is found to be 116.9 g. Find (*a*) the volume of the sample, (*b*) its density and (*c*) its specific gravity.

13. A cube of chalk with porosity 38.4% has an edge length of 1.40 m. The chalk is crushed, closing all the pores, and then reshaped into a cube. What is the edge length of the new cube?

14. Dry surface Bermuda limestone has a porosity of 42.0%. Dry dense Bermuda limestone, porosity 5.00%, has a density of 2.72 g/cm^3. The two limestones have the same grain density. Find the mass of a 2.50 m^3 block of surface Bermuda limestone saturated with water.

15. A cylindrical sample of rock has a diameter of 8.48 cm and a length of 14.6 cm. When dry it weighs 22.8 N; when saturated with water it weighs 28.0 N. (*a*) Find the porosity of the rock. (*b*) What is the volume of the water in the sample when saturated with water?

16. A particular dry oil shale has a density of 2.13 g/cm^3. When saturated with oil of specific gravity 0.600, the rock yields 30.0 gallons of oil per 1.00 ton of saturated rock. Find the porosity of the shale. (1 U.S. fluid gallon = 3786 cm^3; 1 ton = 8896 N.)

17. Mine spoils of porosity n_0 are dumped into a triangular trench to depth H_0, as shown in Fig.(1.6). Over time, the spoils become compacted under their own weight until the porosity is reduced to n. Show that the subsidence ΔH of the surface of the spoils is given by the expression

$$\Delta H = H_0[1 - \sqrt{\frac{1-n_0}{1-n}}].$$

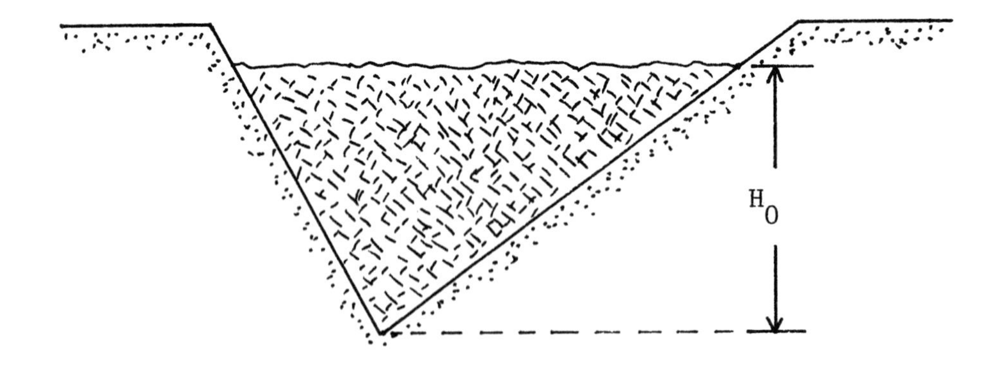

Fig.(1.6) Problem 17

(*Hint*: Draw a diagram for the compacted spoils and then make use of the properties of similar triangles.)

18. A cylindrical sample of rock has a diameter of 6.57 cm and a length of 15.8 cm. The rock has a unit weight of 36.2 kN/m^3. Find the mass of the rock sample.

19. A rock has a density of 2.77 g/cm^3. The grains of the rock have a density of 3840 kg/m^3. Calculate the porosity of the rock.

20. A block of dimension stone has edge lengths 1.13 m, 2.26 m, 1.30 m. When dry the mass of the block is 10,300 kg. The porosity of the stone is 26.4%. Find the mass of the block when it is saturated with liquid mercury.

21. A dry oil shale has a unit weight of 26.3 kN/m^3. When saturated with oil of unit weight 5.80 kN/m^3, the shale has a unit weight of 28.9 kN/m^3. How many gallons of oil can be extracted from 4.72 X 10^6 m^3 of saturated oil shale?

22. A rectangular slab of rock has edge lengths 28.2 cm, 1.46 m, 1.88 m. When dry, the unit weight of the rock is 25.7 kN/m^3. When saturated with oil of unit weight 6.40 kN/m^3, the unit weight of the rock becomes 28.1 kN/m^3. Find the total volume of the pores in the rock.

23. A block of oil shale with a volume of 0.774 m^3 is saturated with 0.311 m^3 of oil. The unit weight of the saturated oil shale is 27.8 kN/m^3. After all of the oil has been driven out of the rock, the unit weight of the rock is 25.2 kN/m^3. Find the specific gravity of the oil.

24. A 2.74 m^3 block of rock saturated with water has a unit weight of 32.4 kN/m^3. The rock has a porosity of 0.276. Find the weight of the rock when dry.

25. A dry oil shale has a unit weight of 25.8 kN/m^3. When saturated with oil of specific gravity 0.650, the shale has a unit weight of 29.3 kN/m^3. How many barrels of oil can be extracted from 7400 m^3 of this saturated oil shale? (1 barrel = 119,300 cm^3.)

26. A dry rock has a density of 2730 kg/m^3 and a grain density of 4.82 g/cm^3. (*a*) Find the porosity of the rock. (*b*) Find the unit weight of the rock.

27. A block of rock saturated with water has edge lengths of 1.20 m, 1.47 m, 1.35 m. The porosity of the rock is 28.4%. Find the volume of water squeezed out of the rock when it is crushed so that all the pores are closed.

28. A 2420 kg block of rock has edge lengths 0.94 m, 1.26 m, 1.15 m. The grain density is 2.87 g/cm^3. Find the porosity of the rock.

29. A block of rock has a volume of 5.30 m^3. When dry its mass is 14,700 kg. When saturated with water its mass is 16,600 kg. Find the porosity of the rock.

30. Mine tailings saturated with water form a triangular prism 15.2 m wide, 15.2 m high and 83.0 m long when dumped on the ground. Later, after the tailings have settled, squeezing out all the water, they form a triangular prism of the same length but now 19.5 m wide and 8.70 m high. Find the porosity of the freshly-dumped tailings.

31. Clay sediment of porosity 48.2% is deposited into a triangular trench to a depth equal to 7.26 m. (*a*) Find the porosity of the clay when it has settled by 54.4 cm. (*b*) Find the greatest possible settlement that can occur due to compaction.

32. Dry mine spoils of porosity 60% are dumped into a triangular trench to a depth equal to 2.87 m. Find the depth of the spoils when they have compacted to a porosity of 30%.

33. Material with a porosity n_0 is stored in a rectangular trench, filling the trench to depth H_0. Show that the porosity n of the material after settling a distance equal to $\frac{1}{2}H_0$ is given by $n = 2n_0 - 1$.

34. A rock with porosity 37.5% and bulk density 1.70 g/cm^3 consists only of orthoclase and pyroxene. Find the abundances of these minerals in the rock.

35. Calculate the porosity of a rock that is 50.0% quartz, 50.0% muscovite and that has a bulk density of 2.00 g/cm^3.

36. Calculate the porosity of a rock that is 50.0% quartz, 50.0% muscovite, and that has a bulk density of 2.00 g/cm^3 when saturated with water.

37. A sample of rock has a volume of 27.6 cm^3. Its composition is 36.2% orthoclase and 17.4% pyrite; the rest is galena. The porosity of the rock is 21.0%. Find the mass of the rock sample if it is saturated in oil of specific gravity 0.600.

38. Derive Eq.(1.24). *Hint*: Draw a diagram; label the slope angle of the line and the quantities y_p and $mx_\mathrm{p} + b$.

39. On a triangular composition diagram for rocks composed of the minerals R, S and T,

locate the regions occupied by rocks with compositions (*a*) 20% R, 40% T; (*b*) 60% S, 10% T; (*c*) 100% R.

40. On a triangular composition diagram for rocks made of the minerals P, Q, R, locate the region containing rocks with compositions in the range 10% → 30% P, 20% → 50% Q.

Chapter 2

Stress

2.1 Unconfined Compressive Strength

Suppose that, from a block of a particular type of rock, a test specimen in the shape of a right circular cylinder (length about twice the diameter) is cut. The values of the length and diameter are recorded. The cylinder is placed on a test bench, standing on one of its circular ends.

A force F is applied perpendicular to the top face, area A, and directed towards it. To ensure that the action of this force is distributed uniformly over the face, the force actually is applied to a plate placed on the cylinder. (The weight of this plate is negligible compared to the force F, and can be ignored, as can the weight of the cylinder itself.) The bench exerts an equal and opposite force F upward on the bottom face (as long as the bench does not collapse!). The situation is shown in Fig.(2.1). No forces are exerted on the curved side of the cylinder: laterally, the cylinder is unconfined. Hence, this procedure is called an *unconfined compression test.*

The test continues by steadily increasing the magnitude of the applied force F. Now, it is common experience that everything is breakable. Therefore, it is expected that when the applied force becomes sufficiently large, the cylinder will break or fail. Call F_{ult} the *ultimate force*, that is, the value of the applied force at which failure of the rock specimen takes place.

Just what is meant by saying that the specimen *fails*, rather than, say, breaks? The word 'break' implies that the cylinder undergoes a complete fracture, separating cleanly into two or more pieces. But, in engineering situations, things need not necessarily reach such an extreme state for the rock mass represented by the cylinder to no longer be able to perform its designed function. Hence, the term 'fail' often is used, and so F_{ult} is the force at which this failure occurs.

This unconfined compression test can be carried out on other cylinders cut from the same parent block of rock, cylinders with different diameters (and, thereby, different cross-sectional areas A), but with the same length to diameter ratio. It seems reasonable to expect that F_{ult} for a larger cylinder will be greater than for a smaller cylinder. Actual tests bear this out. But these tests also indicate that the quantity σ_u, defined by

$$\sigma_u = \frac{F_{ult}}{A}, \tag{2.1}$$

has the same value for all cylinders made from this particular kind of rock. Hence σ_u, called the *unconfined compressive strength* is a property of the rock type itself, with a numerical value independent of the size and shape of the cylinder.

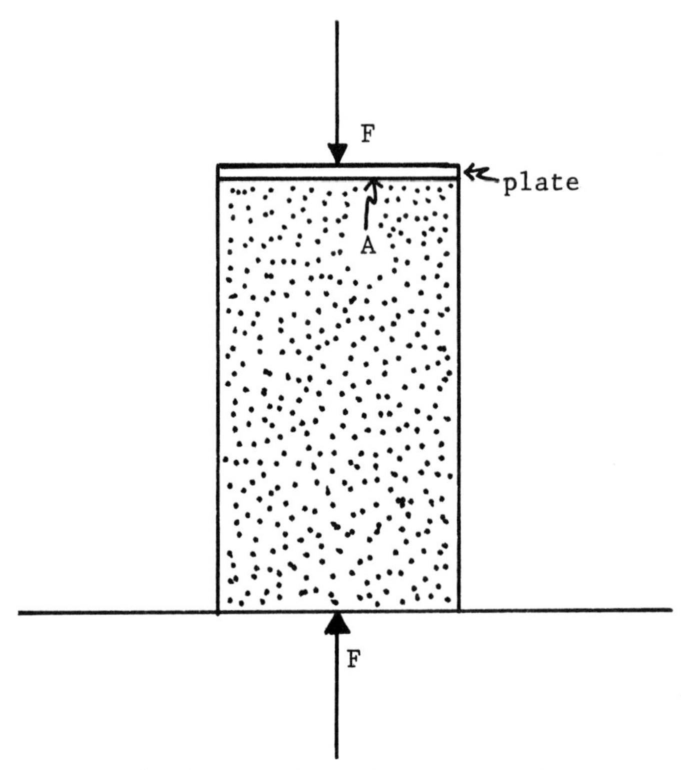

Fig.(2.1) Unconfined Compression Test

In actuality, the value of σ_u does depend, to some extent, on the dimensions and shape of the cylinder. Cylinders with different ratios of diameter to length yield slightly different values of σ_u. Also, the shape of the cross section (circle vs. square, for example) also affects the value of the unconfined compressive strength. However, these variations will be ignored, and the numerical value of the unconfined compressive strength will be taken to depend only on the kind of rock composing the cylinder.

Values of the unconfined compressive strength vary from about 10 MPa for welded volcanic ash (*tuff*) to about 350 MPa for some basalts. (See p. 27 for the unit Pa.) It is important to recognize that there can be wide variations in the physical properties of rocks with the same geologic name but found in different locations. For example, for sandstones σ_u can vary from 70 MPa to 220 MPa; for granite, the range is 140 MPa to 230 MPa, approximately.

2.2 Force and Stress

In Eq.(2.1), the right-hand side has a force F_{ult} divided by the area A on which it acts; this ratio defines σ_{u}. If the force F acting is less than F_{ult}, then F/A is less than σ_{u}. In this more general case, the quantity with symbol σ (no subscript) is defined by

$$\sigma = \frac{F}{A}. \tag{2.2}$$

That is, σ is defined as the force F divided by the area A of the surface on which the force acts. The quantity σ is called *stress.* Equation (2.2) is sometimes read, or verbalized, as "stress equals force per unit area", where the "per unit" stands for division.

In Fig.(2.1), the force F is an *axial force*, since its line of action is parallel to the axis of the cylinder. Also, the force is perpendicular, or normal, to the area A. In addition, the force is a compressive force, as it evidently is directed so as to compress the cylinder along its length. Hence, the stress σ is an *axial compressive stress.*

If the forces F on the ends of the cylinder are reversed in direction, so as to tend to pull the cylinder apart, then the associated stress is an *axial tensile stress.*

From Eq.(2.2), it can be seen that the units of stress are the units of force divided by the units of area. This means that the SI base units of stress are N/m^2. This combination of units is given its own name, the *pascal*, abbreviation Pa. That is, $\text{N/m}^2 = \text{Pa}$.

Now 1 Pa is a very small stress. (Atmospheric pressure is about 100,000 Pa at the Earth's surface.) In engineering situations, it is common to encounter millions of pascals. Hence, the SI prefixes are often employed to simplify the writing:

$$\begin{aligned} 1 \text{ kPa} &= 1 \text{ X } 10^3 \text{ Pa}, \\ 1 \text{ MPa} &= 1 \text{ X } 10^6 \text{ Pa}, \\ 1 \text{ GPa} &= 1 \text{ X } 10^9 \text{ Pa}. \end{aligned}$$

In CGS metric units, force is in dynes and area in cm^2, so the units of stress are dyne/cm^2. (This combination of units is not given any special name, although $1 \text{ X } 10^6 \text{ dyne/cm}^2 = 1$ bar.) In the USCS system (the old British system), the units of stress are lb/in^2 (pounds per square inch, or psi). Neither dyne/cm^2 nor psi units will be used in this book.

Sometimes the term *load* is used. However, load is not a quantity defined in the metric system of units. Rather, it is a descriptive term and can mean, among other things, mass, weight, or stress. When the term load is encountered in a book, technical paper or report, its definition must be ascertained. Sometimes authors are careless and do not spell out their definition of load. But, if a numerical value is given, then the units reveal the meaning of the word. Thus, a load of 500 MPa denotes stress, but a load of 500 kg indicates mass. Because of this ambiguity in the use of load, no special symbol is given for load; rather, the symbol for the quantity corresponding to the use of the word load should be employed.

It is common in engineering to use different terms interchangeably, such as force and stress, mass and weight. The context in which these terms are used, and especially the units when numerical values are given, can resolve an ambiguity in a particular case. Readers must be aware that meanings may be different in the next use, even on the same page of a document. In this book, however, only the proper terminology, corresponding to accepted scientific usage, is adopted.

****

EXAMPLE 1
A load of 1730 N is applied perpendicularly to a square surface of edge length 1.43 cm. Find the applied stress, in MPa.

The load has units of N (Newtons), so in this case load refers to force. Therefore, write F = 1730 N. Since Pa = N/m^2, the area must be expressed in m^2. To do this, either convert the edge length to m and then calculate the area, or calculate the area in cm^2 and convert to m^2. Suppose that the first method is adopted. For a square $A = L^2$, where L is the edge length, so that $A = (0.0143 \text{ m})^2$, $A = 2.045 \text{ X } 10^{-4} \text{ m}^2$. By Eq.(2.2),

$$\sigma = \frac{F}{A},$$

$$\sigma = \frac{1730 \text{ N}}{2.045 \text{ X } 10^{-4} \text{ m}^2},$$

$$\sigma = 8.46 \text{ X } 10^6 \text{ Pa},$$

$$\sigma = 8.46 \text{ MPa}.$$

****

****

EXAMPLE 2
A rectangular block of rock with base edge lengths 7.53 m, 4.16 m rests on a pillar of diameter 12.5 cm, as shown in Fig.(2.2). The unconfined compressive strength of the pillar rock is 62.3 MPa. The unit weight of the block is 29.3 kN/m^3. For what height h of the block is the pillar on the point of collapse? (Ignore the weight of the pillar.)

The pillar is at the point of collapse when the compressive stress on the upper surface of the pillar (the block-pillar contact surface) equals the unconfined compressive strength of the pillar, which is assumed to equal that of the rock making up the pillar. The compressive stress is $\sigma = F/A$. At the point of collapse, $\sigma = \sigma_u$. The area A is the cross-sectional area of the pillar, the area of the surface on which the block rests. In terms of the diameter d of the pillar, $A = \pi d^2/4$. The force F equals the weight W of the block, and this weight is

given by $W = \gamma V$. The volume V of a rectangular block is given by $V = wbh$, where w, b and h are the three edge lengths of the block.
Now put all of this together as follows:

$$\sigma = \frac{F}{A},$$

$$\sigma_{\mathrm{u}} = \frac{W}{A},$$

$$\sigma_{\mathrm{u}} = \frac{\gamma(wbh)}{\pi d^2/4}.$$

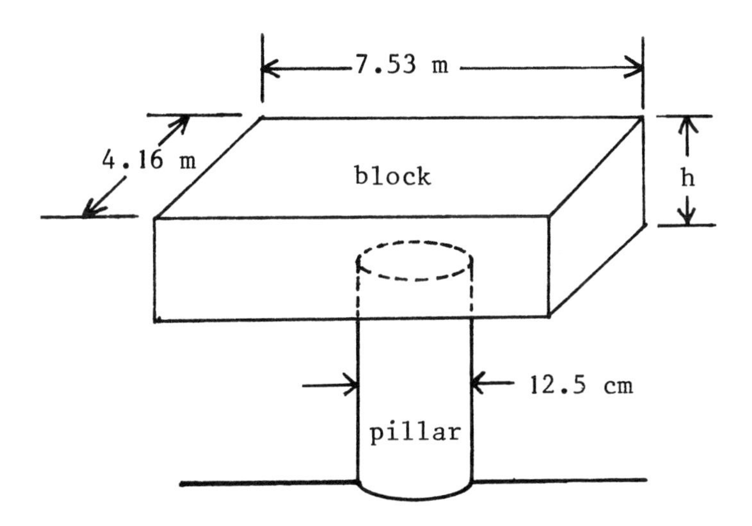

Fig.(2.2) Example 2

In this equation, all quantities are given except h. Solving for h gives

$$h = \frac{\pi d^2 \sigma_{\mathrm{u}}}{4\gamma wb}.$$

Now substitute the data, using SI base units. The diameter d of the pillar is given in cm and must be converted to m. Also, note the SI prefixes in σ_{u} and γ. Numerically, then,

$$h = \frac{\pi(0.125 \text{ m})^2(62.3 \text{ X } 10^6 \text{ Pa})}{4(29.3 \text{ X } 10^3 \text{ N/m}^3)(7.53 \text{ m})(4.16 \text{ m})},$$

$$h = 0.833 \text{ m},$$

$$h = 83.3 \text{ cm}.$$

****

2.3 Stress in the Earth

Rock under stress stores energy. This energy is the work done in deforming the rock. Any engineering project involving the removal or excavation of rock can result in the release of this energy. For reasons of safety, the sudden release of large amounts of energy (*rock bursts*) should be avoided. Therefore, before undertaking construction projects requiring the disturbance of significant quantities of rock, the state of stress in the rock before work begins (the *in situ*, or initial, stress) should be determined.

Although many factors bear on the state of stress at any point in a rock formation, the weight of the overlying materials (the *overburden*) is usually the most important one. It is this contribution to the stress that will be analyzed here.

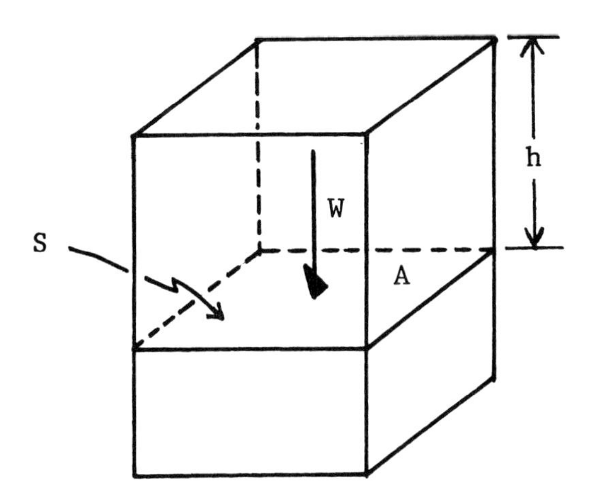

Fig.(2.3) A Rock Column

Consider a vertical column of rock as in Fig.(2.3). The rock has a uniform density ρ and cross-sectional area A. Now evaluate the stress at points on a cross-sectional surface S located at a distance h below the top of the column.

To do this, imagine the column to be made of two parts: the part below S and the part above S. If S is visualized as the top surface of the bottom part, then the force on S is the weight W of the upper part of the column. The associated stress is

$$\sigma = \frac{F}{A},$$

$$\sigma = \frac{W}{A}.$$

But $W = \rho V g$, where V is the volume of the upper part of the column. Since $V = Ah$,

$$\sigma = \frac{\rho(Ah)g}{A},$$

so that

$$\sigma = \rho g h. \tag{2.3}$$

In terms of the rock's unit weight γ,

$$\sigma = \gamma h. \tag{2.4}$$

Equation (2.3) gives the stress acting downward across the surface S. There is also a stress acting vertically upward across S. If S is now thought of as the bottom surface of the upper part of the column, then this stress is seen to be due to the normal force exerted by the lower part of the column on the upper part. For the upper part to be in static equilibrium, this normal force must equal the weight of the upper part. Hence, the state of stress at S can be represented as in Fig.(2.4).

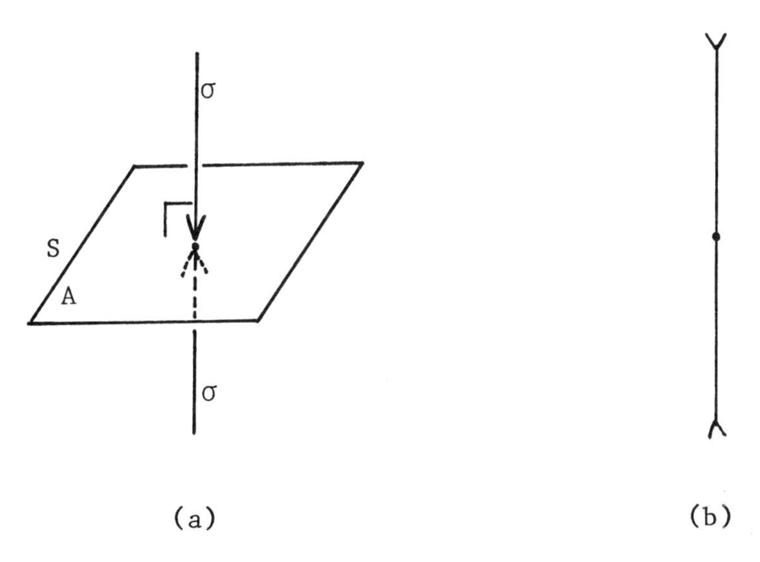

Fig.(2.4) Stress at Surface S in Rock Column

It is important to realize that these equal but oppositely directed stresses do not cancel out. In Fig.(2.4a), the downward directed stress is due to the force exerted by the upper part of the column on the lower part, whereas the upward directed stress is due to the force exerted by the lower part of the column on the upper part. That is, although these stresses act across the same surface, they act on different objects and therefore do not annul each other.

The stress at S is often drawn as in Fig.(2.4b). A line segment, the length of which is proportional to the numerical value of σ, is terminated at each end by an arrowhead. In this case, the arrowhead is inward pointing, suggesting a compressive state of stress.

Now apply these ideas to finding the vertical stress beneath the ground, such as at point P in Fig.(2.5). Point P is at a distance h beneath the surface of the Earth, which is assumed to be locally horizontal. The subsurface rocks have density ρ. But point P can be pictured as lying inside a vertical column of rock, such as the one shown in outline by the dashed lines. Hence, the stress at P is given by Eq.(2.3):

$$\sigma = \rho g h. \tag{2.5}$$

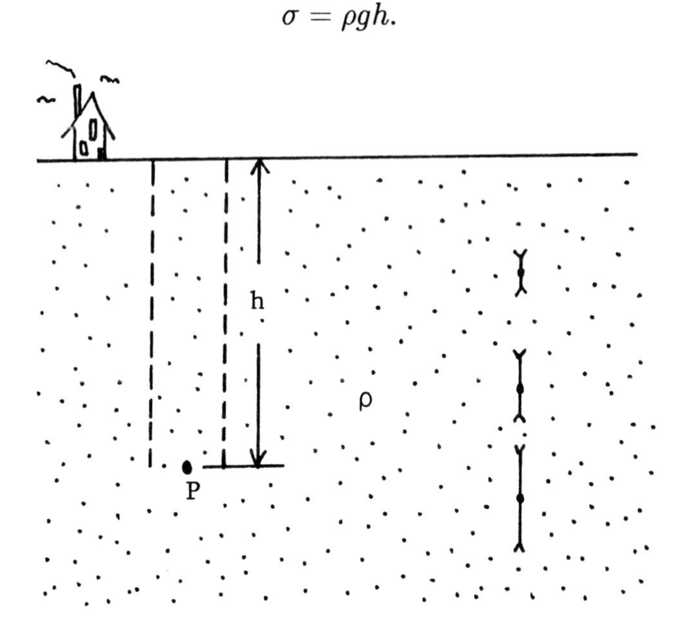

Fig.(2.5) Subsurface Stress

In terms of the unit weight of the rock, the stress is

$$\sigma = \gamma h. \tag{2.6}$$

Note that the SI base units of unit weight γ implied by Eq.(2.6) are Pa/m. In Chapter 1, the units of γ are given as N/m^3. But Pa = N/m^2, so it is evident that Pa/m = N/m^3; the units of unit weight can be expressed in either of these forms.

Now suppose that, beneath the surface, the rocks are arranged in several horizontal layers of different thicknesses and densities (a fairly common geometry), and that the stress is needed to be known at a depth beyond at least that of the first interface between two different layers. For example, suppose that the stress must be calculated at point P in Fig.(2.6).

The vertical stress is due to the total weight of the overburden. But the weight of any object is equal to the sum of the weights of its parts. This implies that Eq.(2.5) can be applied to each layer above P and the results added. That is,

$$\sigma = \sum_i \rho_i g h_i, \tag{2.7}$$

where the subscript i identifies the thicknesses of the various layers of rock above point P. For the layer in which P is embedded, the corresponding h is only the thickness of that part of the layer above P, not the total thickness of the layer.

For example, for the configuration of Fig.(2.6), the stress at P is

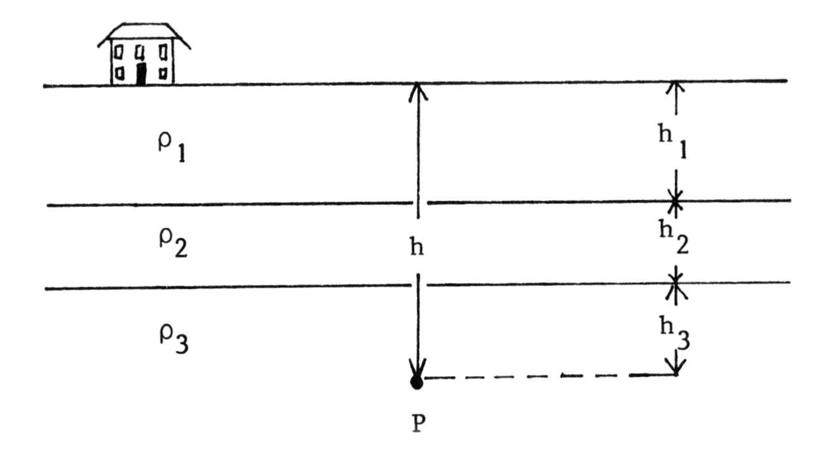

Fig.(2.6) Multilayered Subsurface Rock

$$\sigma = \rho_1 g h_1 + \rho_2 g h_2 + \rho_3 g h_3, \tag{2.8}$$

where

$$h_3 = h - h_1 - h_2.$$

Note that h_3 is not the total thickness of layer 3.

The stress that has been calculated in this section, the vertical stress due to the weight of material lying above, is called the *lithostatic stress.* At most locations, there is also some horizontal stress, but this stress component is ignored in this book. (There is no simple formula, like Eq.(2.6), for the horizontal stress, the value of which may depend on the detailed geologic history of the region, or the effect of current *tectonic* (large scale) processes.)

****

EXAMPLE 3

Figure 2.7 shows a cross section of rock beneath the ground in a certain region. Three horizontal layers or *beds* of rock are shown, together with their dry unit weights. The numbers on the right show the depths of the two interfaces, and also the depth of point P which lies in rock type C. Rock layers A and C cannot be penetrated by water. (*a*) Find the vertical stress at P. (*b*) Water seeps into layer B, saturating the rock. The stress at P rises to 12.00 MPa as the water enters. Find the porosity of rock B.

(*a*) Apply Eq.(2.7), but written in terms of unit weight; that is

$$\sigma = \gamma_A h_A + \gamma_B h_B + \gamma_C h_C.$$

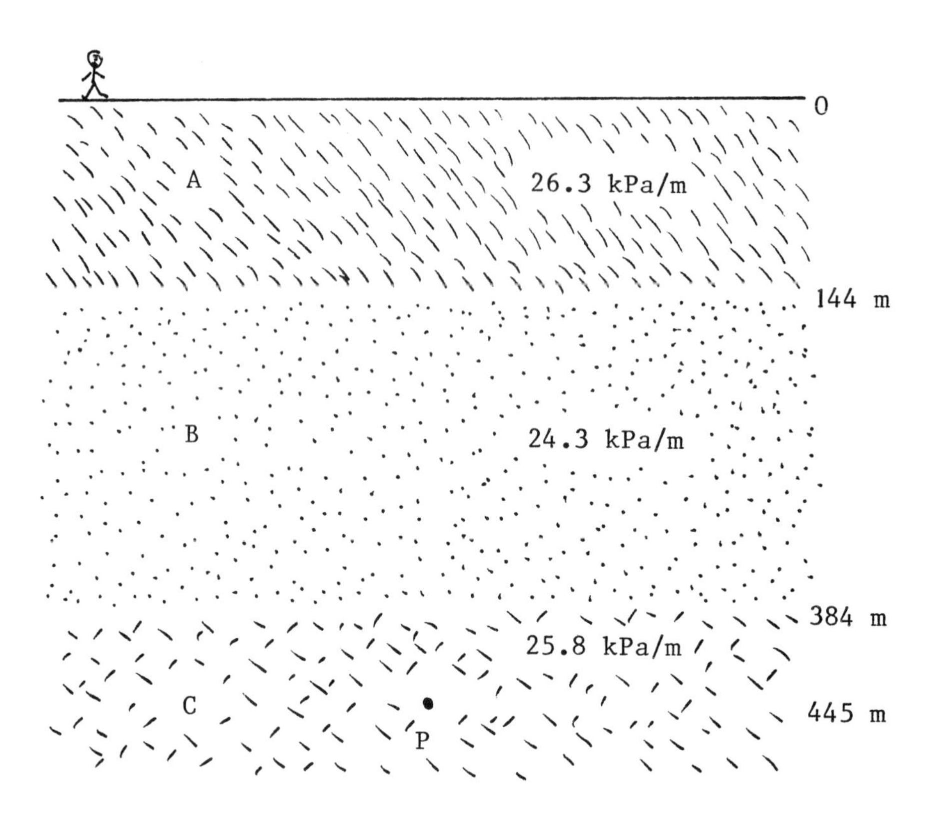

Fig.(2.7) Example 3

Reaching point P from the surface requires passing through the entire thicknesses of layers A and B. Therefore, set $h_{\mathrm{A}} = 144$ m and $h_{\mathrm{B}} = 384$ m $-$ 144 m, $h_{\mathrm{B}} = 240$ m. Finally, a distance $h_{\mathrm{C}} = 445$ m $-$ 384 m, $h_{\mathrm{C}} = 61$ m through layer C must be traversed. Thus,

$$\sigma = (26.3 \text{ kPa/m})(144 \text{ m}) + (24.3 \text{ kPa/m})(240 \text{ m}) + (25.8 \text{ kPa/m})(61 \text{ m}),$$
$$\sigma = 1.119 \text{ X } 10^7 \text{ Pa},$$
$$\sigma = 11.2 \text{ MPa}.$$

(b) The unit weight γ_{B} of layer B as used in part (a) is the dry unit weight. When this layer is saturated with water, the dry unit weight must be replaced with the saturated unit weight $\gamma_{\mathrm{B,sat}}$. By Eq.(1.17) this is

$$\gamma_{\mathrm{B,sat}} = \gamma_{\mathrm{B}} + n\gamma_{\mathrm{w}},$$

where γ_{w} is the unit weight of water. When this saturated unit weight is substituted for the dry unit weight in the equation for σ in (a), an additional contribution of $n\gamma_{\mathrm{w}}h_{\mathrm{B}}$ is made to the stress. This extra stress $\Delta\sigma$ due to the water must be 12.00 MPa $-$ 11.19 MPa, or $\Delta\sigma$ = 0.81 MPa. Therefore

$$\Delta\sigma = n\gamma_{\mathrm{w}}h_{\mathrm{B}},$$
$$810 \text{ kPa} = n(9.8 \text{ kPa/m})(240 \text{ m}),$$
$$n = 0.344.$$

****

2.4 Underground Rooms

Suppose that it is needed to excavate a room (chamber) beneath the surface of the Earth, and that the horizontal dimensions (or one of them) are greater than the distance over which the subsurface rock is *competent*, or self-supporting. It is then necessary to prop-up the roof. One way to do this is to support the roof with pillars (posts, columns). These pillars may be manufactured, of steel for example. Or, they may of material just left behind for this purpose in the excavation, and therefore be formed of the same kind of rock that was removed to form the room. Clearly, for safe construction, it is wise to have at hand a relation between the room's dimensions and depth and the number and size of the pillars required.

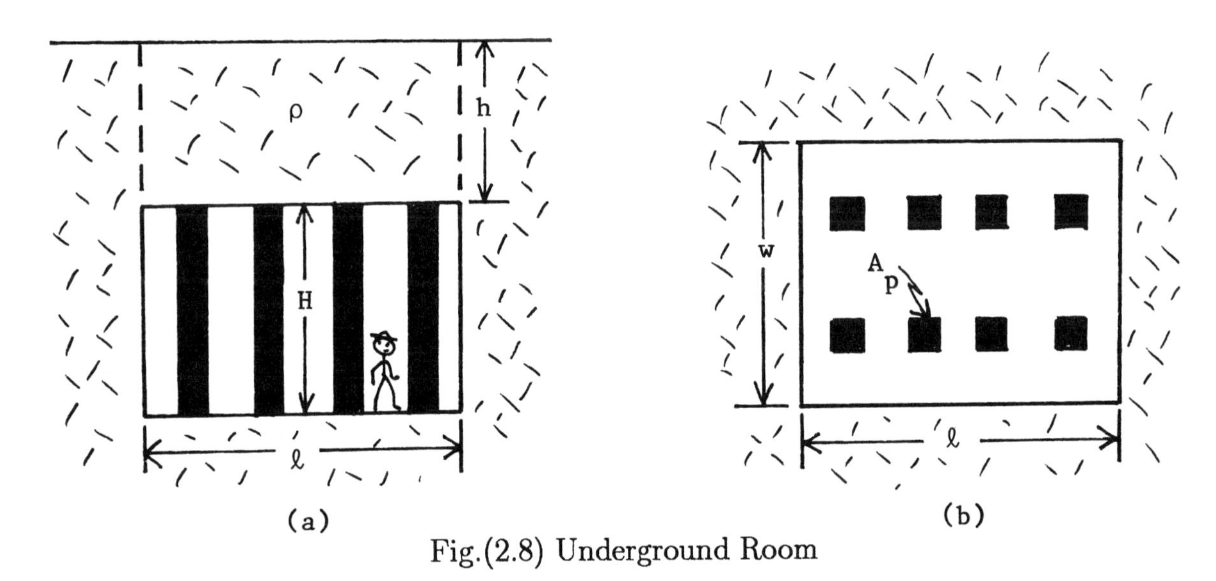

Fig.(2.8) Underground Room

In this analysis, it will be assumed that the stress in the Earth is lithostatic (i.e., is vertical and due to the weight of the overburden). Also, that the pillars are identical, and that the force exerted by the roof of the room on each pillar is the same.

The situation is shown on Fig.(2.8) for a rectangular room. The roof of the room is at a depth h below the surface of the Earth. The subsurface rocks are presumed to be of uniform density ρ. There are n identical pillars, each of area A_{p}, supporting the roof. The area of the room is A; that is, $A = \ell w$, where ℓ and w are the length and width of the roof (and floor). Because effects associated with the length of the pillars are ignored, the height H of the room does not immediately enter.

The weight W of the material between the roof and the surface, the rock between the dashed lines of Fig.(2.8a), acts on the upper cross-sectional surfaces of the pillars. Each pillar exerts a resisting normal force F_{p} vertically upward on this block of rock. For this block of rock (the overburden) to be in equilibrium,

$$nF_{\mathrm{p}} = W. \tag{2.9}$$

Write Eq.(2.9) in terms of the stresses σ_p, the compressive stress on each pillar, and σ, the lithostatic stress in the Earth at the depth h. Since $\sigma_p = F_p/A_p$ and $\sigma = W/A$, then, upon dividing both sides of Eq.(2.9) by the pillar area A_p, the following relations are created:

$$n(\frac{F_p}{A_p}) = (\frac{W}{A})(\frac{A}{A_p}),$$

$$n\sigma_p = \sigma(\frac{A}{A_p}),$$

$$\sigma_p = (\rho g h)(\frac{A}{nA_p}). \tag{2.10}$$

If the pillars support the roof, then σ_p, the compressive stress on each pillar as calculated from Eq.(2.10), will be at most equal to σ_u, the unconfined compressive strength of the pillar rock. That is, $\sigma_p \leq \sigma_u$.

In practice, it is not judicious to have $\sigma_p = \sigma_u$. For then, if (for example) the density of the subsurface rocks have been underestimated, the roof will collapse. The same catastrophe will occur if, instead, the unconfined compressive strength of the pillars is overestimated. Suppose that the pillars are made of rock left behind. An unconfined compressive strength measured on a small sample of rock in the laboratory may not correspond to the value for the same kind of rock in the field. For instance, the presence of *joints* (cracks) in the field rock will lower the effective value of σ_u. Therefore, for the sake of safety, it is desired that $\sigma_p < \sigma_u$, where σ_u is the best available estimate for the unconfined compressive strength of the pillar rock.

By increasing n, the number of pillars, and/or A_p, the cross-sectional area of each pillar, the value of σ_p decreases; see Eq.(2.10). But, a whole room full of pillars defeats the purpose (whatever that is) of constructing the room in the first place. An engineering judgement must be made as to the degree of safety to be built into the design, to reasonably guard against the lack of absolutely precise knowledge of the numerical values of the relevant field parameters.

In this spirit, define the *factor of safety FS* against roof collapse of the underground room by the equation

$$\sigma_u = (FS)\sigma_p. \tag{2.11}$$

The factor of safety FS (a single symbol, not F times S!) tells by what factor the actual pillar stress is smaller than the unconfined compressive strength. Naturally, $FS \geq 1$. Incorporating the factor of safety, Eq.(2.10) becomes

$$\sigma_u = (FS)(\rho g h)(\frac{A}{nA_p}). \tag{2.12}$$

The quantity nA_p represents the total pillar area in a room of floor area A. Recall that the area A is the area obtained from the edge lengths of the room. But for rooms far underground, the supporting pillars of which are simply left behind after excavation, the total pillar area may be a significant fraction of the room area. This means that the useful room area A_{use}, the area on which activities for which the room was built can be performed, is significantly lower than A. In fact,

$$A_{\text{use}} = A - nA_{\text{p}}. \tag{2.13}$$

In effect, each pillar supports an area A/n of the roof, called the *tributary* area A_{trib}:

$$A_{\text{trib}} = \frac{A}{n}. \tag{2.14}$$

The adjective 'tributary' describes the purpose of the pillars in channelling the stress on the roof, area A, into the pillars, total area $nA_{\text{p}} < A$.

If the pillars are formed of material left behind as the room is excavated, then sometimes it is convenient to express the factor of safety in terms of the volume of material that must actually be removed in the excavation.

Suppose that, as in Fig.(2.8), the height of the room is H. The volume $V = AH$ is the volume of the room as calculated from the edge lengths. If there are no pillars, then this is the actual free volume created. The total volume of all the pillars present is $V_{\text{ps}} = n(A_{\text{p}}H)$.

Multiply the right-hand side of Eq.(2.12) by H/H to get

$$\sigma_{\text{u}} = (FS)(\rho g h)(\frac{AH}{HnA_{\text{p}}}),$$

$$\sigma_{\text{u}} = (FS)(\rho g h)(\frac{V}{V_{\text{ps}}}). \tag{2.15}$$

If, before construction begins, the final dimensions of the room are known, then calculations can be done in terms of the volumes themselves. Sometimes, however, the final dimensions are not known, since the room may have to be periodically enlarged as plans change, or as there is increased demand for use of the facility. In this case, it is handy to work with the volume ratios, since these do not change as the room expands.

The volume of material that must be removed V_{rv} is

$$V_{\text{rv}} = V - V_{\text{ps}}.$$

Therefore, the fraction f of material that must be carried away, defined by

$$f = \frac{V_{\text{rv}}}{V}, \tag{2.16}$$

becomes

$$f = 1 - \frac{V_{\text{ps}}}{V},$$

and therefore

$$\frac{V}{V_{\text{ps}}} = \frac{1}{1-f}.$$

Now substitute this result into Eq.(2.15) and solve for f to get

$$f = 1 - (FS)(\frac{\rho g h}{\sigma_u}).$$

The quantity $\rho g h$ is the lithostatic stress at the roof; this is often called the *roof load* σ_{RL}; that is

$$\sigma_{RL} = \rho g h. \tag{2.17}$$

In terms of this roof load,

$$f = 1 - (FS)(\frac{\sigma_{RL}}{\sigma_u}). \tag{2.18}$$

Equation (2.18) gives the fraction of material that must be removed in order to obtain a chamber with a specified factor of safety against collapse.

The type of construction described here is called *room and pillar* or *checkerboard* excavation and is appropriate for rooms with flat roofs. Flat roofs cannot transfer the stress to the sides of the chamber, so that the entire overburden load must be carried by the pillars. If the room is narrow, as in a tunnel, then the roof can be *arched* (i.e., not be flat, but given a quasi-circular profile); this shape of roof can transfer a substantial part of the roof load to the walls, reducing the number of needed pillars.

****

EXAMPLE 4

A chamber is to be constructed 840 m beneath the ground (depth to the roof) for the deposition of toxic waste material. The roof is to be supported by circular pillars, formed from rock left behind in the excavation. This rock has an unconfined compressive strength of 55.0 MPa. The unit weight of the overburden is 23.6 kN/m^3. A factor of safety of 1.80 is required. (*a*) The dimensions of the facility will be increased as more toxic material is trucked in and dumped. As the chamber is expanded, what ratio of total pillar area to useful room area must be maintained? (*b*) For each 125 m^2 of roof area, how many pillars, each of diameter 1.62 m, are needed?

(*a*) In the notation of Eq.(2.12), the data are: $FS = 1.80$, $\sigma_u = 55$ MPa, $h = 840$ m and $\gamma =$ 23.6 kPa/m. Neither the area A of the chamber nor the area A_p of the pillars are given. It looks as though there is one equation with two unknowns! However, this is not really so, for the ratio of the two areas is one unknown. As mentioned above, the final size of a project may not be known during the design or early construction period, but certain ratios will not change if the facility grows. The current problem seeks the ratio nA_p/A_{use}, rather than the ratio nA_p/A that appears in Eq.(2.12). But, by Eq.(2.13),

$$A/nA_p = 1 + (A_{use}/nA_p).$$

Therefore, Eq.(2.12) can be written as

$$\sigma_u = (FS)(\gamma h)(1 + \frac{A_{use}}{nA_p}).$$

This can be solved for the ratio in question to obtain

$$\frac{nA_p}{A_{use}} = \frac{(FS)\gamma h}{\sigma_u - (FS)\gamma h}.$$

Upon substitution of the data into the right-hand side, paying attention to SI prefixes, it is found that $nA_p/A_{use} = 1.85$. This means that the useful room area is only a little more than one-half the total area of the pillars; the chamber is mostly pillars!
(*b*) In this part there are specific dimensions. The area A_p is

$$A_p = \pi d^2/4,$$

$$A_p = \pi(1.62 \text{ m})^2/4,$$

$$A_p = 2.061 \text{ m}^2.$$

With $A = 125$ m^2 and this value of A_p, Eq.(2.13) gives

$$2.061n = 1.85(125 - 2.061n),$$
$$n = 39.37.$$

It might seem appropriate to round this number to 40, since the pillars are presumed to be identical, so that there can only be an integral number of them in a room. But this problem does not say that the roof area of 125 m^2 pertained to one room, only for the number of pillars for each 125 m^2. Only for a final room should n be rounded to an integer. For example, if the completed chamber has a roof area of 1250 m^2, the number of pillars should be (39.37)(10) = 394, and not (40)(10) = 400, as would be obtained if the 39.37 is rounded to 40.

****

2.5 Shear Stress

In Fig.(2.1) the force F is directed perpendicular to and toward the surface of area A to which it is applied. This resulting stress is a compressive stress. Another important stress occurs when the force is parallel to the surface of area A to which it is applied. This stress is called a *shear stress.* See Fig.(2.9).

The numerical value of the shear stress is defined just as it is for a compressive (or tensile) stress: stress = force/area. However, it is necessary to distinguish between compressive stress and shear stress when dealing with forces applied to surfaces. In Eq.(2.2), the symbol σ is used for compressive stress. Extending the notation to include shear stress can be done in two ways: (i) By attaching a two-component subscript to σ, the first component indicating the orientation of the surface and the second the direction of the force; (ii) By using a different symbol. Alternative (i) is favored when treating stress in its full three-dimensional character, but alternative (ii) is adequate, and less tedious, when two-dimensions suffice, as is often the case.

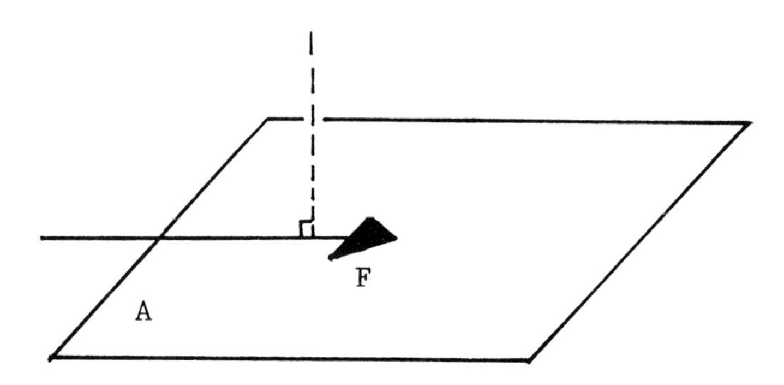

Fig.(2.9) Shear Stress

When alternative (ii) is adopted, as is done here, the symbol τ is often chosen as the symbol for shear stress; hence, the definition of shear stress is written

$$\tau = \frac{F}{A}, \tag{2.19}$$

it being understood that the force is applied parallel to the surface. The SI base units for shear stress are, by Eq.(2.19), N/m^2 or Pa, just as for compressive stress.

On diagrams, a shear stress is often represented by a *harpoon arrow* ($\leftharpoonup$) drawn parallel to the surface on which the shear stress acts and in the direction of the responsible force.

The *shear strength* τ_u of a rock is the value of the greatest shear stress that the rock can sustain without rupture. The value may depend on the particular surface to which the stress is applied.

The shear strength can be measured, easily in principle, by the procedure shown on Fig.(2.10). A block of the rock to be tested is placed into a restraining slot, as shown. A distributed force F is applied perpendicularly to the upper part of one face. This force sets up a shear stress on the internal surface shown by the dashed line. The slot in which the block is placed exerts an equal and opposite force. If ℓ and w are the length and width of the block (the width being into the page), then $\tau = F/(\ell w)$. The force F is increased until, at the value $F_{\rm max}$ shear failure on the plane takes place. The shear strength $\tau_{\rm u}$, on the indicated plane, is given by

$$\tau_{\rm u} = \frac{F_{\rm max}}{\ell w}. \tag{2.20}$$

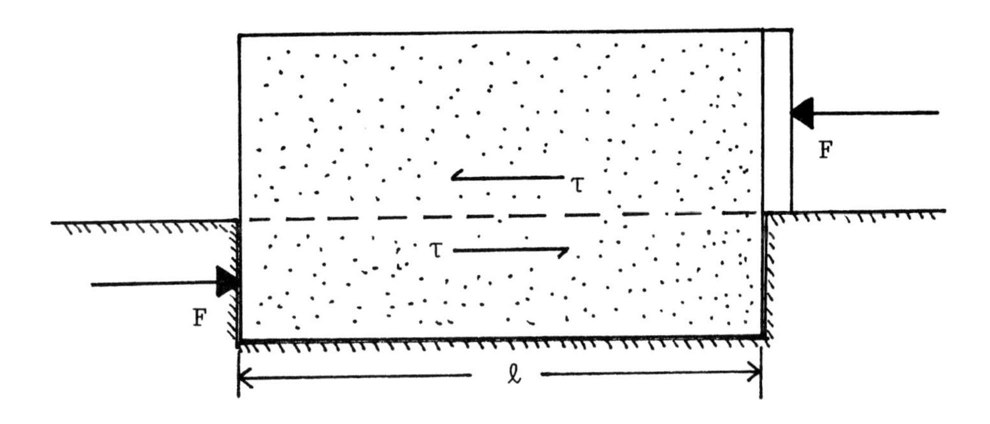

Fig.(2.10) Testing for Shear Strength

Measured values of the parameters describing the strength of rocks (unconfined compressive strength, tensile strength, shear strength) vary over a wide range for different kinds of rock, and for the same kind of rock found at different locations. But, on average, for igneous and metamorphic rock,

$$\sigma_{\rm u}(\text{compressive})/\sigma_{\rm u}(\text{tensile}) \approx 11,$$

$$\sigma_{\rm u}(\text{compressive})/\tau_{\rm u} \approx 6.$$

For sedimentary rock, on average,

$$\sigma_{\rm u}(\text{compressive})/\sigma_{\rm u}(\text{tensile}) \approx 8,$$

$$\sigma_{\rm u}(\text{compressive})/\tau_{\rm u} \approx 4.$$

That is, the compressive strength is the greatest, followed by shear strength. (Note that, unlike axial stress, there is only a single shear stress: reversing the direction of the shear stress across the dashed surface in Fig.(2.10) represents the same stress.) Rocks are weakest under tension. Hence, it is usually important when using rock in construction projects to minimize tensile stresses. For typical numerical values of the strengths, see the problems in this chapter and in Chapter 3.

****

EXAMPLE 5

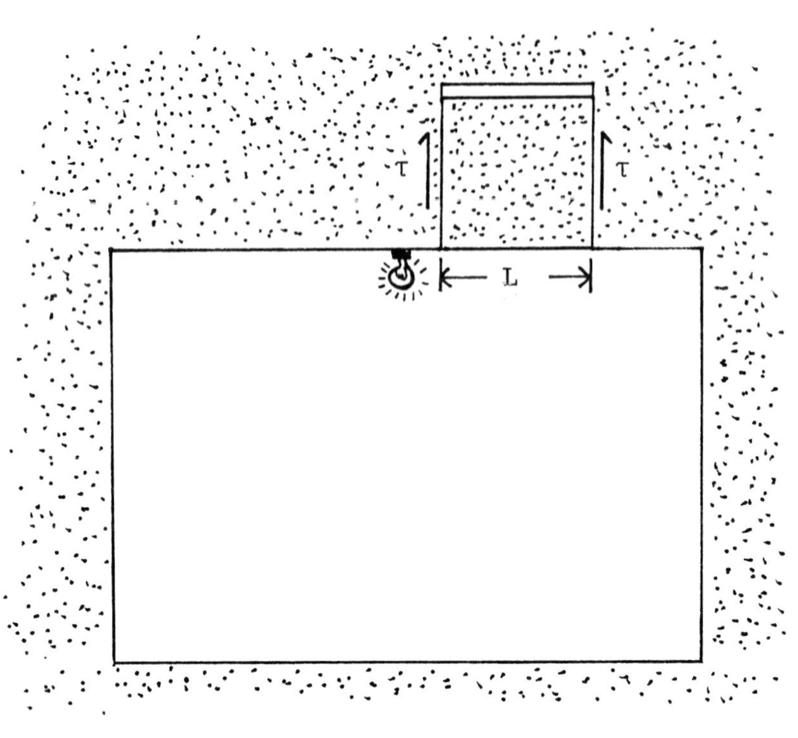

Fig.(2.11) Example 5

Figure (2.11) shows a cubical block of rock in the flat roof of an underground tunnel. The block is a *key block*; that is, it may fall out. The density of the block is ρ and each edge length is L. Due to friction, neighboring rock exerts equal upward shear stresses τ on the two side faces of the block but, due to the presence of small gaps, no stresses on the front, back, and top faces. Show that, if the block does not fall, the shear stress on each side face must be given by

$$\tau = \frac{1}{2}\rho g L.$$

For equilibrium, the total force (not stress) on the block must be zero. The weight W of the block acts vertically downward; the forces due to the shear stresses act upward. If A is the area of each face, then

$$\Sigma F = 0,$$

$$2\tau A - W = 0,$$

$$2\tau(L^2) - \rho g(L^3) = 0,$$

$$\tau = \frac{1}{2}\rho g L.$$

****

2.6 Problems

1. A force of 17,200 N is exerted perpendicularly against a surface of area 0.136 m^2. Find the stress.

2. What force is needed to provide a stress of 726 kPa on a surface of area 2.82 cm^2?

3. A cylinder of rock has a length of 5.82 cm and a diameter of 2.14 cm. Oppositely-directed axial forces of 14.8 kN each are exerted against the ends. Find the stress.

4. What must be the length, in centimeters, of the sides of a square surface on which a force of 41.3 kN supplies a load of 37.2 MPa?

5. A cylinder of rock has length 12.6 cm, density 2.77 g/cm^3 and mass 324 g. The cylinder is under an axial compressive stress of 28.1 MPa. Find the axial force being applied to the cylinder.

6. Minnesota quartzite has an unconfined compressive strength of 629 MPa, but the unconfined compressive strength of Utah quartzite is only 148 MPa. The diameter of the thinnest cylinder of Minnesota quartzite that supports a certain weight is 3.85 cm. Find the diameter of the thinnest cylinder of Utah quartzite that supports the same weight.

7. In a certain region where subsurface rocks have density 3.08 g/cm^3, what is the lithostatic stress at a depth of 4.75 km?

8. Find the unit weight of subsurface rocks if the vertical stress is 9.00 MPa at a depth of 366 m.

9. By how much does the vertical stress at a depth of 1.22 km exceed the stress at a depth of 840 m if the underground rocks have density 2420 kg/m^3?

10. Rocks of density 2.30 g/cm^3 overlay rocks of density 2.87 g/cm^3. The lithostatic stress at a depth of 580 m is 13.7 MPa. Find the depth to the horizontal interface between the two kinds of rock.

11. A block of density 3.10 g/cm^3 and edge lengths 8.30 m, 3.20 m, 4.50 m sits on top of a pillar with a square cross section of edge length 23.0 cm. Find the stress exerted by the block on the top of the pillar.

12. A block of rock with edge lengths 1.63 m, 2.17 m, 1.44 m sits on top of a column that has a cross-sectional area 0.214 m^2. The stress on the top of the column due to the weight of the block equals 743 kPa. Find the unit weight of the rock making up the block.

13. A block of rock with a density of 2330 kg/m^3 and a volume of 7.81 m^3 sits on top of a 3.66 m high column with a square cross section. The compressive stress on the top of the column due to the weight of the block is 4.94 MPa. Find the total surface area of the sides of the column.

14. A room with floor dimensions 55.0 m X 33.0 m is excavated 196 m beneath the surface of the ground (depth to the roof). The density of the overburden is 2.90 g/cm^3. The roof of the room is supported by a single pillar of unconfined compressive strength 46.6 MPa. A factor of safety of 3.00 is desired. Find the useful room area.

15. A rectangular room with floor edge lengths 15.2 m and 22.7 m is to be built 135 m beneath the ground (depth to the roof). The overburden density is 3.22 g/cm^3. The room has a flat roof which is to be supported by pillars. Each pillar has an unconfined compressive strength of 380 MPa and area 0.424 m^2. A factor of safety of 2.30 is needed. (*a*) How many pillars are required? (*b*) Calculate the pillar tributary area.

16. A circular room of diameter 37.1 m and height 5.30 m is excavated at a depth to the floor of 230 m. The roof is supported by 18 identical pillars, each carrying the same load. The pillars are made of material with an unconfined compressive strength of 84.0 MPa and each has a cross-sectional area of 5.738 m^2. The factor of safety is 1.40. Find the unit weight of the overburden.

17. A room 18.0 m X 32.0 m and 12.0 m high is constructed 170 m beneath the ground (depth to the roof). The density of the subsurface rock is 3.30 g/cm^3. The roof is supported by ten identical square pillars, made of material left behind, with an unconfined compressive strength of 23.5 MPa. The factor of safety is 1.50. (*a*) Find the volume of material removed to make the room. (*b*) Assuming that the conditions described actually hold, what is the compressive stress on each pillar?

18. A tunnel 240 m long is built through a small mountain composed of rocks with density 2.21 g/cm^3. The tunnel is 3.80 m wide and 4.30 m high with a flat roof 134 m beneath the top of the mountain. The roof is supported by a single row of posts, each with a diameter of 26.4 cm and made of material with an unconfined compressive strength of 412 MPa; see Fig.(2.12). Find the adjacent post spacing, center to center, in the tunnel to give a factor of

safety against collapse of 1.25. There is a post at each end.

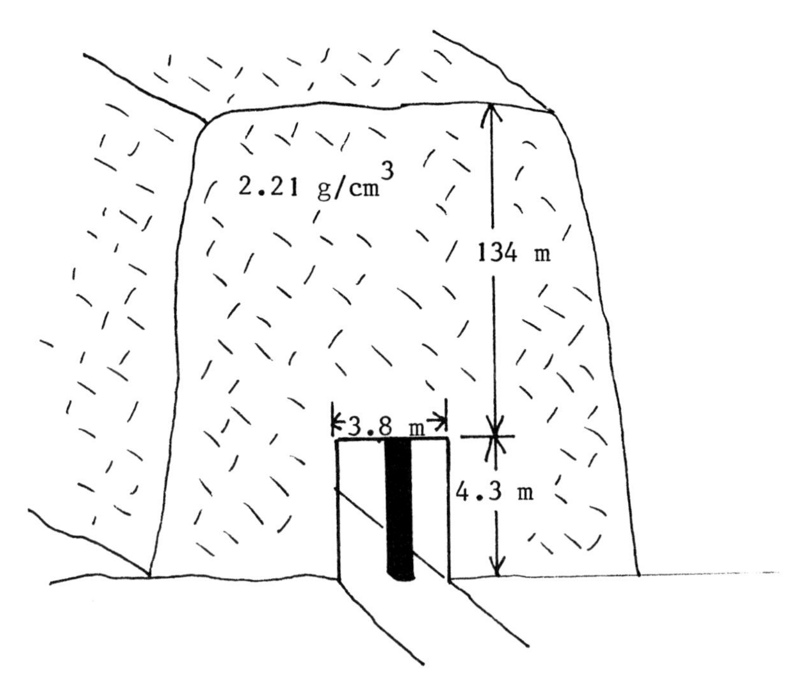

Fig.(2.12) Problem 18

19. Engineering activities can trigger rock failure. In a certain region, because of the presence of rock fractures, failure can arise if the lithostatic stress exceeds only 16% of the unconfined compressive strength of the rock. Suppose that a hill rises at an elevation angle of 28°. The rock has an unconfined compressive strength of 53 MPa and unit weight 27 kPa/m. A horizontal tunnel with *portal* (entrance) at the base of the hill is to be constructed into the hill. At what distance x from the portal will rock failure first be encountered? See Fig.(2.13).

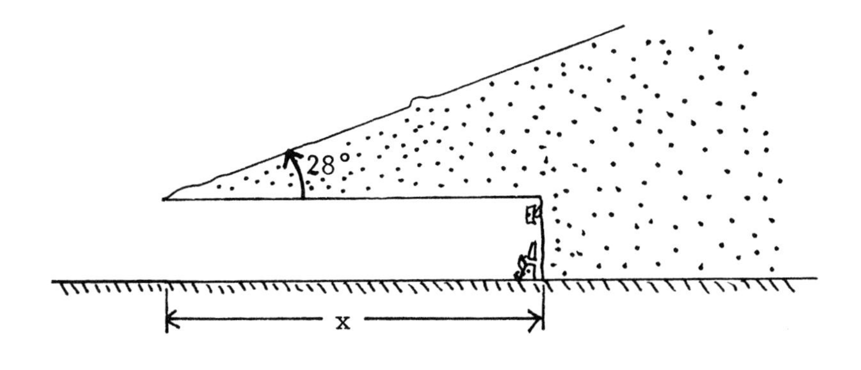

Fig.(2.13) Problem 19

20. Colorado granite has an unconfined compressive strength of 226 MPa and a unit weight of 25.9 kN/m^3. Find the greatest possible height of a free-standing vertical column of this rock. Assume that the column contains no joints or any other defect.

21. At what depth below the ground surface in Fig.(2.14) is the value of the vertical stress equal to 6.08 MPa?

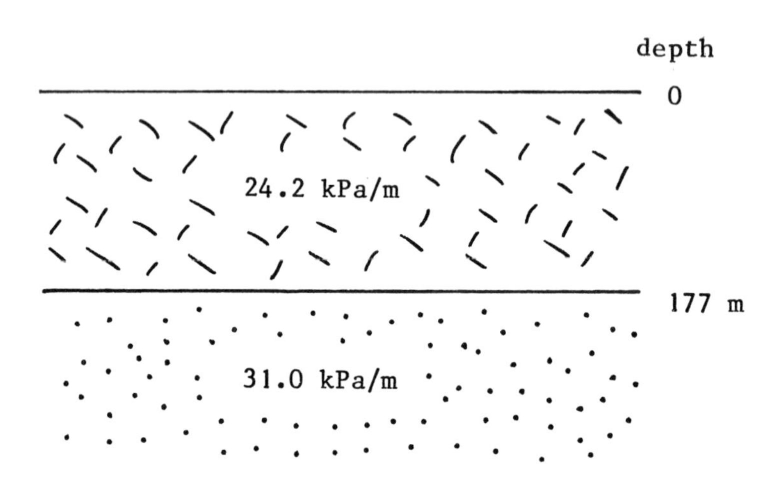

Fig.(2.14) Problem 21

22. Near the town of Igneousville, rocks of density 2.42 g/cm^3 extend from the surface to depth 126 m. Between depths of 126 m and 194 m are rocks of density 2.91 g/cm^3. A third rock layer of density 3.20 g/cm^3 extends from depth 194 m to 230 m. Find the lithostatic stress at depths (a) 167 m and (b) 220 m.

23. A rectangular room with floor dimensions 38.2 m and 19.5 m is excavated at a depth to the roof of 130 m. The density of the overburden is 2200 kg/m^3. The roof of the room is supported by 28 identical pillars, each carrying the same load. Each pillar has an area of 1.36 m^2 and is made of rock with unconfined compressive strength 92.5 MPa. Soon after completion, terrorists begin to destroy the pillars one by one. What is the greatest number of pillars that can be destroyed without causing the roof to collapse? (Assume that the pillars are destroyed from all parts of the room, so that the remaining pillars span regions of the roof over which the roof rock is competent.)

24. The phase diagram of carbon, Fig.(2.15), shows the ranges of temperature and pressure in which carbon will crystallize either as diamond or graphite. What is the minimum depth at which diamonds can form if the local temperature is 1000°C and the subsurface rocks

have density 3.1 g/cm^3?

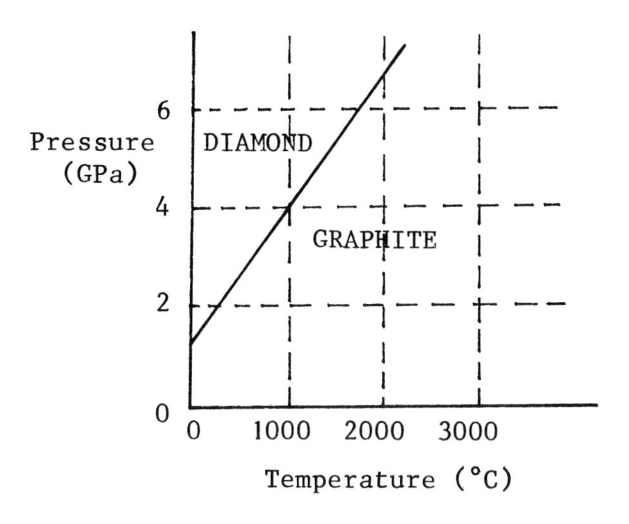

Fig.(2.15) Problem 24

25. Figure (2.16) shows three subsurface layers of rock; the thicknesses of the layers are given, together with the densities of the two upper layers when dry. Under dry conditions, the stress at a depth of 1.20 km is 34.4 MPa. The upper and lower layers have zero porosity, but the middle layer has a porosity of 36.5%. (*a*) Find the stress at a depth of 350 m. (*b*) Find the stress at a depth of 600 m. (*c*) Calculate the density of the rock in the lowest layer. (*d*) Water now seeps into the middle layer, filling all the pores. What now is the stress at a depth of 1.20 km?

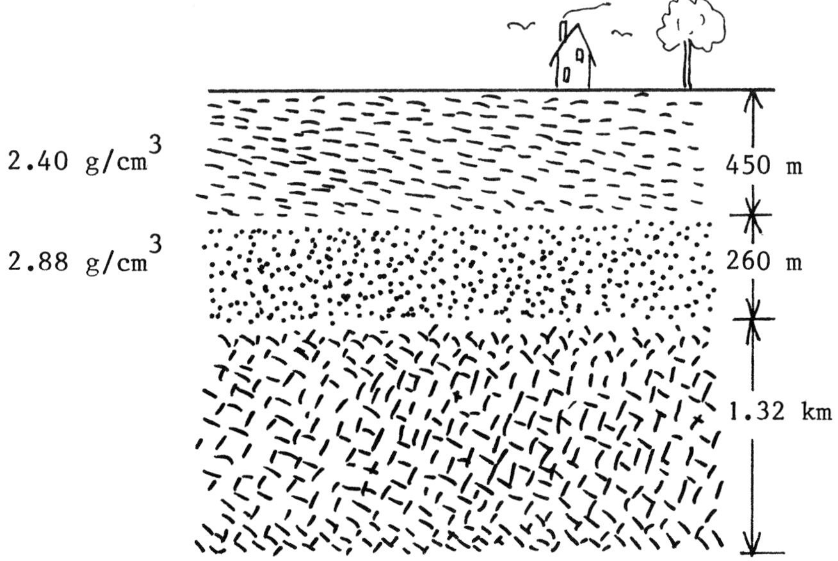

Fig.(2.16) Problem 25

26. Show that, for an underground room in which the pillars just barely support the roof, the roof load σ_{RL} is given by

$$\sigma_{\mathrm{RL}} = (1 - f)\sigma_{\mathrm{u}},$$

where f is the fraction of material removed, and σ_{u} is the unconfined compressive strength of the pillar material.

27. Prove that the factor of safety against roof collapse FS for an underground room satisfies the inequality

$$FS \leq \sigma_{\mathrm{u}}/\sigma_{\mathrm{RL}},$$

where σ_{u} is the unconfined compressive strength of the pillars and σ_{RL} is the roof load.

28. Calculate the roof load of the chamber of Example 4.

29. A flat coal seam 2.40 m thick lies at a depth of 127 m to the top of the seam. The overburden is shale, with density 2.43 g/cm^3; the coal has a density of 1.60 g/cm^3. The coal is being mined, in the room and pillar manner, with the supporting pillars being coal left behind. The unconfined compressive strength of the coal is 20.7 MPa. The floor and roof of the mining chamber are at the bottom and top of the coal seam. The factor of safety during the mining process is 2.00. How far must a tunnel 8.30 m wide advance to recover 3790 metric tons of coal? (1 metric ton = 1000 kg.)

30. Engineers often report stress in improper units, such as kg/cm^2. To how many kPa does 1 kg/cm^2 correspond?

31. A house of weight W is built on a barrier island, where frequent ocean surges are expected. The house is raised on n piles, which penetrate the sand and transfer the load of the house to underlying bedrock. Assume that each pile has the same cross-sectional area A, and that each pile supports the same load. Show that, for a factor of safety FS against collapse of the piles, the number of piles needed is given by

$$n = \frac{(FS)W}{\sigma_{\mathrm{u}}A},$$

where σ_u is the compressive strength of the pile material; see Fig.(2.17).

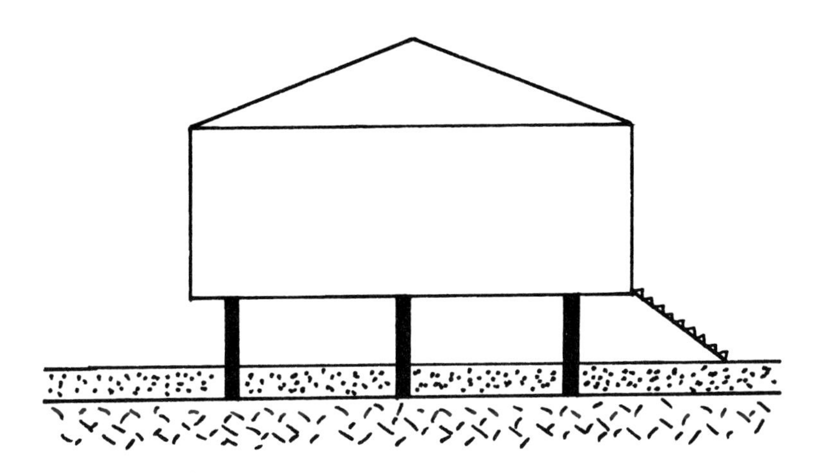

Fig.(2.17) Problem 31

32. In a certain region, rocks of density 2.36 g/cm^3 extend from the surface down to depth 85.0 m. Between depths 85.0 m and 140 m is a different kind of rock. Finally, between depths 140 m and 190 m the density of the rocks is 2.97 g/cm^3. The lithostatic stress at a depth of 180 m is 4.53 MPa. Find the density of the rocks in the middle layer.

33. Explorers seeking a hidden underground tomb walk slowly along a secret tunnel that slopes downward at 25.8°, as shown in Fig.(2.18). The surrounding rock has a unit weight of 24.6 kN/m^3 and unconfined compressive strength 33.7 MPa. The explorers have been warned not to go past the point where the vertical stress equals 25.0% of the unconfined compressive strength of the rocks. How far along the tunnel can the explorers walk before reaching this point?

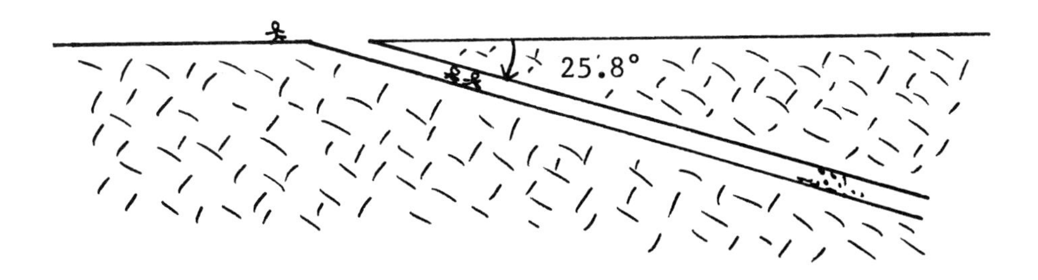

Fig.(2.18) Problem 33

34. A block with edge lengths a, b, c (dimension a normal to page) is about to drop from the roof of an underground tunnel; see Fig.(2.19). The overburden exerts a compressive stress

σ on the top surface of the block; friction with neighboring rock exerts equal upward shear stresses τ on each of the four side faces. Find the unit weight of the block if a = 1.33 m, b = 1.15 m, c = 1.40 m, σ = 882 kPa and τ = 272 kPa.

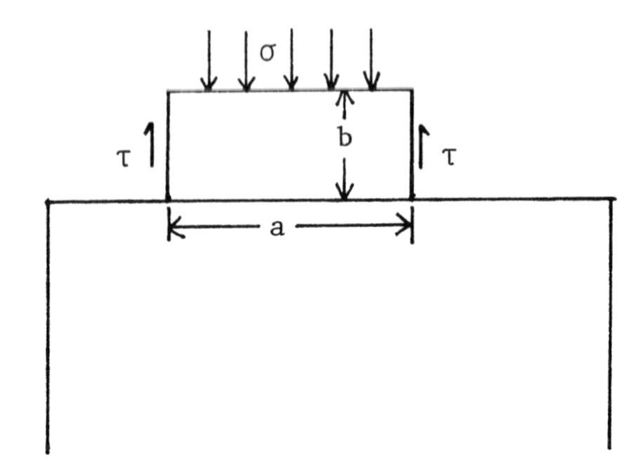

Fig.(2.19) Problem 34

35. The tensile strength of a rock can be measured directly by cementing a metal cap to each end of a cylinder of the rock and applying equal tensile forces F to these caps, which ensure that the force is applied uniformly over the ends of the cylinder; see Fig.(2.20). Diabase rock from New York has a tensile strength of 55.1 MPa, one of the highest tensile strengths of any rock. (a) What tensile force must be applied to a cylinder of this diabase 14.3 cm long and 12.0 cm in diameter to cause failure? (b) Steel has a tensile strength of 365 MPa. Find the diameter of the thinnest steel cylinder that can sustain the tensile force found in (a).

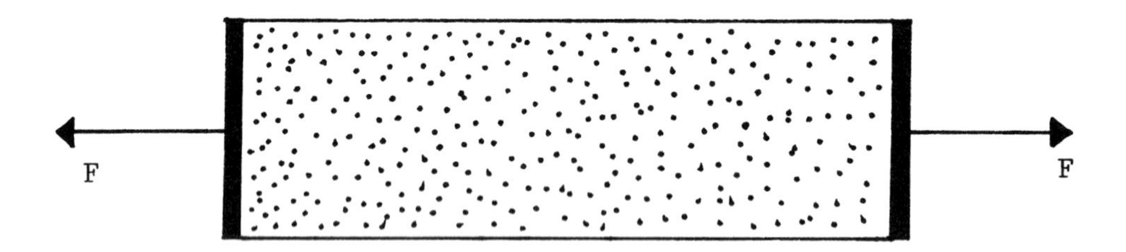

Fig.(2.20) Problem 35

36. The shear strength of steel is 17.2 MPa. A force is applied at 90° to a square steel sheet 1.20 cm thick and 1.84 m in edge length to punch out a hole 1.50 cm in diameter. Find the minimum force needed.

37. At the face of a mountain of granite, large joints (fractures) often develop, the joints being parallel to the face of the mountain. In one such situation shown in Fig.(2.21), a fracture appears at a distance h = 35.0 cm into the face of the mountain. Due to erosion at the foot of the mountain, the outermost sheet of granite, ℓ = 2.55 m in length and w = 1.63 m in width (normal to page), is subject to sliding. The elevation angle of the mountain is θ = 38.0°. The granite has density ρ = 2.64 g/cm^3. Find the smallest shear stress that must exist along the fracture to hold the sheet of granite in place.

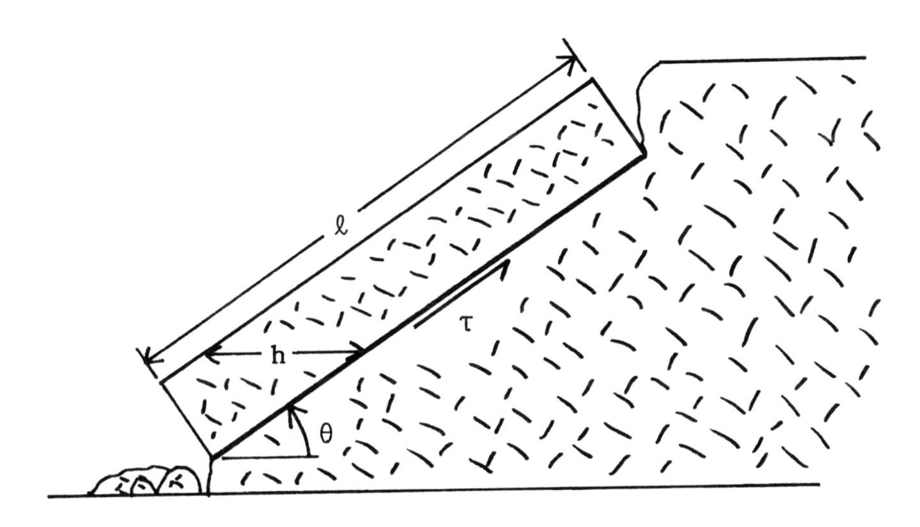

Fig.(2.21) Problem 37

Chapter 3

Rock Deformation

3.1 Uniaxial Load

Chapter 2 describes the unconfined compression test, the unconfined tension test, and the shear test. These tests are performed on rock samples to determine the strength of the rock under the various loading conditions specific to the tests. In each test, the applied force is 'cranked up' to the value needed to trigger rock failure.

In this chapter, the same tests (and one new test) are revisited, but this time in each case the magnitude of the force applied is kept significantly below the value that leads to failure. The focus of attention now is on the change in size and shape of the rock sample, that is, the *deformation* induced by the applied force even though the rock sample does not break.

Start, then, with the unconfined compression and tensile test. The arrangement of the applied forces for the two tests is called a *uniaxial load* because the forces are parallel to the axis of the cylinder. Figure (3.1) shows the two loadings, together with the cylinder when under no load ($F = 0$). Unlike Fig.(2.1), this figure displays the test cylinder and the loading forces only; no supporting equipment (test bench, caps, etc.) are shown.

Figure (3.1a) is of a test cylinder of rock at rest under *no load*; that is, there are no forces being applied to the cylinder. (The force of gravity, the weight of the cylinder, is ignored, as it is much smaller than the forces that must be applied to cause significant deformation.) In this no-load situation, the length of the cylinder is given the symbol L_0, the diameter D_0 and the cross-sectional area A_0.

In Fig.(3.1b) a *uniaxial tensile load* is applied. This means that equal and oppositely directed forces F are simultaneously applied perpendicularly to the ends of the cylinder, the forces being directed so as to tend to pull the cylinder apart. As discussed in Chapter 2, this load is unconfined, since no forces are applied to the sides of the cylinder. As indicated in Fig.(3.1b), the cylinder becomes longer and thinner than the no-load cylinder. The changes in length and diameter are greatly exaggerated on the drawing. It is assumed that the forces are not so large as to bring the cylinder close to rupture.

In Fig.(3.1c) the forces are reversed in direction relative to Fig.(3.1b). In this case, the forces tend to compress the cylinder and therefore this situation is described as a *uniaxial compressive load.* Under such a load, the cylinder of rock becomes shorter and thicker than the no-load cylinder. Again, the deformations are greatly exaggerated in the drawing.

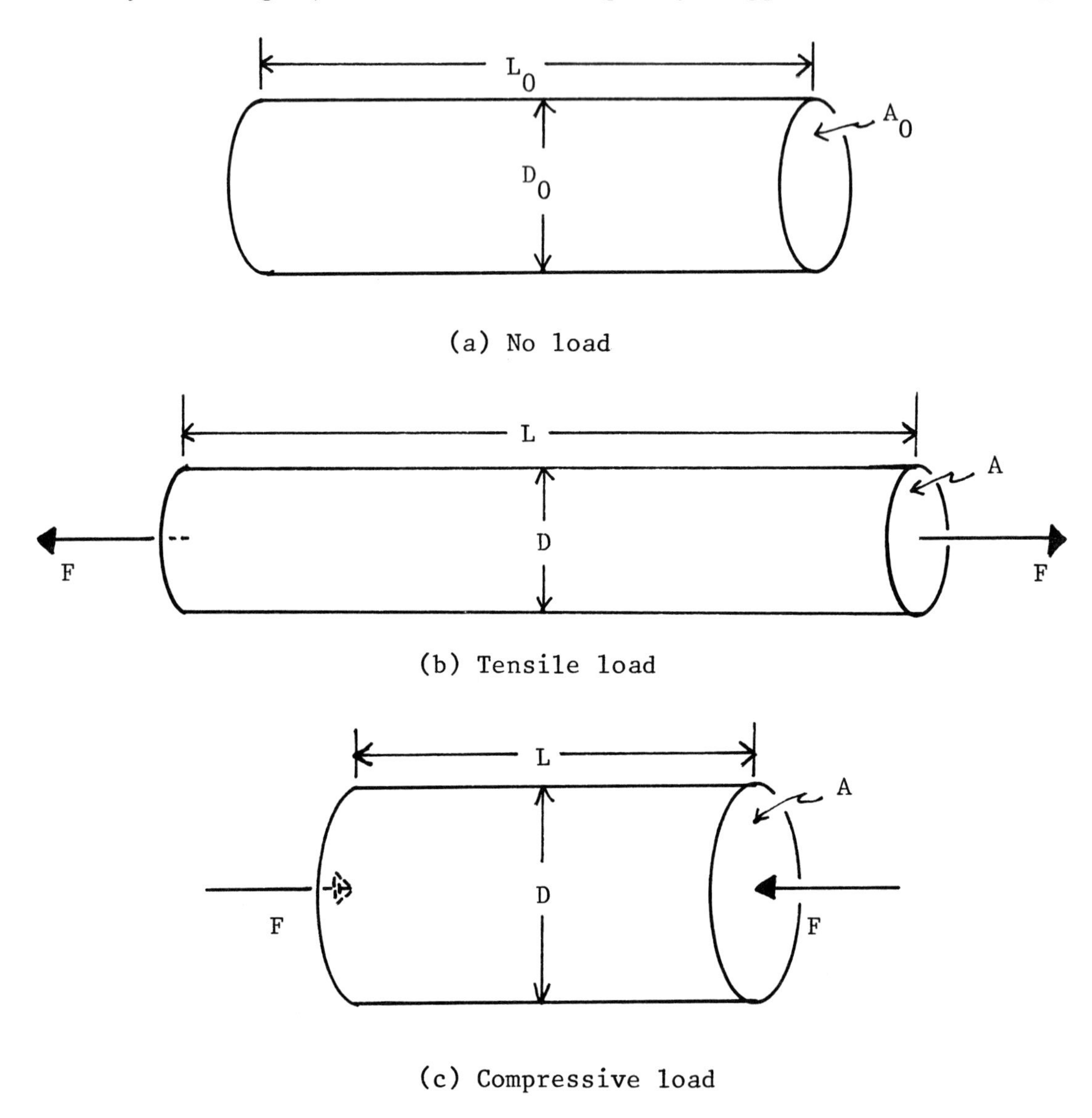

Fig.(3.1) Uniaxial Loads

For both the tensile and compressive load, the length of the cylinder under the load is given the symbol L, the diameter D and the cross-sectional area A (i.e., the same symbols as for the no-load situation, but without the subscripts).

As discussed in Chapter 2, force divided by area equals stress. When the forces are first applied, the cross-sectional area of the cylinder is A_0. But the cylinder eventually assumes a different cross-sectional area A (it takes time for the cylinder to deform). Which area should be used to define the stress? Either the 'initial' area A_0 or the 'final' area A can be used, but the same choice must be maintained throughout a calculation. If the no-load area A_0 is selected, then the stress so calculated is called the *engineering stress*; if the area A under the load is used, then the stress calculated is called the *true stress*. This book uses engineering stress only. That is, the stress σ is defined as

$$\sigma = \frac{F}{A_0}. \tag{3.1}$$

In Chapter 2, the change in cross-sectional area under the load is ignored in discussing the various rock strengths. Hence, in Chapter 2, the area is written as just A, for simplicity in writing.

As a measure of the rock deformation, the *axial strain* ϵ is defined by

$$\epsilon = \frac{L - L_0}{L_0}. \tag{3.2}$$

In terms of the change in length ΔL, where

$$\Delta L = L - L_0, \tag{3.3}$$

the axial strain can be written

$$\epsilon = \frac{\Delta L}{L_0}. \tag{3.4}$$

Engineers often verbalize Eq.(3.4) as "strain equals the fractional change in length".

Note that the denominator in the definition of axial strain is L_0, to be consistent with the denominator A_0 is the definition of stress, Eq.(3.1). [The true strain to be used with true stress F/A is not $\Delta L/L$ but $\ln(\Delta L/L)$. This book works only with engineering stress and strain.]

The SI base unit of stress is the pascal Pa (see Chapter 2). To determine the units of axial strain, observe that the lengths, in this case L and L_0, are to be expressed in the same units (as in any equation). Then Eq.(3.2) shows that the units cancel. Strain is dimensionless: it has no units.

Sometimes, however, strain is written with units, such as m/m, or cm/cm, even though these units cancel. It is difficult to find a reason for this, except tradition. But this is not all. It is not uncommon, in engineering literature, to see strain written with units such as mm/m. Here a rationale can be discerned. It is this: the change in length for real rocks is very small compared with the no-load length. Therefore, the change in length may be expressed in small length units, like mm, whereas the length itself may be in longer units, such as m, yielding mm/m for the units of strain. However, in any equation as written in this book that contains the strain, ϵ must be expressed in dimensionless form.

It is found by actual experiments that, as long as the stress is neither too large nor too small, the stress is directly proportional to the strain; that is

$$\sigma = E\epsilon, \tag{3.5}$$

where E is a fixed number, or constant, for each kind of rock. (Recall that rocks from different localities but of the same geologic type can have different physical properties, and therefore should be considered as different kinds of rock.) This constant of proportionality E is known both as *Young's modulus* and as the *modulus of elasticity*. Since σ has units of Pa and ϵ has no units, Eq.(3.5) indicates that E has units of Pa.

Now, the numerical value of Young's modulus (to pick one of the two names for E) is chosen to be always positive. But the axial strain is negative for a compressive load [see Eq.(3.2) with $L < L_0$]. Hence, for Eq.(3.5) to yield a positive value for E, a negative sign must be attached to σ. Therefore, amend Eq.(3.1) for the stress to read

$$\sigma = \pm\frac{F}{A_0}. \tag{3.6}$$

The plus (+) sign is used for a tensile load, and the minus (−) sign for a compressive load.

****

EXAMPLE 1

An axial compressive load of 37.6 kN is applied to a cylinder of rock with a no-load length of 12.6 cm and a no-load diameter of 4.83 cm. For the cylinder of rock, Young's modulus equals 35.0 GPa. Find the change in length of the cylinder under the load.

The change in length is ΔL and can be found from the strain ϵ; see Eq.(3.4). The strain can be found from Eq.(3.5) and the needed stress is evaluated from Eq.(3.6). Since the units of the load are N (newtons), the term *load* must here refer to the applied force F (see Chapter 2 for a discussion of *load*). For a compressive load, select the minus sign in Eq.(3.6). To obtain the stress in Pa (to agree with the units of Young's modulus E), the no-load diameter must be converted to meters. Start, then, by calculating the stress by Eq.(3.6):

$$\sigma = -\frac{F}{\pi D_0^2/4},$$

$$\sigma = -\frac{37.6 \text{ X } 10^3 \text{ N}}{\pi(0.0483 \text{ m})^2/4},$$

$$\sigma = -2.052 \text{ X } 10^7 \text{ Pa}.$$

Now find the axial strain by Eq.(3.5):

$$\sigma = E\epsilon,$$
$$-2.052 \text{ X } 10^7 \text{ Pa} = (35.0 \text{ X } 10^9 \text{ Pa})\epsilon,$$
$$\epsilon = -5.863 \text{ X } 10^{-4}.$$

Finally, use Eq.(3.4) to calculate the change in length:

$$\Delta L = \epsilon L_0,$$
$$\Delta L = (-5.863 \text{ X } 10^{-4})(12.6 \text{ cm}),$$
$$\Delta L = -7.39 \text{ X } 10^{-3} \text{ cm}.$$

The negative sign indicates that the cylinder is shorter under the load. The change in length could have been found from $\Delta L = L - L_0$ by first finding the length L under the load by Eq.(3.2) and then subtracting the given value of L_0. However, more than four significant figures would have to be carried in the calculation to get 3 sig fig in ΔL, and there would be more steps in the calculation than are really necessary. Hence, this procedure is not recommended.

****

As Fig.(3.1) suggests, a cylinder of rock gets thinner under an axial tensile load and thicker under an axial compressive load. This lateral, or transverse, deformation is expressed by the *transverse strain* ϵ_T, which is defined in strict analogy with the axial strain ϵ. In Eq.(3.7), then, D_0 is the no-load diameter and D is the diameter under the load, so that

$$\epsilon_T = \frac{D - D_0}{D_0}. \tag{3.7}$$

Like the axial strain, the transverse strain is dimensionless.

It turns out that, in the *elastic region*, the range of values of applied stress in which Eq.(3.5) is valid, the ratio ν, defined by

$$\nu = -\frac{\epsilon_T}{\epsilon}, \tag{3.8}$$

has a particular fixed value for each kind of rock. The quantity ν is called *Poisson's ratio.* Since both of the strains ϵ and ϵ_T are dimensionless, ν must be dimensionless also.

Why the minus sign in Eq.(3.8)? Under an axial tensile load, $\epsilon > 0$ since $L > L_0$, but $\epsilon_T < 0$ since $D < D_0$ (the cylinder gets longer but thinner). With the minus sign, Eq.(3.8) yields $\nu > 0$. With an axial compressive load, however, $\epsilon < 0$ since $L < L_0$ and $\epsilon_T > 0$ since $D > D_0$. But again, Eq.(3.8) gives $\nu > 0$. In short, the minus sign in Eq.(3.8) ensures that Poisson's ratio is a positive quantity.

****

EXAMPLE 2 A cylinder of rock with a Poisson's ratio of 0.422 has a no-load length equal to 17.6 cm and a no-load diameter of 6.20 cm. Under an axial compressive load the axial strain is found to be −3.44 mm/m. Find the change in diameter under the load.

The transverse strain can be found from the axial strain and Poisson's ratio. However, first the axial strain must be put into dimensionless form, as follows:

$$\epsilon = -3.44 \text{ mm/m},$$
$$\epsilon = -3.44 \text{ X } 10^{-3} \text{ m/m},$$
$$\epsilon = -3.44 \text{ X } 10^{-3}.$$

From Eq.(3.8),

$$\epsilon_T = -\nu\epsilon,$$
$$\epsilon_T = -(0.422)(-3.44 \text{ X } 10^{-3}),$$
$$\epsilon_T = 1.452 \text{ X } 10^{-3}.$$

Now find the change in diameter by rewriting Eq.(3.7) in a manner analogous to Eq.(3.4), and then solve for the change in diameter to get

$$\Delta D = \epsilon_T D_0,$$
$$\Delta D = (1.452 \text{ X } 10^{-3})(6.20 \text{ cm}),$$
$$\Delta D = 9.00 \text{ X } 10^{-3} \text{ cm},$$
$$\Delta D = 90.0\ \mu\text{m}.$$

In the last line, μm (*micron*) is defined by 1 μm = 1 X 10^{-6} m.

****

3.2 Volumetric Strain

Since the diameter and length of the test cylinder of rock both change under an axial load, it is likely that the volume of the cylinder changes also under the load. For a circular cylinder, the no-load volume V_0 is

$$V_0 = \frac{1}{4}\pi D_0^2 L_0. \tag{3.9}$$

The volume V of the cylinder under the applied load is

$$V = \frac{1}{4}\pi D^2 L. \tag{3.10}$$

Equation (3.10) assumes that the cylinder of rock remains a cylinder under the load; this is observed to be the case as long as the applied stress does not approach the strength of the rock . To express V in terms of V_0, and therefore be able to compare the volumes under no load and under a load, write L in terms of L_0 through the axial strain, and D in terms of D_0 through the transverse strain. Equations (3.2) and (3.7) can be written as

$$L = L_0(1 + \epsilon),$$

$$D = D_0(1 + \epsilon_T).$$

Substituting these two equations into Eq.(3.10) and invoking Eq.(3.9) yields

$$V = \frac{1}{4}\pi[D_0(1+\epsilon_T)]^2[L_0(1+\epsilon)],$$

$$V = \frac{1}{4}\pi D_0^2(1+\epsilon_T)^2 L_0(1+\epsilon),$$

$$V = (\frac{1}{4}\pi D_0^2 L_0)(1+\epsilon_T)^2(1+\epsilon),$$

$$V = V_0(1+\epsilon_T)^2(1+\epsilon),$$

$$V = V_0(1+\epsilon+2\epsilon_T+2\epsilon_T\epsilon+\epsilon_T^2+\epsilon_T^2\epsilon). \tag{3.11}$$

In actual engineering situations, the strains are small. Strains larger than about 0.001 seldom are encountered before failure of the rock. [Thus, the changes in the dimensions of the cylinders shown in Fig.(3.1) are greatly exaggerated.] Hence, at most

$$\epsilon \approx 1 \text{ X } 10^{-3},$$
$$\epsilon_T \approx 1 \text{ X } 10^{-3},$$
$$\epsilon_T\epsilon \approx 1 \text{ X } 10^{-6},$$
$$\epsilon_T^2 \approx 1 \text{ X } 10^{-6},$$
$$\epsilon_T^2\epsilon \approx 1 \text{ X } 10^{-9}.$$

In practical situations, the last three terms in Eq.(3.11) are negligible, so set them equal to zero. Equation (3.11) now reduces to

$$V = V_0(1+\epsilon+2\epsilon_T).$$

This equation can be rearranged, using Eq.(3.8), as follows:

$$V = V_0 + V_0(\epsilon+2\epsilon_T),$$

$$\frac{V-V_0}{V_0} = \epsilon+2\epsilon_T,$$

$$\frac{V-V_0}{V_0} = \epsilon(1+2\frac{\epsilon_T}{\epsilon}),$$

$$\frac{V-V_0}{V_0} = \epsilon(1-2\nu). \tag{3.12}$$

The quantity on the left of Eq.(3.12) looks like a volume strain [compare with Eq.(3.2) for the axial strain]; this fractional change in volume is called the *volumetric strain*, or the *dilatation*. This book does not give a special symbol for the volumetric strain, but leaves Eq.(3.12) as it stands.

Now examine this special case: suppose that a rock is *incompressible* . The volume of a test cylinder of an incompressible rock does not change under a load, so that $V = V_0$. This does not mean that the length and diameter do not change. However, since the volumetric strain is zero, Eq.(3.12) reduces to

$$0 = \epsilon(1 - 2\nu).$$

Since $\epsilon \neq 0$,

$$1 - 2\nu = 0,$$

so that

$$\nu = \frac{1}{2}.$$

However, no material is truly incompressible, so $\nu < 1/2$. Also, lateral strains are not zero in unconfined uniaxial loading, so $\nu = 0$ does not occur. The range of values of Poisson's ratio for ordinary rocks is $0 < \nu < \frac{1}{2}$.

The quantities E (Young's modulus or modulus of elasticity) and ν (Poisson's ratio) are two *elastic moduli* of the rock. The elastic moduli are of fundamental engineering importance since their values determine the amount of deformation of the rock under an applied load.

****

EXAMPLE 3 A cylinder of rock under no load has a volume of 4844.00 cm^3. Under an applied axial load the volume becomes 4845.13 cm^3. For this rock Poisson's ratio equals 0.416. Find the axial strain under the load. Is the load compressive or tensile?

Since both the no-load volume V_0 and the volume under the load V are expressed in the same units (cm^3), these units will cancel when they are substituted into the left side of Eq.(3.12), so there is no need to change from cm^3 to the SI base unit m^3. Hence, Eq.(3.12) gives

$$\frac{V - V_0}{V_0} = \epsilon(1 - 2\nu),$$

$$\frac{4845.13 - 4844.00}{4844.00} = \epsilon[1 - 2(0.416)],$$

$$\epsilon = 1.39 \text{ X } 10^{-3}.$$

Since $\epsilon > 0$, the load is tensile.

****

3.3 Shear Load

In uniaxial loading, the forces F are applied at 90° to the surfaces on which they act. Now consider *shear load*, the forces being applied parallel to the surfaces on which they act.

Figure (3.2) shows a cylinder of rock under no load (dashed lines) and under a shear load (solid lines). At its base the cylinder is fixed, so that the base cannot move. The cylinder has length L and cross-sectional area A. The diagram is of a rectangular cylinder, rather than the circular cylinder used in discussing axial load: the shape of the cross section of the cylinder makes no diffference to the results in either case.

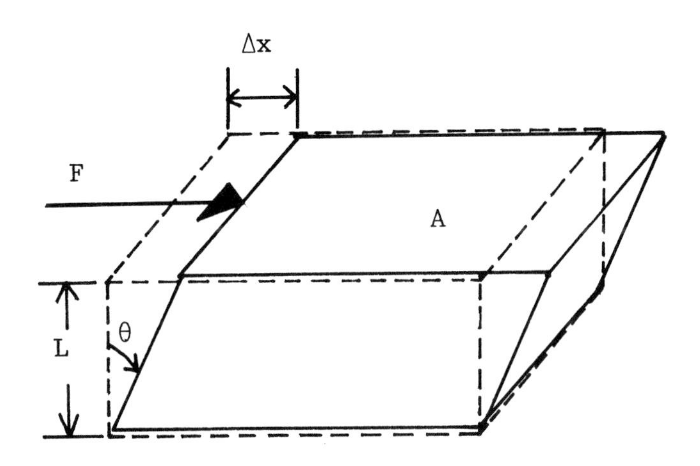

Fig.(3.2) Shear Load

In shear loading, a force F is applied on, and parallel to, a surface of the rock; in Fig.(3.2), the force is applied to the top surface. Since the base of the cylinder is fixed, in the ground, say, there must be an equal and opposite force F exerted on the bottom face by the ground; this force is not shown in Fig.(3.2). Under the action of the force shown, the top surface of the rock cylinder will be pushed aside a small distance Δx. If bending is avoided, the sides of the cylinder remain straight and the cross-sectional area A does not change, unlike the deformation in axial loading. The cylinder under load makes an angle θ with the vertical. The shape of the cylinder changes; under the load it is no longer a right cylinder, but is oblique.

As described in Chapter 2, the shear stress associated with the shear force is defined as force per unit area, just as for axial stress. That is, the shear stress τ is defined by

$$\tau = \frac{F}{A}, \tag{3.13}$$

where the force F is parallel to the surface on which it acts. The units of shear stress are $\mathrm{N/m^2}$, or Pa, just as for axial stress. Unlike axial loading, there is no need to use + and − signs for the shear stress, since the situation in Fig.(3.2) is unchanged if the direction of the force is reversed (as though the figure is viewed from behind). The shear stress is taken to be positive.

The *shear strain* associated with the shear stress is defined as the deflection angle θ of the loaded cylinder, as shown in Fig.(3.2). This angle is expressed in radians (rad). From the right triangle in Fig.(3.2), it is seen that $\tan\theta = \Delta x/L$. Now, shear strains encountered in actual engineering situations are very small (as are axial and transverse strains in axial loading). That is, θ is a very small angle. For a very small angle expressed in radians, $\tan\theta \approx \theta$. Therefore, for all practical purposes, the shear strain θ is given by

$$\theta = \frac{\Delta x}{L}. \tag{3.14}$$

As with uniaxial loading, experiments performed on cylinders of rock in which strains resulting from applied stresses are measured show that, provided that the stresses do not approach the shear strength, the stress is proportional to the strain. That is, an equation similar to Eq.(3.5) for uniaxial loading can be written for shear loading. It would be very convenient mathematically if the proportionality constant had the same numerical value for both loading modes, but this is not the case. Using the symbol G for the new proportionality constant, experiment shows that for shear loading

$$\tau = G\theta. \tag{3.15}$$

The proportionality constant G is called the *shear modulus*; it is also known as the *modulus of rigidity*. Each type of rock has a particular value of the shear modulus G. The units of G are the units of τ divided by units of θ. The unit of τ are Pa; θ expressed in radians has no units. Hence, the units of G are Pa (as are the units of E).

In Chapter 1, G is used as the symbol of specific gravity. There is no standardization of symbols in engineering geology, but G is a common choice for shear modulus. Therefore, when the symbol G is encountered, the context in which it is being used, and/or the units (if any), will reveal its identity as either specific gravity or shear modulus.

It is often convenient to work with the applied force F and linear deflection Δx, rather than the angular deflection θ. For this purpose, Eqs.(3.13), (3.14), (3.15) can be combined to obtain the relation

$$\frac{F}{A} = G(\frac{\Delta x}{L}). \tag{3.16}$$

****

EXAMPLE 4 Granite from Quincy, Massachusetts has a shear modulus of 34.5 GPa. A block of Quincy granite with edge lengths 0.722 m, 1.200 m, 1.330 m rests on a horizontal surface with its shortest dimension perpendicular to the surface. The base is fixed. A horizontal force of 22.8 MN is applied to the top surface of the block. Find (a) the linear deflection and (b) the angular deflection, in degrees, of the block.

(a) The surface to which the force is applied has area $A =$ (1.200 m)(1.330 m), so that $A =$ 1.596 m^2. The distance L is the edge length perpendicular to the surface on which the force acts; in this case, $L = 0.722$ m. Equation (3.16) gives

$$\frac{F}{A} = G(\frac{\Delta x}{L}),$$

$$\frac{22.8 \text{ X } 10^6 \text{ N}}{1.596 \text{ m}^2} = (34.5 \text{ X } 10^9 \text{ Pa})(\frac{\Delta x}{0.722 \text{ m}}),$$

$$\Delta x = 0.299 \text{ mm}.$$

(*b*) Use Eq.(3.14) to get the angular deflection in radians, and then convert to degrees. Since π rad $= 180°$,

$$\theta = \frac{\Delta x}{L},$$

$$\theta = \frac{0.299 \text{ X } 10^{-3} \text{ m}}{0.722 \text{ m}},$$

$$\theta = 4.141 \text{ X } 10^{-4} \text{ rad},$$

$$\theta = (4.141 \text{ X } 10^{-4} \text{ rad})(180°/\pi \text{ rad}),$$

$$\theta = 0.0237°.$$

****

3.4 Bulk Loading

Uniaxial and shear loading cannot be applied to ideal liquids, that is, those devoid of viscosity or friction, since it is not possible to have a free-standing column of liquid. For liquids, the moduli E and ν have no meaning, and $G = 0$. However, both solids and liquids can undergo bulk loading.

In uniaxial and shear loading on a rock cylinder, equal and opposite forces are applied on only two opposite faces of the cylinder. In *bulk loading*, forces are applied at 90° to the entire surface of the cylinder. Futhermore, the resulting stress has the same value at all points on the surface, and is either all compressive or all tensile. Since the force is applied perpendicular to the surface, σ is used as the symbol for the stress. Bulk loading is nearly always compressive. Sometimes, instead of stress σ, the term *pressure* p is used when discussing bulk loading. In fact, pressure is the preferred term when the load is applied to a liquid.

Figure (3.3a) shows a rectangular cylinder under no load; the volume of the cylinder is V_0. In Fig.(3.3b) a compressive bulk loading is applied: equal compressive normal stresses are applied to each of the six faces of the cylinder. Since the load surrounds the cylinder (there are no unconfined surfaces), the cylinder is compressed in all directions, resulting in a cylinder with smaller volume V. As in Figs.(3.1) and (3.2), this deformation is exaggerated in Fig.(3.3).

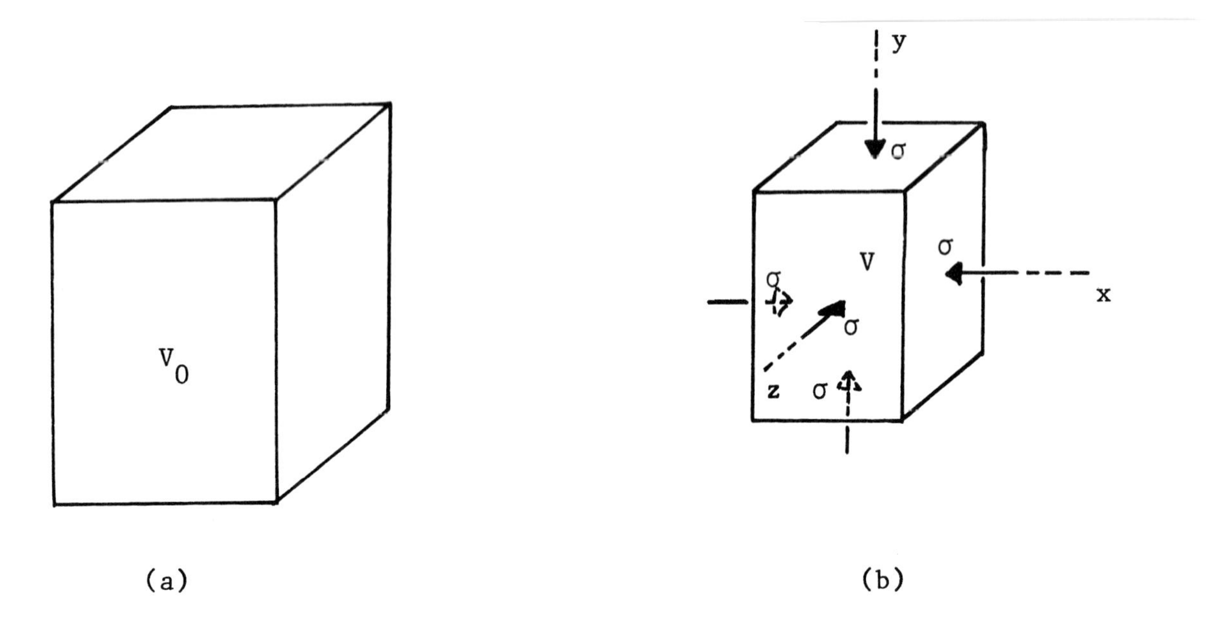

Fig.(3.3) (a) No load. (b) Bulk Load.

The strain associated with bulk loading is, therefore, a volumetric strain $\Delta V/V_0$. Note that Eq.(3.12) cannot be used here, for that equation applies only to uniaxial loading. As with uniaxial and shear loading, experiment shows that the applied stress, if not too large, is proportional to the resulting strain. Expressed mathematically, this means that

$$\sigma = -k(\frac{\Delta V}{V_0}), \tag{3.17}$$

where

$$\Delta V = V - V_0. \tag{3.18}$$

The constant of proportionality k is called the *bulk modulus* of the material making up the cylinder. Since the strain $\Delta V/V_0$ has no units, k must have the same units as stress; i.e., Pa.

A feature of Eq.(3.17) not found in uniaxial or shear loading is the negative sign. For compressive bulk loading, σ is taken to be positive. But in this (common) case, $\Delta V < 0$, since the cylinder shrinks. The presence of the minus sign ensures that k is positive.

The bulk modulus of water at 20°C and atmospheric pressure is 2.18 GPa. For some values for rock, see the examples and problems.

Sometimes, in place of the bulk modulus k, the *compressibility* K is used. The compressibility is defined by

$$K = \frac{1}{k}, \tag{3.19}$$

and has units of Pa^{-1}. In this book the bulk modulus, rather than the compressibility, is used.

****

EXAMPLE 5 Vermont marble has a bulk modulus of 71.9 GPa and a density of 2.70 g/cm^3. A block of Vermont marble has edge lengths 1.22 m, 1.08 m, 1.43 m in place at a depth of 130 m in a quarry. The block is excavated and removed to the surface. By how much does the volume of the quarried block exceed the volume of the block in situ? Assume that the lithostatic stress acts on all surfaces of the block in place in the quarry.

The increase in volume of the block when the stress is relieved in the mining process is the same as the decrease in volume that would occur if a similar block at the surface is subjected in the laboratory to a bulk stress equal to the lithostatic stress at a depth of 130 m in rocks of the given density. That is, set σ in Eq.(3.17) equal to the lithostatic stress given by Eq.(2.5) to obtain

$$\rho g h = -k\left(\frac{\Delta V}{V_0}\right).$$

Since $V_0 = 1.884\ \text{m}^3$, this equation gives

$$(2700\ \text{kg/m}^3)(9.8\ \text{m/s}^2)(130\ \text{m}) = -(71.9 \text{ X } 10^9\ \text{Pa})\left(\frac{\Delta V}{1.884\ \text{m}^3}\right),$$

$$\Delta V = -90.1\ \text{cm}^3.$$

This is the decrease in volume as the stress is applied to a block originally under no load. If the stress is removed from a block under load, as when a block is quarried, then the increase in volume is equal to 90.1 cm^3. Care must be taken in such mining operations to relieve the stress slowly; a sudden release of stress could result in a dangerous rock burst as the rock responds.

****

3.5 Four Elastic Moduli

Up to this point, four elastic moduli have been introduced, each describing a deformation of a rock under a specific loading mode. These moduli are, with their units,

E (Young's modulus; modulus of elasticity), Pa,
ν (Poisson's ratio), no units,
G (shear modulus; modulus of rigidity), Pa,
k (bulk modulus), Pa.

Now it turns out that these four moduli can be grouped into two sets of two each. Specifically, E and ν constitute one set, and G and k the other. The two sets are related: given E and ν, the values of G and k can be calculated, and vice versa.

For example, consider the bulk loading of Fig.(3.3b). The resulting deformation is the same as the accumulated deformation due to three successive uniaxial loadings, one each along the x, y, z axes. For each of these individual uniaxial loadings, Eq.(3.12) does apply. For instance, due to a uniaxial loading along the x axis, the resulting change in volume ΔV_x is given by

$$\Delta V_x = \epsilon_x(1 - 2\nu)V_0. \tag{3.20}$$

If the cylinder of rock then undergoes a uniaxial loading along the y axis, and then a uniaxial loading along the z axis, the successive changes in volume are given by

$$\Delta V_y = \epsilon_y(1 - 2\nu)V_0, \tag{3.21}$$

$$\Delta V_z = \epsilon_z(1 - 2\nu)V_0. \tag{3.22}$$

It might be objected that the V_0 in Eq.(3.21) should be replaced with $V_0 + \Delta V_x$, since the uniaxial loading in the x direction has changed the volume of the cylinder before application on the uniaxial load in the y direction. Similarly, the V_0 in Eq.(3.22) should be replaced with $V_0 + \Delta V_x + \Delta V_y$. But these substitutions, if made, would introduce a quadratic term, $\epsilon_x\epsilon_y$, in the strains when, say, Eq.(3.20) is substituted for ΔV_x, and $V_0 + \Delta V_x$ is substituted for V_0 in Eq.(3.21). It has previously been remarked that such quadratic terms are too small to be of importance in actual situations. Hence, leave Eqs.(3.21) and (3.22) as written above.

The total change in volume due to the three uniaxial loadings is

$$\Delta V = \Delta V_x + \Delta V_y + \Delta V_z.$$

Using Eqs.(3.20), (3.21), (3.22), this can be expressed as

$$\Delta V = (\epsilon_x + \epsilon_y + \epsilon_z)(1 - 2\nu)V_0. \tag{3.23}$$

By Eq.(3.5), write

$$\sigma_x = E\epsilon_x,$$

$$\sigma_y = E\epsilon_y,$$

$$\sigma_z = E\epsilon_z.$$

But in bulk loading, the stresses in the three directions are equal; that is

$$\sigma_x = \sigma,$$

$$\sigma_y = \sigma, \tag{3.24}$$

$$\sigma_z = \sigma.$$

But this means that the strains also are equal:

$$\epsilon_x = \epsilon,$$

$$\epsilon_y = \epsilon,$$

$$\epsilon_z = \epsilon,$$

where

$$\epsilon = \frac{\sigma}{E}.$$

Putting all this together, Eq.(3.23) becomes

$$\frac{\Delta V}{V_0} = 3(\frac{\sigma}{E})(1 - 2\nu). \tag{3.25}$$

Before comparing Eq.(3.25) with Eq.(3.17), a minus sign must be attached. This is because in uniaxial loading, compressive stress is taken to be negative, but in bulk loading a compressive stress is considered positive. Doing this, and then setting Eqs.(3.17) and (3.25) equal will lead to

$$-\frac{\sigma}{k} = -3(\frac{\sigma}{E})(1 - 2\nu),$$

$$k = \frac{E}{3(1 - 2\nu)}. \tag{3.26}$$

Thus, the value of k can be calculated from the values of E and ν.

The shear modulus can also be calculated from E and ν. The derivation of the equation for G is more tedious than the derivation above of the equation for k and is omitted here. The equation itself is

$$G = \frac{E}{2(1 + \nu)}. \tag{3.27}$$

To summarize: the values of the four moduli are not independent. Given E and ν, the values of k and G can be found from Eqs.(3.26) and (3.27).

Of course, Eqs.(3.26) and (3.27) can be inverted; that is, rewritten in a form suitable for the case where the values of G and k are known, and it is E and ν that need to be found. To invert the equations, solve for E in Eq.(3.26) and substitute into Eq.(3.27) to get

$$G = \frac{3k(1 - 2\nu)}{2(1 + \nu)}.$$

Only elementary algebra skills are needed to solve this equation for ν; the result is

$$\nu = \frac{3k - 2G}{6k + 2G}. \tag{3.28}$$

Finally, substitute Eq.(3.28) into Eq.(3.26) and rearrange to obtain

$$E = \frac{9kG}{3k + G}. \tag{3.29}$$

****

EXAMPLE 6 Limestone from the town of Solenhofen in Germany has a Poisson's ratio of 0.276 and a Young's modulus equal to 63.0 GPa. Find the expected values of the other moduli.

The other moduli are the shear modulus G and the bulk modulus k. The values $\nu = 0.276$ and $E = 63.0$ GPa are given, so that Eqs.(3.26) and (3.27) are appropriate. For the shear modulus, use Eq.(3.27):

$$G = \frac{E}{2(1+\nu)},$$

$$G = \frac{63.0 \text{ GPa}}{2(1+0.276)},$$

$$G = 24.7 \text{ GPa}.$$

The bulk modulus is found from Eq.(3.26):

$$k = \frac{E}{3(1-2\nu)},$$

$$k = \frac{63.0 \text{ GPa}}{3[1-2(0.276)]},$$

$$k = 46.9 \text{ GPa}.$$

****

3.6 Limiting Cases

The equations connecting the elastic moduli, Eqs.(3.26), (3.27), (3.28), (3.29), can be used to obtain constraints on the numerical values that the moduli can take. One such constraint has already been found for Poisson's ratio: from Eq.(3.12) it has been concluded that

$$0 < \nu < \frac{1}{2}, \tag{3.30}$$

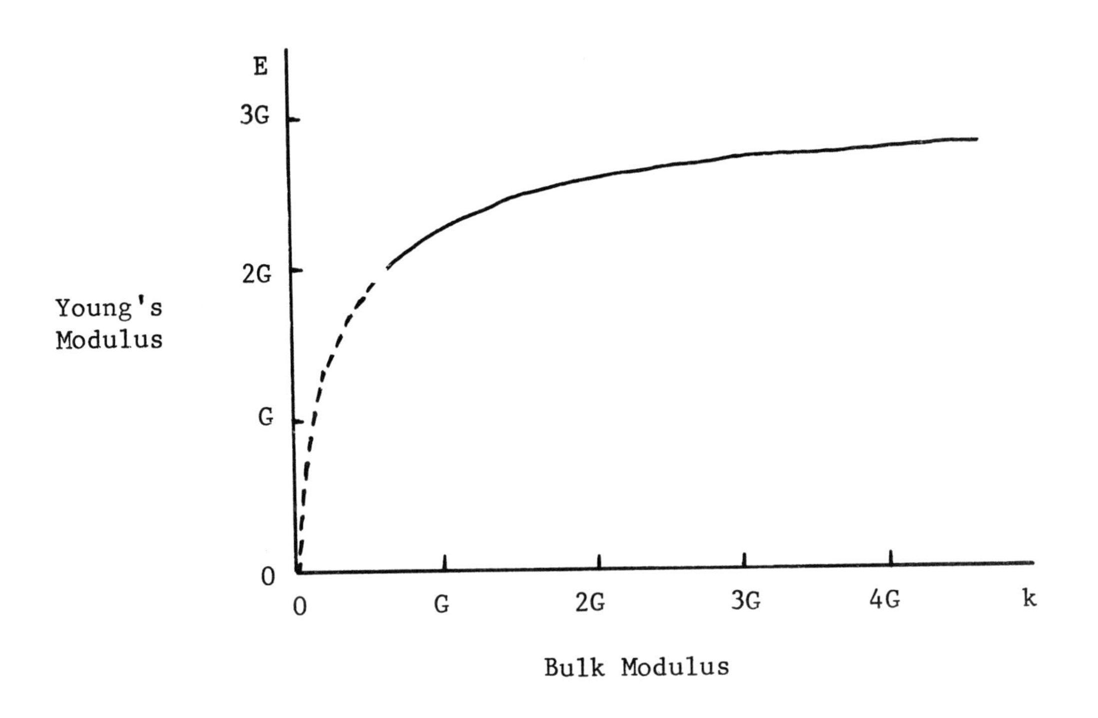

Fig.(3.4) Young's Modulus As a Function of Bulk Modulus

Now consider the bulk and shear moduli as they appear in Eq.(3.28). Since $\nu > 0$, Eq.(3.28) predicts that

$$3k - 2G > 0,$$

$$k > \frac{2}{3}G. \tag{3.31}$$

Substitute $k = \frac{2}{3}G$ into Eq.(3.29); the resulting equation yields $E = 2G$. But $k > \frac{2}{3}G$. Does this mean that $E > 2G$ or $E < 2G$? To answer this question, evaluate the derivative of E with respect to k, holding G fixed, to find

$$\frac{\partial E}{\partial k} = (\frac{3G}{3k + G})^2.$$

Since the derivative can be expressed as a real quantity squared, this ensures that

$$\frac{\partial E}{\partial k} > 0, \tag{3.32}$$

and therefore $k > \frac{2}{3}G$ implies that $E > 2G$.

From "Eq.(3.31)" (not really an equation but an inequality), it is evident that $k > \frac{2}{3}G$. Is there an upper limit to the value of k relative to G to go with this lower limit? Look at Eq.(3.28) and substitute $G = 0$, to find that $\nu = \frac{1}{2}$. Hence, if $k \gg G$, then $\nu \approx \frac{1}{2}$. Does this reveal anything about E? Turn to Eq.(3.29). If $k \gg G$, then

$$3k + G \approx 3k,$$

and Eq.(3.29) gives $E = 3G$. Equation (3.32) then implies that, in reality, $E < 3G$. Earlier, it was found that $E > 2G$. Hence, relative to G, the possible values of Young's modulus E are restricted to the range

$$2G < E < 3G. \tag{3.33}$$

The behaviour of E as a function of k is shown on the graph of Fig.(3.4). Only the solid part of the line represents physically realistic, as opposed to mathematically possible, values.

"Equations (3.30), (3.31), (3.33)" are examples of limiting cases. Such relations may or may not be useful in specific situations. For instance, if a new kind of rock is discovered, or a new material fabricated, and the measured values of the elastic moduli do not satisfy these inequalities, then it may be concluded either that errors were made in the lab, or that the rock is not isotropic, homogenous, and elastic (as assumed in this chapter), or both.

3.7 Problems

1. A cylinder of rock with a Young's modulus of 72 GPa is put under an axial compressive load of 180 MPa. The no-load length of the cylinder is 22 cm. By how much does the length change under the load?

2. A cylinder of rock with a no-load length of 7.826 cm has a length of 7.844 cm under an axial tensile load. The transverse strain under the same load is $-9.500 \text{ X } 10^{-4}$. Calculate Poisson's ratio for this rock.

3. A cylinder of rock with a no-load length of 77.73 cm has an axial strain of $-2.600 \text{ X } 10^{-3}$ when under an axial compressive load. Find the length under the load.

4. Calculate the tensile stress on a cylinder of rock with a no-load diameter of 14.3 cm under a load of 38.6 kN.

5. A slate from Michigan has a tensile strength of 25.5 MPa. Find the greatest tensile force that can be applied to a cylinder of slate with a diameter of 16.0 cm.

6. A cylinder of rock is subjected to an axial tensile load of 41.20 kN. The no-load length of the cylinder is 23.360 cm. The no-load diameter is 3.881 cm and the diameter under the load is 3.879 cm. For this rock, Young's modulus is 26.30 GPa. Calculate (*a*) the stress on the cylinder, (*b*) the change in length under the load, (*c*) the transverse strain, and (*d*) Poisson's ratio of the rock.

7. A cylinder of rock has a no-load diameter of 12.2 cm and a no-load length of 236.0 cm. Young's modulus for this rock is 53.7 GPa and Poisson's ratio equals 0.421. The rock is subjected to an axial compressive load of 159 MPa. Find (*a*) the change in length under the load, and (*b*) the change in diameter.

8. A cylindrical sample of rock has the following dimensions: no-load length 1.600 m, no-load diameter 8.30 cm; length under load 1.576 m, diameter under load 8.36 cm. The axial load is 622 kN. Find (*a*) Poisson's ratio, and (*b*) Young's modulus.

9. An Indiana limestone has a Young's modulus equal to 27.0 GPa. Under an axial compressive load the axial strain on a cylinder of this limestone is −1.80 mm/m. Find the axial stress on the cylinder.

10. A cylinder of rock has a no-load length of 27.60 cm and a no-load diameter of 6.83 cm.

For this rock, Young's modulus equals 25.4 GPa and Poisson's ratio is 0.276. The cylinder is put under an axial tensile load of 52.7 kN. Find the changes in (*a*) length, and (*b*) diameter under the load.

11. A cylindrical column of rock has a length of 4.633 m and a diameter of 5.738 cm when under no load. But, under an axial compressive load of 72.6 MPa, the length decreases to the value 4.628 m and the diameter becomes 5.741 cm. Find (*a*) Poisson's ratio for the rock forming the column, and (*b*) the modulus of elasticity of the rock.

12. Under an axial compressive load of 462.0 kN, the diameter of a cylindrical sample of rock increases from 7.559 cm to 7.562 cm. The no-load length of the cylinder is 89.62 cm. For this rock, Poisson's ratio is 0.304. Find (*a*) the change in length of the cylinder under the load, (*b*) Young's modulus of the rock, and (*c*) the change in volume under the load.

13. Under no load a cylinder of rock has a volume of 435.8 cm^3. For this rock, Poisson's ratio equals 0.327. Under an axial load, the axial strain is $-2.56 \text{ X } 10^{-3}$. Find the change in volume under the load.

14. A cylinder of sandstone has a no-load length of 4.826 m and a no-load diameter equal to 63.65 cm. Poisson's ratio is 0.3440 and Young's modulus has the value 21.37 GPa. Under an applied compressive load the length becomes 4.819 m. Calculate (*a*) the change in diameter under the load, and (*b*) the value of the applied force.

15. A cylindrical rock column under no load has a length of 3.522 m and a diameter equal to 5.73 cm. Under an axial compressive load of 39.5 kN, the length becomes 3.510 m. Poisson's ratio is 0.320. Find the modulus of elasticity.

16. A granitic rock cylinder has a no-load length equal to 14.30 cm and a no-load diameter of 4.63 cm. Under an axial compressive load of 51.7 kN, the length decreases by 0.162 mm. The rock has a Poisson's ratio of 0.322. Find (*a*) the change in diameter under the load, and (*b*) Young's modulus of the rock.

17. Some "exotic" materials have a negative value of Poisson's ratio: they get thinner under an axial compressive load! The minimum negative value of Poisson's ratio is for a material that, under an applied axial load, maintains the same ratio of length/diameter (i.e., keeps the same shape). Show that, in this case, (*a*) $\nu = -1$, and (*b*) $\Delta V/V_0 = 3\epsilon$ for small axial strains.

18. Find a formula for density strain $\Delta\rho/\rho_0$ under uniaxial loading.

19. Calculate the shear stress needed to give a rock with a shear modulus of 37.5 GPa a shear strain equal to 0.280°.

20. A 158 cm high column of limestone stands vertically with its base fixed to the ground. The column has a cross-sectional area of 1.26 m^2. The limestone has a shear modulus of 17.3 GPa and a Young's modulus of 41.3 GPa. (*a*) Find the horizontal force that must be applied to the top surface of the column in order to obtain a 1.40 mm horizontal deflection of this surface. (*b*) Find Poisson's ratio of the limestone.

21. A particular sandstone has a Young's modulus equal to 76.5 GPa and a Poisson's ratio equal to 0.170. Find the confining pressure needed to reduce the volume of a 1.820 m^3 block of this sandstone by 12.0 cm^3.

22. Calculate the compressibility of a rock with a shear modulus of 31.6 GPa and a Young's modulus of 78.2 GPa.

23. A block of rock in situ at a depth of 1.60 km measures 1.40 m, 1.22 m, 2.74 m in edge lengths. The density of the subsurface rocks is 2.84 g/cm^3. The rock has a bulk modulus of 87.0 GPa. Find the change in volume of the block when brought to the surface.

24. Gabbro rock from French Creek, Pennsylvania has a bulk modulus of 88.5 GPa and a Young's modulus of 104.3 GPa. Find the pressure needed to compress this rock from a cube with edge length 38.72 cm to a cube with edge length 38.16 cm.

25. Show that the values of the elastic moduli for the limestone from the town of Solenhofen in Germany (see Example 6) fall within the theoretically required ranges, where applicable.

26. A rock is said to be *perfectly elastic* if its bulk modulus equals its modulus of elasticity. (*a*) Find the corresponding value of Poisson's ratio. (*b*) Express the modulus of elasticity in terms of the modulus of rigidity.

27. A quartzitic sandstone has a bulk modulus of 41.7 GPa and a shear modulus with a value of 42.8 GPa. Find the values of the other moduli.

28. For a rock whose value of Young's modulus is in the middle of its allowable range relative to the shear modulus, show that Poisson's ratio has the value in the middle of its allowable range.

Chapter 4

Stability of Rock Slopes

4.1 Non-Engineered Slopes

The Earth's solid surface is not everywhere flat. Hills, mountains, and valleys, for example, are local features that incorporate hilly topography. The ever-present force of gravity, acting on material resting on an inclined surface, tends to induce landslides, rock slides, avalanches, etc. These events present immediate hazards to nearby life and property. Hence, it is usually a serious concern to see to it that a local arrangement of earth materials on a slope does not pass from a static to a dynamic state. Engineering intervention may be necessary if the risk of sliding is significant. Just how to assess this risk, and then to reduce it, is the theme of this chapter.

Examine, then, a block of intact rock resting on an inclined surface or slope. *Intact* rock means that the rock has the strength to resist rupture under the applied forces, and that if these forces set the block in motion, the rock will slide, or at least begin to slide, as a single entity or block. It is assumed that there is no rock mass immediately downslope from the block to prevent sliding (the block *daylights into free space*).

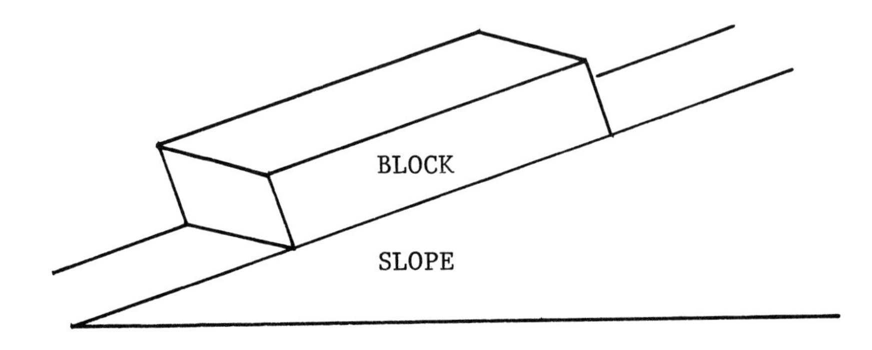

Fig.(4.1) Block on Slope

In Fig.(4.1) a block of rock is shown resting on a slope. This sketch is an idealization of an actual field situation; for example, the actual block may not be a perfectly rectangular solid in shape, and there may be other blocks of rock in close proximity upslope, or to the sides. It is also assumed that the slope itself is an intact block, and that all the rock is dry.

The block is at rest. The immediate query is: How stable is the block in this position? Can the possibility of the block sliding at some time in the future be quantitatively evaluated?

To answer these questions, it is necessary to examine the forces acting on the block. These forces are conveniently divided into two groups: *driving forces*, forces tending to move the block down the plane, and *resisting forces*, forces tending to hold the block in place.

Begin with the driving forces. One driving force that is always present, as already mentioned, is the force of gravity due to the Earth (minus the block); this force is represented by the weight W of the block. It is not feasible to measure W directly by placing the block on a scale, but the weight can be calculated if the mass density ρ and volume V of the block can be ascertained. The density could be found by chipping off a representative piece of the block of manageable size, and then finding the volume and density as indicated in Chapter 1. The weight itself is found either from $W = \rho g V$ or, using the unit weight γ instead of ρ, from $W = \gamma V$.

The weight force acts at the *center of gravity* of the block and is directed vertically down. For a uniform rectangular block, the center of gravity is at the middle of the block, i.e., at the point where the body diagonals intersect. For other shapes, the location of the center of gravity may be more difficult to find.

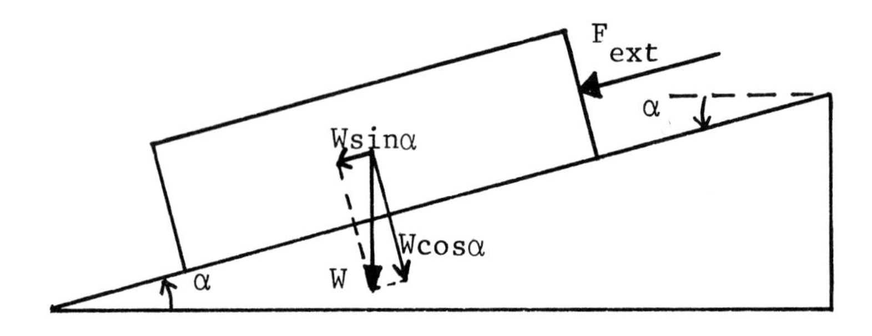

Fig.(4.2) Driving Forces

The weight force is shown on Fig.(4.2). Only the component $W \sin \alpha$, however, is directed down the slope. The angle α is the angle made by the inclined surface of the slope with the horizontal. As shown on Fig.(4.2), it may be thought of as an *elevation angle* above the horizontal, or as a *dip angle* below the horizontal.

Gravity cannot be "switched-off", so the $W \sin \alpha$ driving force is ever present. But there may, or may not, be other driving forces acting. For instance, although for clarity the block is drawn as an isolated entity on the slope, it may in fact be separated from an uphill block only by a narrow fracture, or joint. If the fracture is empty, no other driving force is produced. But if water seeps into the joint, an additional driving force can be generated in several ways. First, there is the hydrostatic force itself. Second, if the water freezes, the accompanying expansion produces strong forces that will be exerted on the rock adjacent to the joint. Third, even if the water does not freeze, it can still expand if the temperature rises (thermal expansion). If the water is confined, the forces produced will act to split the rock even more.

The sum of these driving forces other than gravity is represented by the symbol F_{ext}, for extra driving force. Also, to keep the analysis relatively simple, it is presumed that F_{ext} is directed parallel to, and down, the slope; see Fig.(4.2). Under these conditions, the total driving force DF (single symbol) is given by

$$DF = W \sin\alpha + F_{\text{ext}}. \tag{4.1}$$

Turn now to resisting forces. The first of these to be considered is the force of static friction f_s. This is the familiar force that resists the start of sliding of an object over a surface on which the object is initially at rest, but on which an external force is acting so as to induce sliding. The numerical value of f_s on such an object that nevertheless remains at rest is precisely the value that makes the net force on the object equal to zero, in accordance with Newton's second law ($\Sigma F = ma$ with $a = 0$). If the magnitude of the external force is increased, the value of f_s will also increase to keep the net force equal to zero, and thereby maintain the object at rest.

But the value of f_s cannot increase forever: there is an upper limit to the magnitude of the friction force. Experiments show that this maximum possible value of the friction force, $f_{s,\max}$, in any particular situation, is given by Coulomb's law:

$$f_{s,\max} = \mu_s R. \tag{4.2}$$

On the right-hand side of Eq.(4.2), μ_s is the coefficient of static friction. This is a dimensionless number, the numerical value depending on the composition and condition of the two surfaces that are in contact. The R in Eq.(4.2) is the sum of the normal forces on the object. (The word *normal* is not used as the opposite to *abnormal*, but rather means *perpendicular*.) A normal force is a force on the object that is directed perpendicular to the surface of contact.

Sometimes, in place of the coefficient of friction μ_s, an *angle of friction* ϕ is employed. The angle of friction is related to the coefficient of friction by

$$\tan\phi = \mu_s. \tag{4.3}$$

The choice of using either ϕ or μ_s in the treatment of friction is entirely personal preference. If ϕ is employed, then Eq.(4.2) becomes

$$f_{s,\max} = R\tan\phi. \tag{4.4}$$

Regardless of whether ϕ or μ_s is used, if the value of f_s needed to keep the object at rest comes to exceed $f_{s,\max}$, sliding will occur.

Now apply these considerations to the block of rock on the slope. Under the conditions so far specified, the normal force R must equal the component of the force of gravity that acts normal to the slope, for there is no acceleration normal to the slope, even if the block slides. From Fig.(4.2), it is evident that this component is $W \cos\alpha$. Hence,

$$f_{s,\max} = W \cos\alpha \tan\phi. \tag{4.5}$$

****

EXAMPLE 1 With only friction acting as the resisting force, angle of friction ϕ, and no extra driving force present, find the angle of the steepest slope on which the block can remain at rest.

The driving force is $DF = W \sin\alpha$ and the resisting force is f_s. If the block remains at rest, then

$$\Sigma F = 0,$$

$$W \sin\alpha - f_s = 0,$$

$$W \sin\alpha = f_s.$$

Now $W \sin\alpha$ is larger in value for steeper slopes, since α is greater. Therefore, to keep the block at rest on steeper slopes, f_s must be larger also. The steepest slope on which the block can remain at rest is that for which f_s has increased to its maximum possible value $f_{s,\max}$. By Eq.(4.5), for this steepest slope, the last equation gives

$$W \sin\alpha_{\max} = W \cos\alpha_{\max} \tan\phi.$$

Divide this equation (both sides, of course) by $\cos\alpha_{\max}$. But, for any angle θ, $\sin\theta / \cos\theta = \tan\theta$. Hence, the equation reduces to

$$\tan\alpha_{\max} = \tan\phi,$$

$$\alpha_{\max} = \phi.$$

This result provides a physical interpretation to the angle of friction ϕ. With only friction acting as a resisting force, and no driving force present except gravity, ϕ equals the elevation angle of the steepest slope on which the block can remain at rest. This greatest slope angle for the block to be at rest is sometimes called the *angle of repose*, especially if the block and slope are made of the same material.

****

Another resisting force that occurs naturally in rocks (and soils) is the *cohesion force* F_{coh}. The force of cohesion is a resisting force the value of which is independent of the value of the normal force R. Unlike the friction force, the force of cohesion is directly proportional to the area of contact A between the block and the slope. The force is written as

$$F_{\text{coh}} = cA, \tag{4.6}$$

where the constant of proportionality c is called the *cohesion stress.* Like ϕ (and μ_s), the numerical value of c depends on the composition and condition of the two surfaces in contact. The area A in Eq.(4.6) is shown shaded in Fig.(4.3). The direction of F_{coh} is, like f_s, parallel to the slope and upward (opposite to the direction of the driving force $W \sin \alpha$), opposing the tendency of the block to slide down under gravity.

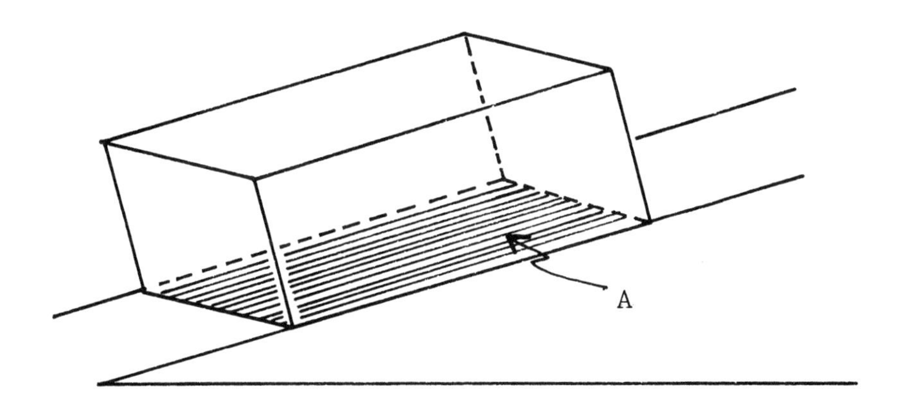

Fig.(4.3) Contact Area

The forces of friction and of cohesion are naturally occurring resisting forces. With both acting, the total resisting force RF_{acting} is

$$RF_{\text{acting}} = f_s + cA. \tag{4.7}$$

If the block remains at rest,

$$\Sigma F = 0,$$

$$DF - RF_{\text{acting}} = 0,$$

$$RF_{\text{acting}} = DF, \tag{4.8}$$

in accordance with Newton's second law $\Sigma F = ma$, with $a = 0$ for a system that remains at rest.

Now suppose that in Eq.(4.7) the acting friction force f_s is replaced with $f_{s,\max}$, the maximum possible value of f_s. Then Eq.(4.7) yields the largest possible value of the total resisting force; this maximum value is given the symbol RF; that is,

$$RF = f_{s,\max} + cA, \tag{4.9}$$

$$RF = W \cos\alpha \tan\phi + cA, \tag{4.10}$$

the last step by Eq.(4.5).

The ratio of the maximum possible value of the total resisting force to the driving force is called the *factor of safety against sliding*, symbol FS; that is,

$$FS = \frac{RF}{DF}. \tag{4.11}$$

Substituting Eqs.(4.1) and (4.10) into Eq.(4.11) yields

$$FS = \frac{W \cos\alpha \tan\phi + cA}{W \sin\alpha + F_{\text{ext}}}. \tag{4.12}$$

Fig.(4.4) House at Risk

The numerical value of the factor of safety indicates how safe the block is against the tendency to slide down the slope. For example, suppose that the purchase of the house shown in Fig.(4.4) is being contemplated; i.e., a house located at the bottom of a hill, say, on which a slab of rock is resting. The builder, or real estate agent, imply that there is no need to worry about the slab suddenly sliding down the hill into the house, because the factor of safety, calculated by Eq.(4.12), has the value 1.12. This means that the maximum available resisting force RF has the value $RF = 1.12(DF)$. For the slab to remain at rest, it is only necessary that $RF = 1.00(DF)$, or $FS = 1$. It seems that there is $0.12(RF)$ of "extra" resisting force, and so no concern need be felt about the presence of the slab lurking uphill.

However, bear in mind that the values of some of the terms in Eq.(4.12) can vary in the course of time, sometimes in very little time. For instance, if the ground becomes saturated with water, the values of $\tan\phi$ and c can decrease considerably, lowering the value of RF. If an uphill tension crack forms and fills with water which later freezes, an extra driving force is produced, increasing the value of DF. Both of these events force the factor of safety to smaller values. As soon as FS diminishes to $FS = 1$, the slab will slide on the slightest disturbance. The fact that this has not yet happened may not be a reason for complacency.

****

EXAMPLE 2 A rectangular block of rock, density 2.90 g/cm^3 and edge lengths 17.0 m, 2.30 m, 8.47 m, rests on a 16.0° incline, as shown in Fig.(4.5). An extra driving force of 734 kN, acting parallel to and down the incline, will just start the block sliding. The angle of friction between block and incline is 7.00°. Find the cohesion stress on the block.

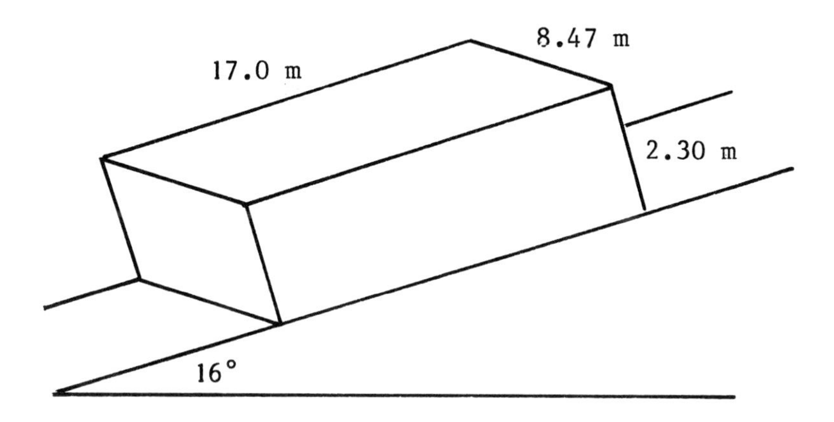

Fig.(4.5) Example 2

The weight of the block is calculated from

$$W = \rho V g$$

Hence,

$$W = (2900\ \mathrm{kg/m^3})[(17\ \mathrm{m})(8.47\ \mathrm{m})(2.3\ \mathrm{m})](9.8\ \mathrm{m/s^2}),$$
$$W = 9.412\ \mathrm{X}\ 10^6\ \mathrm{N}.$$

From Fig.(4.5), the area of contact A between block and slope is seen to be

$$A = (17\ \mathrm{m})(8.47\ \mathrm{m}),$$
$$A = 144.0\ \mathrm{m^2}.$$

With the extra driving force, the block just starts sliding; this implies that $FS = 1$. Use Eq.(4.12), with all quantities in SI base units, to get

$$FS = \frac{W\cos\alpha\tan\phi + cA}{W\sin\alpha + F_{\mathrm{ext}}},$$

$$1 = \frac{(9.412\ \mathrm{X}\ 10^6\ \mathrm{N})\cos 16^\circ \tan 7^\circ + c(144\ \mathrm{m^2})}{(9.412\ \mathrm{X}\ 10^6\ \mathrm{N})\sin 16^\circ + 734\ \mathrm{X}\ 10^3\ \mathrm{N}},$$
$$c = 15.4\ \mathrm{kPa}.$$

****

4.2 Slope Stress

The condition for stability on a non-engineered slope with no extra driving force, as expressed in Eq.(4.12) with $F_{\text{ext}} = 0$, often is written in terms of stresses rather than forces. To do this, set $F_{\text{ext}} = 0$ and divide numerator and denominator by the contact area A, to obtain

$$FS = \frac{(W \cos\alpha/A)\tan\phi + c}{(W \sin\alpha/A)}.$$

Since all of the forces in Eq.(4.12), the resisting forces of friction and cohesion and the driving force due to gravity, act parallel to the incline, and hence to the contact area A, the associated stresses are shear stresses, so that the factor of safety can be expressed as

$$FS = \frac{\tau_{\text{R}}}{\tau_{\text{D}}},$$

where τ_{R} is the total resistive shear stress and τ_{D} is the driving shear stress. There is a compressive stress σ acting across the contact area on the block: it is due to the reaction of the slope surface to the component of the weight perpendicular to the surface, i.e., it is due to the normal force. Since this normal force is $W \cos\alpha$, the compressive stress, in magnitude, is

$$\sigma = W \cos\alpha/A.$$

It follows that the values of the shear stresses τ_{R} and τ_{D} are related to that of the compressive stress:

$$\tau_{\text{R}} = \sigma \tan\phi + c,$$

$$\tau_{\text{D}} = \sigma \tan\alpha.$$

(The last equation follows by noting that $\tan\alpha = \sin\alpha/\cos\alpha$.) Hence, in terms of stress,

$$FS = \frac{\sigma\tan\phi + c}{\sigma\tan\alpha}. \tag{4.13}$$

If $c = 0$ (no cohesion, or cohesion ignored), the factor of safety and the angles of the slope and of friction are related by

$$\begin{aligned} FS &< 1 \text{ if } \phi < \alpha, \\ FS &= 1 \text{ if } \phi = \alpha, \\ FS &> 1 \text{ if } \phi > \alpha. \end{aligned}$$

The situation with $c \neq 0$ is sometimes presented graphically. In Fig.(4.6), the shear stresses τ_{R} and τ_{D} are plotted on the ordinate (the "y axis") and the compressive stress σ on the abscissa (the "x axis"). The driving shear stress plots as a straight line through the origin with slope α, and the resisting shear stress plots as a straight line with σ-intercept c and slope ϕ.

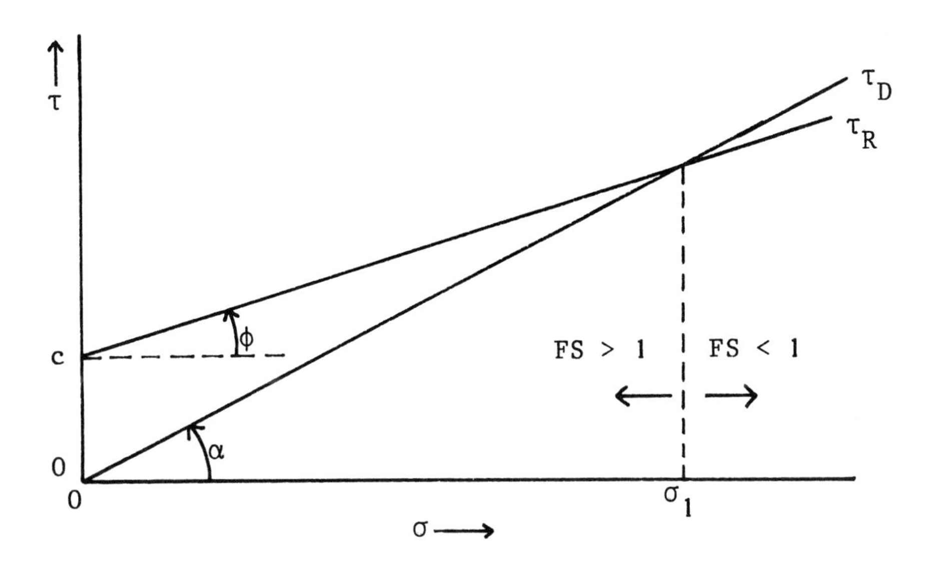

Fig.(4.6) Stresses on a Non-Engineered Slope

Figure (4.6) is drawn with $\phi < \alpha$. (The situation with $\phi > \alpha$ always gives $FS > 1$.) The two lines representing the shear stresses intersect at a value of the compressive stress labelled σ_1. For configurations with $\sigma < \sigma_1, FS > 1$ and for $\sigma > \sigma_1, FS < 1$.

In evaluating the utility of Fig.(4.6), it should be borne in mind that σ hides a dependence on the angle of dip α ($\sigma = W \cos\alpha / A$), so that Eq.(4.13), on which Fig.(4.6) is based, does not explicitly display all the dependence of FS on α. For this reason, it seems more clear-cut to evaluate the factor of safety in terms of forces, as is done in Eq.(4.12).

4.3 Engineered Slopes

Suppose that the factor of safety, calculated from Eq.(4.12), is not sufficiently large to provide confidence that the block will not slip under conditions that are expected to vary. Then it may be necessary to stabilize the block by some engineering expedient, and thereby avoid having to rely completely on the naturally occurring resisting forces. Two related stabilizing techniques are analyzed in this section.

The first of these techniques is the installation of rock bolts. Rock bolts are solid rods, usually of steel, driven through the block into the slope. They are generally installed at 90° to the slope. This means that if the block tries to slide, it will exert a force on the bolt that is parallel to a cross section of the bolt. By Newton's third law, the bolt exerts an equal and opposite force on the block. This force will also be parallel to a cross section of the bolt and therefore is a shear force. The force acts to resist the sliding of the block.

The largest possible value F_B of this force exerted by the bolt on the block is given by

$$F_B = \tau_B A_B, \tag{4.14}$$

where τ_B is the shear strength of the material of which the bolt is made, and A_B is the cross-sectional area of the bolt. As described in Chapter 2, the shear strength represents the greatest shear stress that can be applied, in this case to the bolt, without causing the bolt to lose all meaningfull resistance to the applied shear force. If n identical bolts are installed, the net effect is to add nF_B to the resisting force RF in Eq.(4.10).

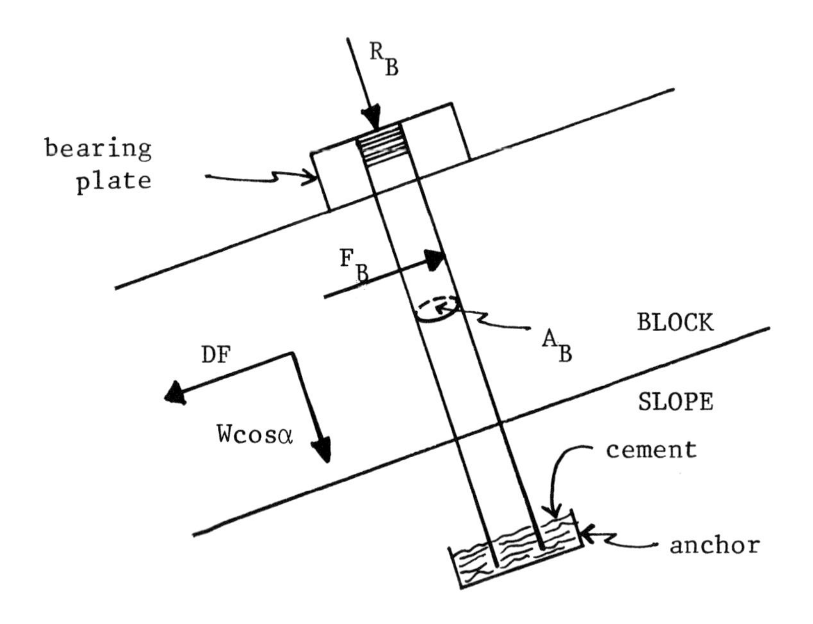

Fig.(4.7) Installed Rock Bolt

If the base of the bolt is cemented into the slope, then the bolt can be tightened. This action squeezes the block and slope together. Since each bolt is installed perpendicular to the slope, the force exerted due to the tightening is also perpendicular, or normal to, the slope. Call this force R_B, the normal force each tightened bolt exerts on the block. This force can be expressed as

$$R_B = \sigma_B A_B, \tag{4.15}$$

where σ_B is the stress on the bolt due to the tightening and, as before, A_B is the cross-sectional area of the bolt. On the bolt, the stress is one of axial tension, since the block and slope want to "spring apart", and the tightened bolt resists this. Hence, the bolt cannot be tightened beyond the tensile strength of the bolt material.

The force R_B itself is not a resisting force on the block, since it is exerted normal to the slope. The effect of R_B is to increase the normal force and therefore to increase the maximum friction force available. By Eq.(4.2), $f_{s,\max} = \mu_s R$. The part of R due to gravity,

$W \cos\alpha$ is always present. If there are n bolts, all tightened to the same tension, then the total normal force becomes

$$R = W \cos\alpha + nR_{\mathrm{B}}. \tag{4.16}$$

Hence,

$$f_{\mathrm{s,max}} = \mu_{\mathrm{s}}(W \cos\alpha + nR_{\mathrm{B}}),$$

or

$$f_{\mathrm{s,max}} = (W \cos\alpha + nR_{\mathrm{B}}) \tan\phi. \tag{4.17}$$

Therefore, by Eqs.(4.14), (4.15), and (4.17), the maximum resisting force available with n rock bolts installed perpendicular to the slope is

$$RF = (W \cos\alpha + n\sigma_{\mathrm{B}} A_{\mathrm{B}}) \tan\phi + cA + n\tau_{\mathrm{B}} A_{\mathrm{B}}. \tag{4.18}$$

It is assumed that the bolts are identical and that all have been tightened to the same tension σ_{B}. The driving force on the block is unaffected by the rock bolts. Combining the last equation with Eqs(4.1) and (4.11) gives for the factor of safety

$$FS = \frac{(W \cos\alpha + n\sigma_{\mathrm{B}} A_{\mathrm{B}}) \tan\phi + cA + n\tau_{\mathrm{B}} A_{\mathrm{B}}}{W \sin\alpha + F_{\mathrm{ext}}}. \tag{4.19}$$

Often in engineering situations, it is important to know the number of bolts needed to achieve a desired factor of safety. With this in mind, Eq.(4.19) can be solved for n with the result

$$n = \frac{FS(W \sin\alpha + F_{\mathrm{ext}}) - W \cos\alpha \tan\phi - cA}{A_{\mathrm{B}}(\sigma_{\mathrm{B}} \tan\phi + \tau_{\mathrm{B}})}. \tag{4.20}$$

****

EXAMPLE 3

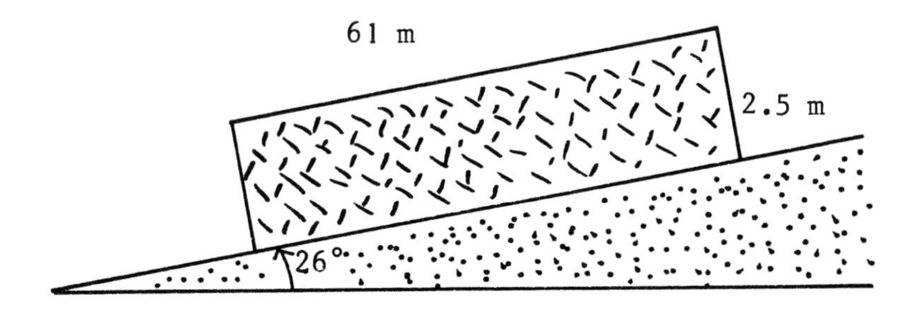

Fig.(4.8) Example 3

The slab shown in Fig.(4.8) has a width of 13 m; its density is 3.2 g/cm^3. The angle of friction between the slab and the slope is 20° and cohesion equals 75 kN/m^2. Rock bolts are installed but not tightened. Each bolt has an area of 6.2 cm^2 and shear strength 740 MPa. A factor of safety of 3.0 is desired. How many rock bolts are needed?

Calculate the weight of the slab from $W = \rho V g$; the result is $W = 62.2$ MN. The angle of friction is $\phi = 20°$ and the angle of the slope, from Fig.(4.8) is $\alpha = 26°$. With cohesion present, $c = 75 \text{ X } 10^3$ N/m^2, the contact area between slab and slope must be calculated. This area is $A = (61 \text{ m})(13 \text{ m})$, $A = 793 \text{ m}^2$. Since the bolts are not tightened, $\sigma_B = 0$. The bolt cross-sectional area is $A_B = 6.2 \text{ X } 10^{-4} \text{ m}^2$ and their shear strength is $\tau_B = 740 \text{ X } 10^6$ Pa. (The SI prefixes cannot be overlooked.) Write Eq.(4.20) with $\sigma_B = 0$, and also with $F_{ext} = 0$, since no extra driving force is mentioned. Using SI base units, then, and with $FS = 3$, the result is

$$n = \frac{W[(FS)\sin\alpha - \cos\alpha\tan\phi] - cA}{A_B\tau_B},$$

$$n = \frac{(62.2 \text{ X } 10^6 \text{ N})(3\sin 26° - \cos 26° \tan 20°) - (75 \text{ X } 10^3 \text{ Pa})(793 \text{ m}^2)}{(6.2 \text{ X } 10^{-4} \text{ m}^2)(740 \text{ X } 10^6 \text{ N})},$$

$$n = 4.3.$$

Of course, there cannot be 4.3 identical bolts. Suppose that "safety first" is the work philosophy; then install $n = 5$ bolts.

****

Another method to prevent sliding is to stitch the block to the slope. The technique is similar to the use of rock bolts. A hole is drilled through the block perpendicular to, and into, the slope. Some cement is poured into the bottom of the hole. Then, instead of a bolt, a cable is inserted. One end of the cable is secured by the cement, when hardened. At the outer surface of the block, a cap and nut secure the other end of the cable. The nut is then tightened.

When the cable is tightened, the block and slope are pulled together. Hence, the effect is the same as tightening a rock bolt. However, a cable, being flexible, is presumed to offer no shearing resistance to the block. If the cable bends after installation, then the block must have moved and this indicates sliding.

The formula for the factor of safety in stitching can be derived from the corresponding formula for rock bolts, Eq.(4.19) by: (*i*) replacing σ_B, bolt tension, with σ_C, cable tension; (*ii*) replacing A_B, bolt cross-sectional area, with A_C, cable cross-sectional area; (*iii*) deleting the term $n\tau_B A_B$. Hence, with n identical stitches, each tightened to tension σ_C, the factor of safety against sliding is given by

$$FS = \frac{(W\cos\alpha + n\sigma_C A_C)\tan\phi + cA}{W\sin\alpha + F_{ext}}. \qquad (4.21)$$

****

EXAMPLE 4

A rectangular block of rock is stitched to a 27.0°-slope. The block has unit weight 30.4 kN/m^3 and edge lengths 10.8 m, 12.6 m, 2.10 m. The angle of friction between block and slope is 18.0°. Ignore cohesion. The stitching cable has cross-sectional area 7.50 cm^2 and is tightened to tension 620 MPa. (*a*) How many stitches are needed to provide a factor of safety against sliding equal to 2.00? (*b*) With the stitches installed, find the smallest extra driving force that will cause the block to slip.

(*a*) The weight of the block is $W = \gamma V$, $W = 8.687$ MN. The angle of friction is $\phi = 18°$ and the angle of the slope is $\alpha = 27°$. Since cohesion is to be ignored, set $c = 0$. The tension in the cable is $\sigma_{\mathrm{C}} = 620$ MPa and the cable area is $A_{\mathrm{C}} = 7.5 \text{ X } 10^{-4}\ \mathrm{m}^2$. No extra driving force is mentioned in this part, so put $F_{\mathrm{ext}} = 0$. Also, $FS = 2$. Use Eq.(4.21) to find

$$FS = \frac{[W \cos\alpha + n\sigma_{\mathrm{C}} A_{\mathrm{C}}] \tan\phi}{W \sin\alpha}.$$

Substitute the data (in SI base units, of course):

$$2 = \frac{[(8.687 \text{ MN}) \cos 27° + n(620 \text{ MPa})(7.5 \text{ X } 10^{-4} \text{ m}^2)] \tan 18°}{(8.687 \text{ MN}) \sin 27°}.$$

Solve for n. Note that since Pa $=$ N/m^2, then MPa $=$ MN/m^2. The result is $n = 35.56$, so that, practically speaking, $n = 36$.
(*b*) With the stitches installed, put $n = 36$. To find the smallest extra driving force needed to cause slipping, set $FS = 1$. In this part, $FS \neq 0$, but $c = 0$ still. Eq.(4.21) now yields

$$FS = \frac{[W \cos\alpha + n\sigma_{\mathrm{C}} A_{\mathrm{C}}] \tan\phi}{W \sin\alpha + F_{\mathrm{ext}}},$$

$$1 = \frac{[(8.687 \text{ MN}) \cos 27° + (36)(620 \text{ MPa})(7.5 \text{ X } 10^{-4} \text{ m}^2)] \tan 18°}{(8.687 \text{ MN}) \sin 27° + F_{\mathrm{ext}}},$$

$$F_{\mathrm{ext}} = 4.01 \text{ MN}.$$

****

4.4 Roadcuts

In the discussion so far, the block has been drawn as a rectangular solid in shape. However, nature seldom forms blocks of rock with such a simple shape. Also, engineering projects often create blocks of a more irregular shape.

For example, suppose that a road under construction must pass for part of its length through hilly or mountainous terrain. Figure (4.9a) shows a cross section of part of a mountain before excavation of the road. The hillside is composed of sedimentary rock. Sedimentary rock often forms in parallel layers (*beds*). The planes of demarcation between contiguous beds are notoriously weak in resisting driving forces that induce sliding. The planes could also represent a system of joints in other kinds of rock. In any event, in Fig.(4.9a), the layers of rock are inclined at angle α with the horizontal and dip directly toward the projected roadway. Any one of these planes separating layers of rock is potentially a plane along which a block (the rock above the plane) can slide.

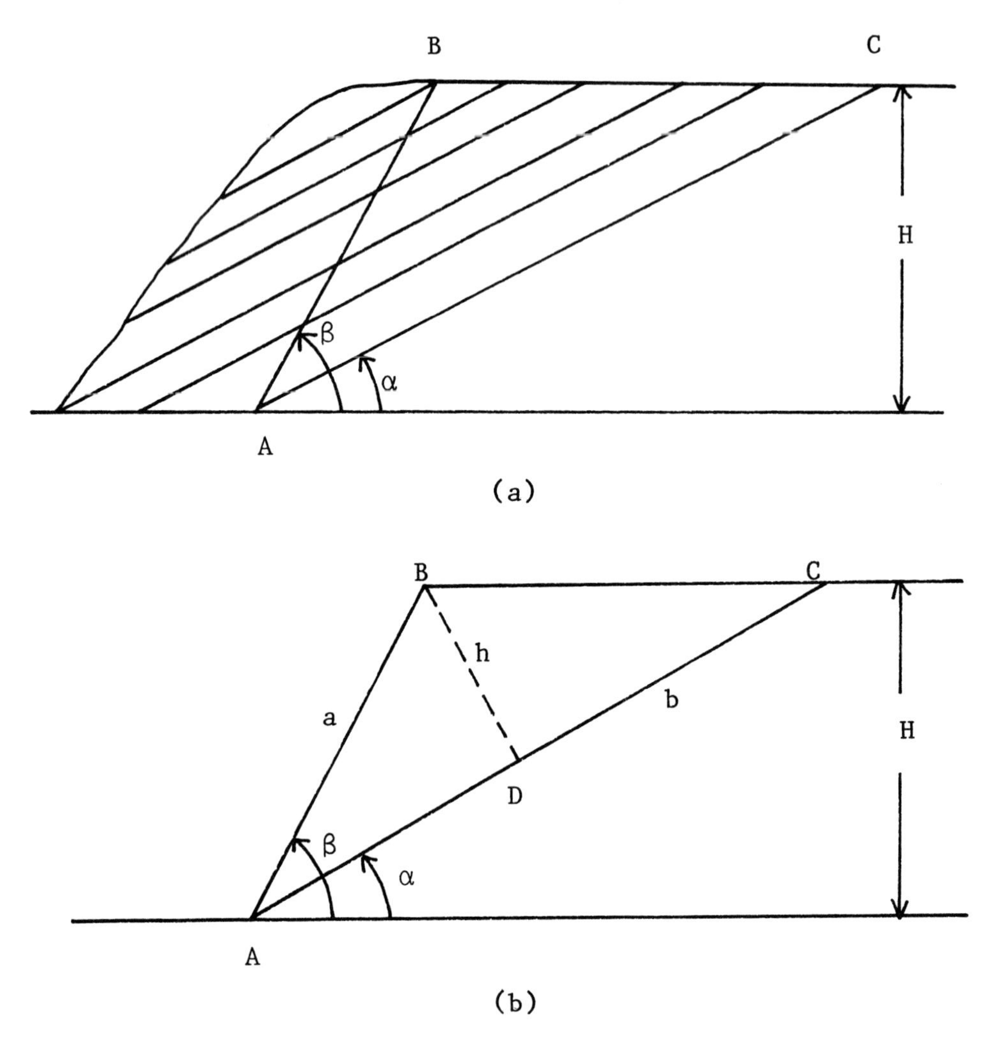

Fig.(4.9) Roadcut

In Fig.(4.9b) the hill is shown with the section that must make way for the road removed. The cut AB is made at an angle β ($\beta > \alpha$) with the horizontal. Only one of the weak bedding planes along which sliding is possible is shown. Specifically, this is the plane with its lower edge through A at the road. It is assumed, however, that the road exerts no supporting force on the block at its *toe* at A. The top of the hill is horizontal.

Of course, Fig.(4.9) is a cross section of a three-dimensional construction: the road extends for a length L perpendicular to the page. The bedding planes also are perpendicular to the page.

The geometry of the roadcut affects the analysis of slope stability only through the calculation of the weight of the block. The weight is given either by $W = \rho V g$ or from

$W = \gamma V$. Previously, the block volume V is calculated as the product of the three edge lengths of the block. However, the block in Fig.(4.9) is not rectangular. Rather, it is a cylinder of length L and triangular cross section ABC.

The volume of a cylinder is the product of the length and the cross-sectional area. Therefore, if A_x is the area of the triangle ABC, then

$$V = A_x L, \tag{4.22}$$

where the area A_x is to be expressed in terms of the height H of the hill, the angle α of the plane and the angle β of the cut.

Write a for the distance AB and b for the distance AC. The area of triangle ABC is

$$A_x = \frac{1}{2} bh, \tag{4.23}$$

where $h =$ BD is the "height" of the triangle perpendicular to "base" AC. Now

$$\sin\alpha = \frac{H}{b},$$

$$b = \frac{H}{\sin\alpha}. \tag{4.24}$$

Similarly,

$$a = \frac{H}{\sin\beta}. \tag{4.25}$$

From triangle ABC and Eq.(4.25),

$$\sin(\beta - \alpha) = \frac{h}{a},$$

$$h = (\frac{H}{\sin\beta}) \sin(\beta - \alpha). \tag{4.26}$$

Substituting Eqs.(4.24) and (4.26) into Eq.(4.23) gives for the area

$$A_x = \frac{1}{2}[\frac{H}{\sin\alpha}][\frac{H}{\sin\beta}\sin(\beta - \alpha)],$$

$$A_x = \frac{1}{2}H^2\left(\frac{\sin\beta\cos\alpha - \cos\beta\sin\alpha}{\sin\alpha\sin\beta}\right),$$

$$A_x = \frac{1}{2}H^2(\cot\alpha - \cot\beta).$$

With the volume of the block given by Eq.(4.22), $V = A_x L$, the weight of the block is

$$W = \rho(A_x L)g,$$

$$W = \frac{1}{2}\rho L H^2 g(\cot\alpha - \cot\beta). \tag{4.27}$$

If cohesion must be accounted for, then an expression is needed for the contact area. By Eq.(4.24), this is

$$A = bL,$$

$$A = HL\csc\alpha. \tag{4.28}$$

The length L may not be the complete length of the cut. The types of rock, the angle of the bedding planes, and their orientation, may vary along the length of the cut. In that case, the length L is the length of an intact block along which these parameters do not change significantly.

It may appear that the block ABC cannot slide, as it seems to be held in place against the roadway at A. But it must be remembered that Fig.(4.9) is an idealization, or approximation, to the actual situation. The block is not perfectly triangular in outline, with absolutely straight edges; A, B, and C are not really mathematical points. The rock mass probably contains many other fractures that will yield if sliding begins. In short, it cannot be considered that the block is geometrically "locked in" at A.

Also, sliding could begin along any of the other bedding planes shown on Fig.(4.9a), for which the associated block "daylights into free space"; there is nothing, even in the idealized construction discussed, to prevent a slide if $FS < 1$. Equations (4.27) and (4.28) can be used for such a block, once the new vertical thickness H of the block is evaluated.

To find a convenient formula for the factor of safety for a non-engineered roadcut, with no extra driving force, in terms of the angles rather than the weight of the block, substitute Eq.(4.27) and (4.28) into Eq.(4.12) to obtain

$$FS = \frac{\tan\phi}{\tan\alpha} + \left(\frac{2c}{\rho g H}\right)\frac{1}{\sin^2\alpha[\cot\alpha - \cot\beta]}. \tag{4.29}$$

Now $\tan\phi/\tan\alpha$ is the value of the factor of safety if $c = 0$; i.e., either if there is no cohesion or cohesion is to be ignored. See Eq.(4.12). Call this value of the factor of safety FS_0:

$$FS_0 = \frac{\tan\phi}{\tan\alpha}. \tag{4.30}$$

Also, define a quantity B by

$$B = \frac{c}{\frac{1}{2}\rho g H}. \tag{4.31}$$

The quantity B is the ratio of the cohesion stress to the vertical stress at one-half the greatest depth in the block; B is dimensionless. With these substitutions, Eq.(4.29) becomes

$$FS = FS_0 + \frac{B}{\sin^2\alpha[\cot\alpha - \cot\beta]}. \tag{4.32}$$

Sometimes it is usefull to calculate, before making a cut, just what angle of cut is needed to achieve a desired factor of safety. For this purpose, Eq.(4.32) can be rearranged to read

$$\cot\beta = \cot\alpha - \frac{B}{(\Delta FS)\sin^2\alpha}, \tag{4.33}$$

where

$$\Delta FS = FS - FS_0, \tag{4.34}$$

FS being the factor of safety desired with cohesion taken into account. If the slope is stabilized with rock bolts or stitches, then Eqs.(4.27) and (4.28) must be substituted into the appropriate equation for the factor of safety of the engineered slope.

****

EXAMPLE 5

For a roadcut like that shown in Fig.(4.9), the vertical thickness of the hill is 16.5 m and the dip angle of the bedding plane is 35.0°. The angle of friction is 31.0° and the cohesion stress equals 38.4 kPa. The unit weight of the rock is 23.7 kPa/m. Find the factor of safety for a vertical cut.

The data are: $H = 16.5$ m, $\alpha = 35.0°$, $\phi = 31.0°$, $c = 38.4$ kPa, $\gamma = 23.7$ kPa/m. For a vertical cut, $\beta = 90°$, $\cot\beta = 0$. First, calculate the quantity B; since $\gamma = \rho g$, Eq.(4.31) yields

$$B = \frac{2c}{\gamma H},$$

$$B = \frac{2(38.4\ \text{kPa})}{(23.7\ \text{kPa/m})(16.5\ \text{m})},$$

$$B = 0.1964.$$

If cohesion was zero, the factor of safety, by Eq.(4.30), would be

$$FS_0 = \frac{\tan 31°}{\tan 35°},$$

$$FS_0 = 0.8581.$$

Equation (4.32) now yields

$$FS = 0.8581 + \frac{0.1964}{\sin^2 35°[\cot 35° - \cot 90°]},$$

$$FS = 1.28.$$

****

This discussion on roadcuts assumes that the angle β of the cut is greater than the dip angle α of the planes of potential slip. What about $\beta < \alpha$? A sketch like Fig.(4.9) with $\beta < \alpha$ shows that the block so formed cannot slide, as it is constrained by the horizontal ground surface. This begs the question: Why not make all cuts with $\beta < \alpha$? The figure shows that this would require the removal of much larger amounts of material than in the case of $\beta > \alpha$, so that it may be simply impractical in terms of impact on the road environment.

4.5 Topples

The blocks whose stability with respect to sliding have been examined in the preceding sections possess one characteristic in common: they are much "longer" than they are "higher". This makes them relatively immune to "tipping over" or *toppling.* Contrast the two blocks shown in Fig.(4.10). Block A will not tip, but block B might, even if sliding does not take place.

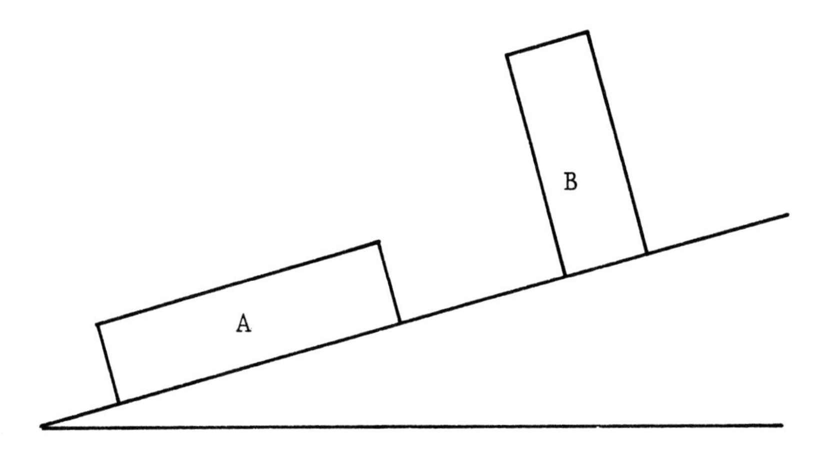

Fig.(4.10) To Tip or Not

It is important to determine the conditions that lead to toppling; clearly it is hazardous to be beneath a block of rock that is susceptible to such a maneuver, as it is to be down-slope from a block of rock with an inclination to slide.

In the preceding sections, the conditions for sliding are analyzed by examining the forces acting on the block, because sliding is a translational motion to which Newton's laws of motion apply.

Toppling is a rotational motion, and rotation is usually analyzed by applying Newton's laws as rewritten in the form convenient for rotational situations: that is, by considering the torques acting on the object. A brief tutorial on torque follows.

See Fig.(4.11). A force F acting on an object at a point P exerts a torque τ about origin O, which lies at a perpendicular distance r from the line of action of F, with τ given by

$$\tau = rF. \tag{4.35}$$

The symbol τ is already used as the symbol for shear stress; it is also used here as the symbol for torque because it is the symbol very commonly used in technical literature.

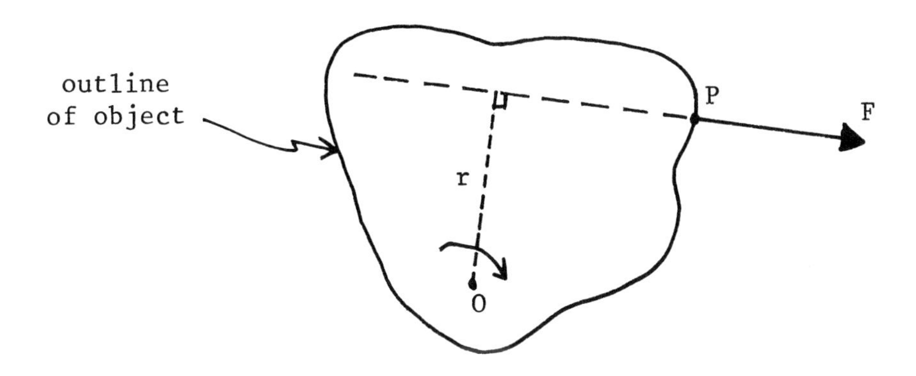

Fig.(4.11) Definition of Torque.

The distance, or line, from O to the line of action of F, the line segment whose length is r, is called the *lever arm* or *moment arm* of the force F about origin O. Torque, itself, is sometimes referred to as *moment*.

The SI base units of torque are, as seen from Eq.(4.35), units of force times units of distance; to wit, N·m. This combination is called a *Joule* (J) in the SI system. But the unit Joule is generally not used with torque; rather, the unit for torque is generally written out as N·m.

There is a *sense* (direction) associated with torque. In Fig.(4.11), the force F, acting alone at point P on the object, pivoted at O, tends to induce a rotation of the object in the clockwise sense, as indicated by the arrow at O. Clockwise torques are taken as negative; counterclockwise torques are positive.

Newton's second law written for rotation reads

$$\Sigma\tau = I\alpha, \tag{4.36}$$

where α now is the angular acceleration of the object, I is its rotational inertia about an axis through O, and $\Sigma\tau$ is the sum of all the torques acting on the object. In examining the conditions for toppling, only the sense of the angular acceleration need be determined, not its numerical value. Therefore, the numerical value of the rotational inertia will not be needed.

To see how Eq.(4.36) is applied, consider the specific case of a uniform, rectangular block of rock on a slope of inclination angle α. (Since a numerical value of angular acceleration will not be calculated, from this point on α will only be used for slope angle. Use of the same symbol for different quantities is not uncommon in engineering writing.) It is assumed here, as in all cases in this chapter, that the block is not susceptible to sliding before toppling.

The block has base length b, height h, and width w (dimension perpendicular to paper); see Fig.(4.12). To see if the block will topple (*i*) suppose that it does topple, (*ii*) draw the block slightly displaced toward toppling, (*iii*) examine the forces and their associated torques, (*iv*) determine the sense of the net torque and therefore of the angular acceleration induced, (*v*) see if the angular acceleration takes the block back to its original position.

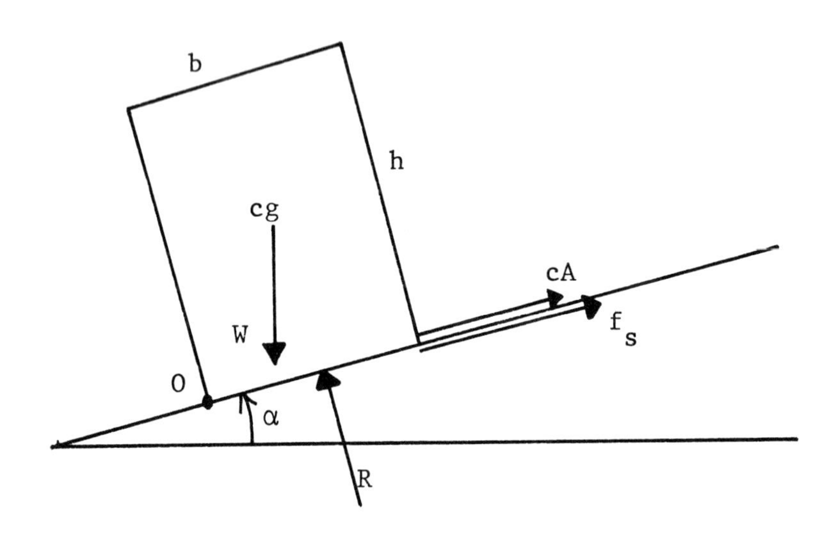

Fig.(4.12) Block before Tipping.

The forces on the block are its weight W, acting vertically down from the center of gravity cg of the block, the normal force R exerted by the slope, and the resisting forces of friction f_s and of cohesion cA. These are shown on Fig.(4.12).

In Fig.(4.13), the block is shown slightly tipped. Since it is presumed that the block does not slide, the axis of rotation about which the block tips passes along the bottom front edge O of the block. Since now there is only a very narrow region of contact between block and slope, namely along the edge through O, the normal force R and the friction force f_s must act there. The cohesion force is proportional to the area of contact; since this area virtually goes to zero in tipping, the cohesion force vanishes and therefore is not shown.

The lines of action of both R and f_s pass through O; hence, each contributes zero torque about O, for their lever arms are zero. The only force that does produce a torque is the weight W, drawn vertically down from the center of gravity. The crucial question, then, is just where, relative to O, the line of action of W passes. Two possible locations for the center of gravity are shown on Fig.(4.13). For position 1, the weight W passes O on the "inside" of the base of the block; this induces a clockwise torque which tends to return the block to its original position. On the other hand, for a center of gravity in position 2, the line of action of the weight W falls "outside" the base of the block; the associated torque about O tends to rotate the block farther away from its original position; i.e., it topples over.

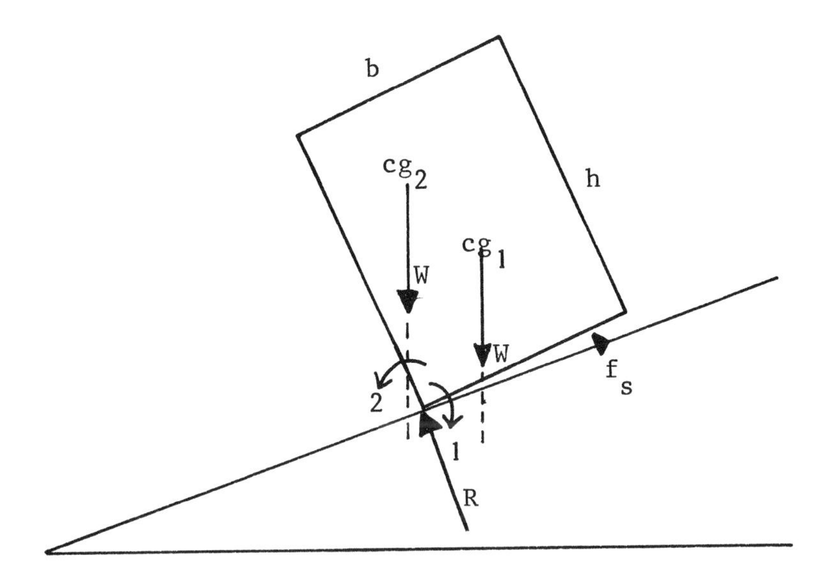

Fig.(4.13) Tipping Block

The two factors that govern toppling, then, are the location of the center of gravity of the block, and the value of the inclination angle of the slope. The steeper the slope the more likely it is that the line of action of W will fall outside the block, indicating toppling. If the line of action of the weight passes through O itself, then the angle of the slope is the largest angle for which the block does not topple, or the smallest angle at which it does (there is no practical difference between these two interpretations).

Figure (4.14) shows a rectangular block at this largest possible slope angle for which toppling does not occur. The center of gravity of a uniform rectangular solid is at the "center"; that is, at the intersection of the body diagonals. For a rectangular block, a line normal to the surface (dashed line) is parallel to the sides of the block. This means that the angle between W and the normal also is α, the slope's dip angle. Therefore, the greatest angle α consistent with stability against toppling is given from

$$\tan\alpha = \frac{b/2}{h/2},$$

$$\tan\alpha = \frac{b}{h}. \tag{4.37}$$

This equation applies for rectangular blocks. A different relation may hold for rocks of a different shape.

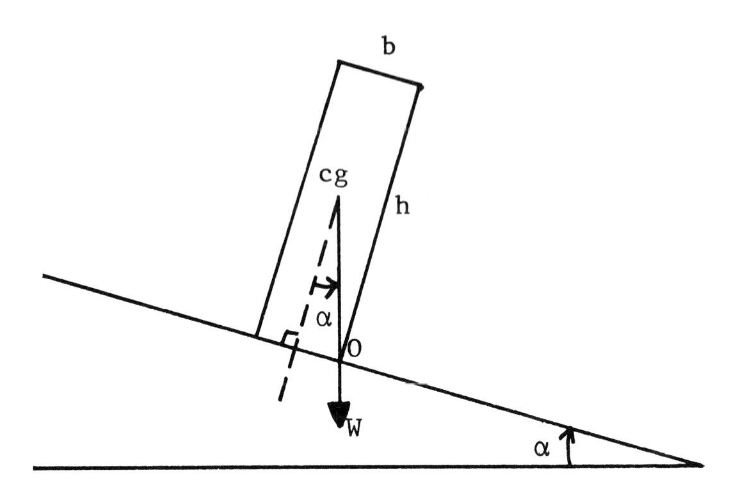

Fig.(4.14) Block in Critical Condition

****

EXAMPLE 6

Due to chemical weathering, a block of granitic rock has eroded to a uniform hemisphere of radius 2.63 m. Find the elevation angle of the steepest slope on which it will not tip. (Assume that the rock does not slide.) For a hemisphere of radius R, the center of gravity is at a distance $\frac{3}{8}R$ above the center of the base.

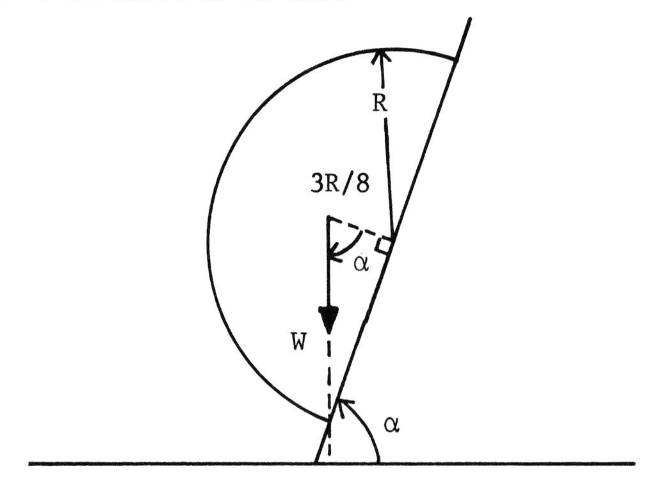

Fig.(4.15) Example 6

Figure (4.15) shows the rock on the point of toppling, since the line of action of the weight W passes through the lowest point on the base perimeter. The slope angle is given by

$$\tan \alpha = \frac{R}{\frac{3}{8}R},$$

$$\alpha = 69.4^\circ.$$

Evidently, a very steep slope is needed to induce toppling. Unless the resisting forces of friction and/or cohesion are exceptionally large, the "block" will slide on shallower slopes.

****

4.6 Problems

1. The rectangular slab of rock of density 3.40 g/cm^3 shown in Fig.(4.16) has edge lengths $a = 28.0$ m, $b = 19.0$ m, $c = 3.60$ m, and is resting on a 17.0°-incline. Ignore friction. The slab will slide if it is disturbed even very slightly. Find the cohesion stress between slab and incline.

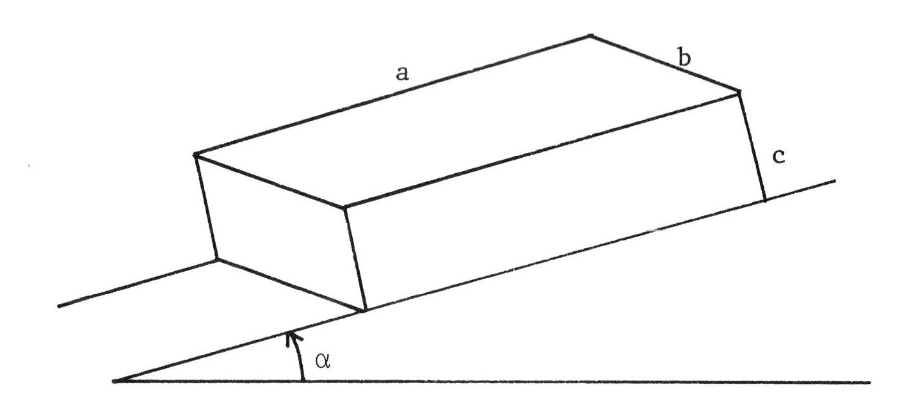

Fig.(4.16) Problems 1 and 2

2. The slab shown in Fig.(4.16) has density 3.15 g/cm^3 and edge lengths $a = 17.6$ m, $b = 9.30$ m, $c = 1.80$ m. The cohesion stress between the slab and the 21.0°-slope is 8.56 kN/m^2. Tests show that the slab will slide if disturbed in the slightest manner. Find the angle of friction between slab and slope.

3. A rectangular block with dimensions 7.92 m, 4.81 m, 1.27 m has a unit weight of 26.4 kN/m^3. It rests on an incline with elevation angle 22.5° with its shortest dimension perpendicular to the incline. Cohesion between block and incline is 23.0 kN/m^2. Ignore friction. Find the factor of safety against sliding.

4. A 17.0 m, 5.80 m, 2.40 m block with density 2600 kg/m^3 rests on an 18.0°-slope with its shortest dimension normal to the slope. The angle of friction between block and slope is 12.0° and cohesion equals 17.0 kPa. (a) Find the factor of safety against sliding. (b) Find the smallest extra driving force that will trigger sliding.

5. The weight of the slab shown in Fig.(4.17) is 22.3 MN. The angle of friction between slab and slope is 8.50° and the factor of safety against sliding is 1.60. Find the force of cohesion

between slab and slope.

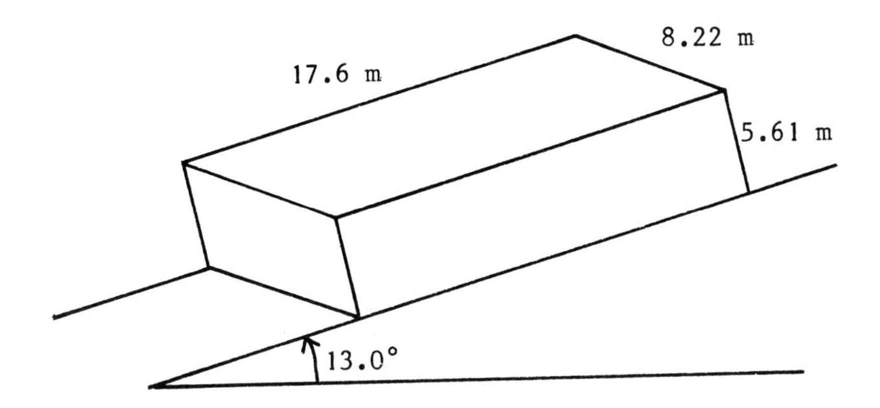

Fig.(4.17) Problem 5

6. A slab of rock has edge lengths 8.30 m, 12.0 m, 1.92 m and a density of 2.96 g/cm^3. It sits on a slope with its shortest dimension perpendicular to the slope. Between slab and slope the coefficient of friction equals 0.411 and the cohesion stress is 47.5 kPa. Tests show that an extra driving force of 5.18 MN will just start the slab sliding. Find the dip angle of the slope. (*Hint*: Write the equation for the factor of safety as a quadratic equation in $\cos\alpha$.)

7. A rock bolt has a cross-sectional area of 6.60 cm^2 and a shear strength of 450 MPa. What shearing force will just rupture the bolt?

8. A block of rock, density 3270 kg/m^3, has edge lengths 14.6 m, 6.31 m, 7.20 m. It is to be bolted to a vertical rock face with 25 identical, loosely-installed rock bolts. It is desired that, if cohesion is ignored, the factor of safety will be 2.00. The bolts have a shear strength of 633 MPa. Find the diameter of the bolts.

9. A slab of rock of mass 2.30 X 10^6 kg is to be bolted to a 31.0°-incline. Friction and cohesion are to be ignored. Each rock bolt has area 4.80 cm^2 and shear strength 510 MPa. Find the minimum number of loosely-installed rock bolts needed.

10. A rectangular block of rock with dimensions 1.22 m, 3.71 m, 1.83 m and of density 2.86 g/cm^3 is to be secured to a vertical rock face with a single rock bolt, as shown in Fig.(4.18). The bolt has a shear strength of 340 MPa. The angle of friction is 25.0°; cohesion is to be

ignored. The bolt is not tightened. Calculate the minimum diameter bolt needed.

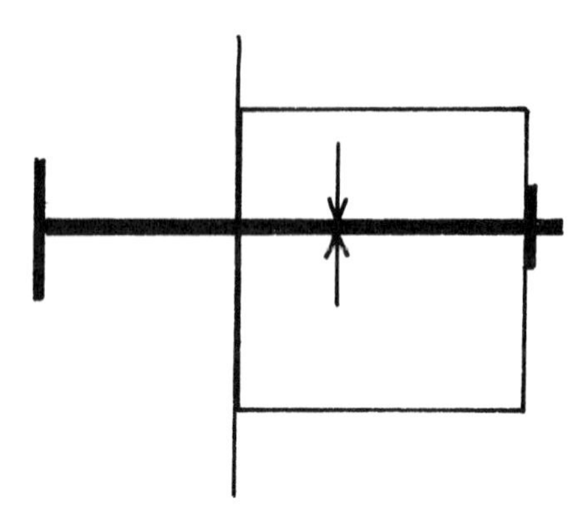

Fig.(4.18) Problem 10

11. A slab resting on a 13.0°-incline has a weight of 48.8 MN. Workers have installed 17 identical rock bolts. Each bolt has a shear strength of 870 MPa and area 14.0 cm^2. Ignore friction and cohesion. The workers did not tighten the bolts. (*a*) Find the factor of safety against sliding. (*b*) Find the minimum extra driving force that will cause the block to slip.

12. The slab shown in Fig.(4.19), mass 2.84 X 10^5 kg, will drop if the vertical joint ruptures. The contact area is 38.0 m^2. Between slab and cliff face the coefficient of friction is 0.300 and cohesion 73.3 kPa. To keep the slab from dropping, stitches are installed. Each cable has area 8.42 cm^2 and is tightened to tension 410 MPa. How many stitches are needed to get a factor of safety of 1.50?

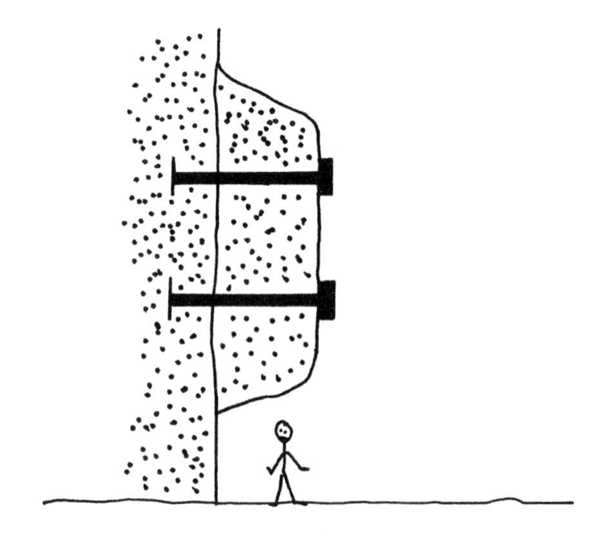

Fig.(4.19) Problem 12

13. A slab with edge lengths 12.0 m, 59.5 m, 2.82 m rests on an incline with its shortest dimension normal to the incline. The elevation angle of the slope is 28.5° and the density of the slab is 3.16 g/cm^3. The angle of friction between slab and incline is 18.0° and cohesion equals 84.2 kN/m^2. Rock bolts are installed but not tightened. Each bolt has an area of 7.30 cm^2 and shear strength 550 MPa. A factor of safety of 2.75 is desired. How many rock bolts are needed?

14. A slab of rock sits on a rock surface that dips at 16.0°. The unit weight of the slab is 28.7 kN/m^3, and its edge lengths are 17.4 m, 8.27 m, 5.58 m. It rests on one of its faces with the largest surface area. The angle of friction is 11.2° and cohesion is 20.7 kPa. (*a*) How many loosely-installed rock bolts are needed to get a factor of safety of at least 1.63? The bolts have shear strength 474 MPa and area 7.30 cm^2. (*b*) With the bolts installed, find the extra driving force that will bring the slab to the verge of slipping.

15. A block of rock weighing 320 MN is to be bolted to a 19.0°-slope. The angle of friction is 14.0°; cohesion has been destroyed by nearby rock blasting. Rock bolts of area 9.34 cm^2 and shear strength 472 MPa are installed. (*a*) How many bolts, not tightened, are needed for a factor of safety of 1.20? (*b*) To what tension must the bolts in (*a*) be tightened to increase the factor of safety by 0.200?

16. A rectangular slab of rock is to be secured to a 32.0°-incline by stitches. The slab of rock has density 2700 kg/m^3 and edge lengths 1.80 m, 7.60 m, 8.20 m. The coefficient of friction between slab and incline is 0.364. Ignore cohesion. The stitching cable has area 9.40 cm^2 and is tightened to tension 320 MPa. (*a*) Find the minimum number of stitches needed. (*b*) Suppose that nine stitches actually are installed. What is the factor of safety that results?

17. A block of rock with dimensions 28.0 m, 13.0 m, 5.30 m rests on a 24.0°-slope with its shortest dimension normal to the slope. The density of the block is 2.90 g/cm^3. Between slab and slope the cohesion stress equals 25.0 kPa and the coefficient of friction is 0.250. A factor of safety of 1.45 is needed. (*a*) If this is to be obtained with loosely-installed rock bolts, how many are required? Each bolt has area 9.20 cm^2 and shear strength 760 MPa. (*b*) The supplier is out of bolts, so the job must be done with stitches. The cable has area 5.30 cm^2 and each stitch is tightened to the tensile strength of 380 MPa. How many stitches are needed?

18. Figure (4.20) shows the cross section of a road cut into the side of a mountain. The line AA is a weak bedding plane along which sliding is possible. The block B, 18.6 m wide, directly above a stretch of the road is separated from uphill rock by a tension crack T normal to AA. The dip angle of the bedding plane is 19.2° and the coefficient of friction between the block B and the bedding plane is 0.390. The density of the block is 2.88 g/cm^3. Ignore

cohesion. (*a*) Show that the block does not slide. (*b*) Water seeps into the tension crack and freezes, exerting a driving force on the block. What value of this force will trigger a slide?

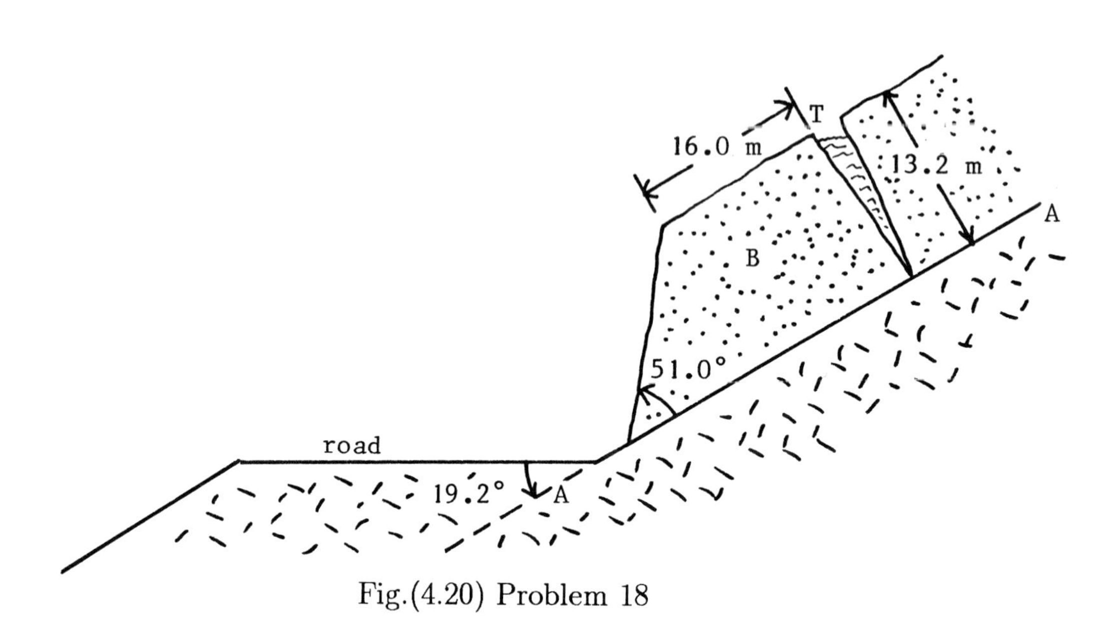

Fig.(4.20) Problem 18

19. A slab of weight 28.0 MN sits on a 23.0°-incline. The angle of friction between slab and incline is 15.0°; cohesion equals zero. (*a*) How many tightened rock bolts are needed to get a factor of safety of at least 1.50? The bolt specifications are: shear strength 230 MPa; area 5.80 cm^2; tightened to tension 86.4 MPa. (*b*) Due to an oversight, only 20 of the needed bolts actually are tightened; what is the real factor of safety?

20. Show that the extra driving force needed to trigger sliding of a block at rest is given by

$$F_{\text{ext}} = W \sin\alpha (FS_{\text{actual}} - 1),$$

where FS_{actual} is calculated with an integral number of bolts or stitches.

21. A block weighing 28.4 MN sits on a 19.0°-slope. The angle of friction is 13.0°. Cohesion equals 430 kPa. Tests show that the factor of safety against sliding is 1.70. Find the contact area betwen block and slope.

22. A block rests on a slope of dip angle α. The angle of friction is ϕ. No engineering steps have been taken to secure the block. Show that, if $\alpha > \phi$, the smallest value of the cohesion stress that keeps the block from sliding is given by

$$c = \sigma(\tan\alpha - \tan\phi),$$

where σ is the stress on the block due to the normal force.

23. Water just fills an isolated pore in a rock. Both the water and the rock are initially at temperature 20°C, but the temperature soon rises to 30°C. Find the stress exerted by the water on the rock due to the thermal expansion. The coefficient of volume expansion of water is 2.55 X 10^{-4} $(C°)^{-1}$, and of the rock is 4.80 X 10^{-5} $(C°)^{-1}$.

24. Suppose that in the roadcut of Fig.(4.9), the depth H is 14.6 m, the bedding angle 26.0° and the cut angle 48.5°. The density of the rock is 2550 kg/m^3. Due to heavy rain the block slides on to the road surface over a 94.0 m length. How many metric tons of rock must be removed to clear the roadway?

25. In Example 5, what angle of cut would yield a factor of safety of (*a*) 2.00, and (*b*) 3.00?

26. Sketch a graph of the factor of safety FS versus roadcut angle β for the range $90° \geq \beta > \alpha$. Pick values of ϕ and α, with $\phi < \alpha$, so that cohesion is necessary for stability. To get a value of B, use $c \approx 300$ kPa, $\gamma \approx 25$ kPa/m and $h \approx 10$ m. A graphing calculator may be handy, but not necessary.

27. A section of thruway is constructed through sandstone beds that dip at 41.5° directly toward the thruway. A section of the roadcut, 19.2 m high, is made at an angle of 70.0° to the horizontal. The angle of friction between the beds of the sandstone is 28.0°, and the density of the rock is 2.24 g/cm^3. Tests show that the factor of safety of the slope thus created is 1.35. Find the cohesion stress in the sandstone.

28. In the thruway roadcut of Problem 27, the factor of safety against sliding is 1.35. It is thought wise to secure the slope with a sufficient number of rock bolts so as to increase the factor of safety to 2.00. The bolts are standard steel rock bolts with diameter 2.54 cm and shear strength 348 MPa. Due to a manufacturing defect, the bolts cannot be tightened. How many bolts are needed for a roadcut 100 m long in the same kind of rock as described in Problem 27?

29. A 56.0°-roadcut is made in sedimentary rock. The bedding planes are parallel and evenly spaced from, and including, the edge E of the block. See Fig.(4.21). The bedding planes dip at 27.0°. Bedding plane I passes through the toe of the cut, just where you are standing. The angle of friction between bedding planes is 20.0°. Tests show that the factor of safety against the entire block sliding as a unit along plane I is 1.60. (*a*) Find the factor of safety against sliding of the entire block above plane II along plane II. (*b*) Similarly, find the factor

of safety against sliding of the block abovc plane III along plane III.

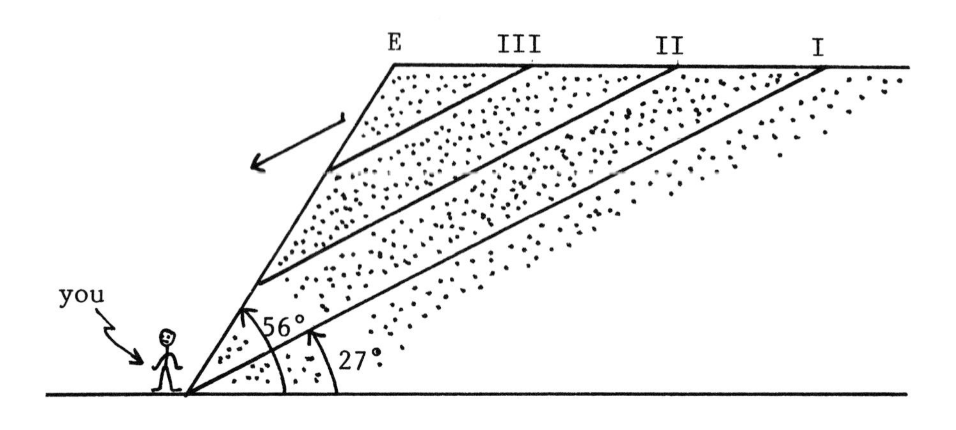

Fig.(4.21) Problem 29

30. A 57.0°-roadcut 11.2 m deep and 75.0 m long is made in granitic rock of density 2.77 g/cm^3. A set of parallel joints in the rock dip at 43.5° directly toward the cut. Tests show that cohesion equals 23.6 kPa and the angle of friction is 19.0°. (*a*) Find the factor of safety. (*b*) How many stitches, each of diameter 2.34 cm and under tension 512 MPa, must be installed to raise the factor of safety to 1.50?

31. Which, if any, of the rectangular blocks in Fig.(4.22) will topple? Assume no sliding.

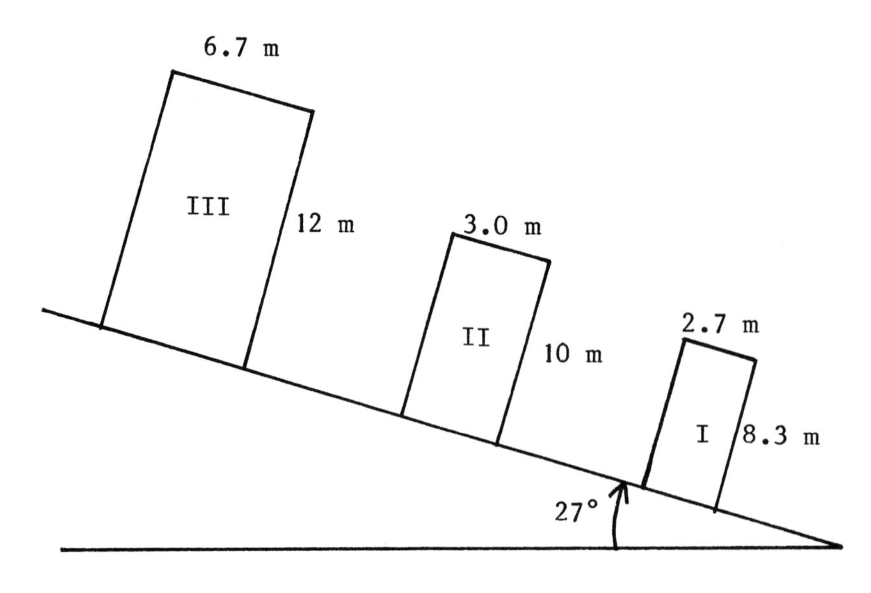

Fig.(4.22) Problem 31

32. A block has the shape of a right-triangular prism, as shown on Fig.(4.23). The block has height 13.0 m and is 11.0 m wide. It rests on a 35.0°-slope. Find the largest possible value of the angle θ of the block so that it will not topple. Assume no sliding.

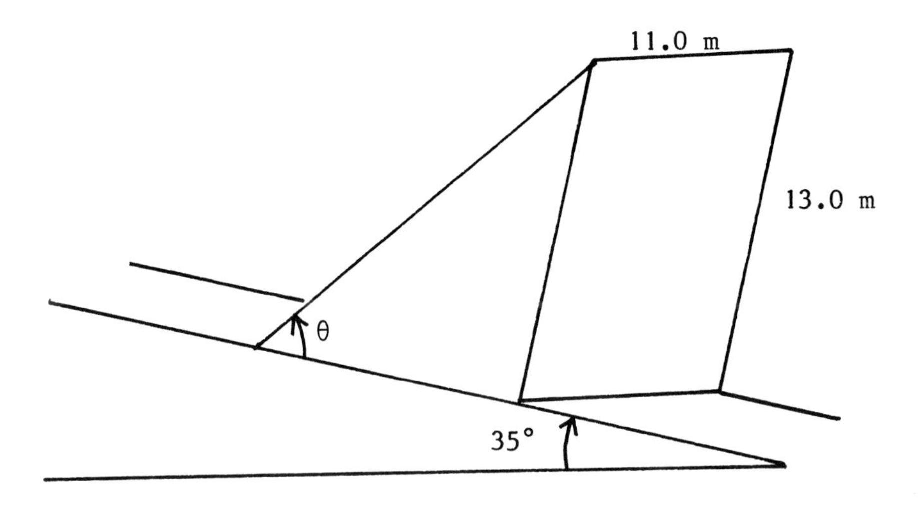

Fig.(4.23) Problem 32

Chapter 5

Soils

5.1 Soil Density

The term 'soil' does not have a single definition, agreed upon by all. To the general public, soil is a material that can support plant life. But to a geologist, soil is an end-product of the weathering of rocks, a material so degraded by the elements, so to speak, that it can be scooped-out virtually by hand. It is the geological definition that is adopted in this chapter.

Even though soil so defined is composed of the remnants of severely weathered rock, the physical properties of soil are very different from those of intact rock. The differences arise from the nature of the grains, and the amount of water in the soil.

Like rock, soil contains open spaces within the solid matrix. In rock studies, the solid part is called the matrix or grains, and the open spaces pores. In describing soil properties, the open spaces are commonly called *voids* , and the solid part just that: the *solids.* The voids are either 'empty' (i.e., contain only air), or are either partially or completely filled with a liquid, most probably water. In this chapter, it will be assumed that the liquid is indeed water.

Water content is such an important parameter in determining the physical properties of soil that it is not a sufficiently good approximation to describe the soil as either 'dry' (contains no water) or 'saturated' (all voids completely filled with water), as was done with rock; see Chapter 1. Rather, with soil, the degree of saturation must be considered.

In describing the basic properties of rock, the porosity is used to specify quantitatively the relative volume of pores. In soil studies, it is more common to use, instead of porosity, a quantity called the *void ratio* e, defined by

$$e = \frac{V_{\mathrm{v}}}{V_{\mathrm{s}}}, \tag{5.1}$$

where V_{v} is the total volume of all the voids within the boundary of the soil sample, and V_{s} is the total volume of all the solids.

By way of a reminder, the porosity is defined by $n = V_{\mathrm{v}}/V$, where V is the total volume of the sample. Since $V = V_s + V_v$, it follows that

$$n = \frac{V_v}{V_s + V_v},$$

$$n = \frac{V_v/V_s}{1 + (V_v/V_s)},$$

$$n = \frac{e}{1+e}. \tag{5.2}$$

Equation(5.2) allows for evaluation of the porosity given the void ratio. To perform the reverse task, invert Eq.(5.2) to get

$$e = \frac{n}{1-n}. \tag{5.3}$$

The porosity can have a value in the range $0 \leq n < 1$. The corresponding range for the void ratio is $0 \leq e < \infty$. Like porosity, void ratio is often expressed as a percent, but must be in decimal form when used in formulas.

Soil can be compacted through a decrease in the void ratio due to the weight of overlying material, just as rock can be similarly compacted. With rock, the compaction is usually written in terms of the porosity n, but soil engineers generally use void ratio e.

Turning now to expressing the degree to which the voids are filled with water, define the *fraction of saturation* f by

$$f = \frac{V_w}{V_v}, \tag{5.4}$$

where V_w is the total volume of all the water in the soil sample; of course, this water is found in the voids. The volume units cancel, so f has no units. For a dry sample $V_w = 0$ so that $f = 0$. For a saturated sample, all the voids are completely filled, so that $V_w = V_v$, giving $f = 1$. The fraction of saturation is sometimes expressed as a percent.

As with rock, the bulk density ρ (often simply called density) is defined by

$$\rho = \frac{m}{V}, \tag{5.5}$$

where m is the mass of a sample of the soil and V is the volume of the sample. This bulk density must be expressable in terms of the densities of the soil components, solids, water and air, and their relative amounts through the void ratio and fraction of saturation. To find the expression, write

$$\rho = \frac{m_s + m_w}{V_s + V_v}, \tag{5.6}$$

where m_s is the total mass of solids and m_w is the total mass of water in the sample. (The mass of air is ignored by virtue of the very small density of air compared with that of water and of the solid part of the soil.) Writing ρ_s and ρ_w for the densities of the solids and of water, respectively, and applying Eq.(5.1) to eliminate the volume V_s of the solid part, Eq.(5.6) becomes

$$\rho = \frac{\rho_s V_s + \rho_w V_w}{V_v + V_v/e},$$

$$\rho = \frac{\rho_s V_s + \rho_w V_w}{V_v(1 + 1/e)}.$$

Now use Eq.(5.4) to eliminate V_v in favor of V_w to find

$$\rho = \frac{\rho_s V_s + \rho_w V_w}{(V_w/f)(1 + 1/e)}. \tag{5.7}$$

Equations (5.1) and (5.4) can be combined to obtain the volume of solids:

$$V_s = \frac{V_w}{ef}. \tag{5.8}$$

Finally, substitute Eq.(5.8) into (5.7) and rearrange a little to arrive at the desired expression relating the densities:

$$\rho = \frac{\rho_s + ef\rho_w}{1 + e}. \tag{5.9}$$

An alternative parameter to f to describe the degree of saturation is the *water content* wc (single symbol). The water content usually is defined as the ratio of the weight of water W_w to the weight of solids W_s, i.e, by

$$wc = \frac{W_w}{W_s}. \tag{5.10}$$

The weight units cancel, so wc is dimensionless.

Since each weight in Eq.(5.10) can be written as the product of the appropriate density, volume and gravity, the water content can be written as

$$wc = \frac{\rho_w V_w}{\rho_s V_s}, \tag{5.11}$$

i.e., as the mass ratio of water to solids.

Applying Eq.(5.8)

$$wc = (ef)(\frac{\rho_w}{\rho_s}). \tag{5.12}$$

Solving Eq.(5.12) for ρ_w and substituting into Eq.(5.9) to obtain an expression for the bulk density in terms of the water content yields

$$\rho = \rho_s\left(\frac{1 + wc}{1 + e}\right). \tag{5.13}$$

Over the entire spectrum of all possible soils, the water content, like the void ratio, has the mathematical range $0 \leq wc < \infty$. However, for any particular soil with a specific value of the void ratio, the maximum possible value of the water content is that given by Eq.(5.12) with $f = 1$ (completely saturated).

****

EXAMPLE 1

(*a*) Calculate the void ratio for a soil that, when 70.0% saturated with water, has a bulk density of 2.06 g/cm^3. The solid part of the soil has a density of 2.92 g/cm^3. (*b*) Find the water content of the soil.

(*a*) The given data are: $f = 0.700$, $\rho = 2.06$ g/cm^3, $\rho_\mathrm{w} = 1.00$ g/cm^3, $\rho_\mathrm{s} = 2.92$ g/cm^3; the quantity to be found is e. Use Eq.(5.9); cancelling the common density units leads to

$$\rho = \frac{\rho_\mathrm{s} + ef\rho_\mathrm{w}}{1+e},$$

$$2.06 = \frac{2.92 + e(0.7)(1)}{1+e},$$

$$2.06 + 2.06e = 2.92 + 0.7e,$$

$$e = 0.632.$$

(*b*) With the void ratio now known, the water content wc can be found directly from Eq.(5.12). Again, the common density units cancel, leaving

$$wc = (ef)(\frac{\rho_\mathrm{w}}{\rho_\mathrm{s}}),$$

$$wc = (0.632)(0.700)(\frac{1}{2.92}),$$

$$wc = 0.152.$$

Equation (5.13) could have been used instead.

****

Two particular values of the water content are especially important in characterizing the physical properties of a soil. The *liquid limit* LL is the smallest value of the water content for which the soil behave like a viscous liquid; i.e., as a liquid with internal friction (*viscosity*). The *plastic limit* PL is the smallest value of the water content for which the soil behaves like a plastic material; i.e., one in which deformations induced by an applied load remain after the load is removed. The liquid limit is greater than the plastic limit: $LL > PL$. The difference is called the *plasticity index* PI:

$$PI = LL - PL. \tag{5.14}$$

All of these quantities can be expressed as a decimal, or as a percent.

It must be borne in mind that soil is a very complex material, and its properties cannot be completely described by giving the values of a few parameters such as PL, LL, or others that are not described here.

5.2 Soil Texture

The particles of solid matter in a particular soil can have a wide distribution of sizes. Fortunately, it is not necessary to determine the size of every particle in a sample of soil to anticipate its engineering properties. Rather, it is usually sufficient to specify the relative number of particles in certain ranges of size. (Sometimes, the term "diameter" is used instead of "size" although, strictly speaking, diameter has a meaning only for a spherical particle; soil grains are actually irregular in shape.)

Different government and private agencies use different size ranges, but they roughly follow the scheme presented here in Table (5.1), where D stands for particle size.

Particle Size Range	Name of Group
$D > 2$ mm	gravel
2 mm $> D >$ 0.05 mm	sand
0.05 mm $> D >$ 0.002 mm	silt
0.002 mm $> D$	clay

Table 5.1 Soil Texture Classification

The choice of the term *clay* for particles less than 0.002 mm in size is unfortunate since "clay" also denotes a certain class of minerals, specifically aluminum silicates that contain hydrated water (the individual water molecules are chemically bonded to mineral molecules). One of these clay minerals, montmorillonite (also known as smectite) can contain from one up to more than ten water molecules per mineral molecule; this results in a wide variation in the volume of the mineral particles depending on the water content. Not all clay minerals are expansive. However, these minerals form small solid particles that roughly (only roughly) fall in the size range labelled as clay.

The size of a soil particle is sometimes expressed on a logarithmic scale often used in describing the size of particles in sedimentary rock. This is the *Wentworth*, or *phi*, scale in which the size of the particle is specified by a dimensionless number phi, symbol ϕ, defined by the equation

$$\phi = -\log_2 D, \tag{5.15}$$

where D is the size, or diameter, of the particle in millimeters. Most calculators do not have a button dedicated to logarithms to the base 2, so for calculational purposes it is better to write Eq.(5.15) as

$$D = 2^{-\phi}, \tag{5.16}$$

so that

$$\ln D = -\phi \ln 2,$$

$$\phi = -\frac{\ln D}{\ln 2}. \tag{5.17}$$

The minus sign is used in the definition of the phi scale because most (not all) of the particles in soil are small, leading thereby to positive values of ϕ. [Logarithms to the base 10 could be used in place of natural logarithms in Eq.(5.17).]

****

EXAMPLE 2
A small cobble has a diameter of 6.27 cm. Find its size on the Wentworth scale.

The size must be specified in mm, so $D = 62.7$ must be used in Eq.(5.17). The units are suppressed, but the value of D must correspond to mm. By Eq.(5.17),

$$\phi = -\frac{\ln D}{\ln 2},$$

$$\phi = -\frac{\ln 62.7}{\ln 2},$$

$$\phi = -5.97.$$

Alternatively, $\phi = -(\log 62.7)/(\log 2)$ could be used.

****

The size distribution of the particles smaller than 2 mm in size is labelled a soil's *texture.* The gravel component ($D > 2$ mm) is omitted in describing texture. This leaves the three particle size groups of sand, silt and clay particles; see Table (5.1). With three groups a triangular diagram, like that used in describing the mineral composition of rocks [Section (1.8)], can be employed to display the size distributions.

In the triangular rock mineral composition diagram, the composition of a point in the triangle is found by projecting from the point to the three perpendicular bisectors of the triangle, meeting them at 90°. Each bisector indicates the composition of the mineral at the apex of that bisector, with 100% at the apex and 0% at the intersection of the bisector with the side of the triangle opposite the apex.

By contrast, soil texture triangular diagrams usually display the percent compositions along the sides of the triangle, rather than along the bisectors. From a point inside the triangle, projections are made to each side along lines parallel to the sides of the triangle. These lines do not meet the sides at 90°.

Figure (5.1) illustrates the difference. The soil texture components are sand (S), clay (C) and Silt (I). By either method, at each apex the composition is 100% of the component at that apex. A point falling on a side of the triangle is made up only of the two components marked at the ends of that side. [For example, a point on the "base" of the triangle of Fig.(5.1) is composed of sand and silt only.]

There are two ways to label the sides. On the base, the composition could be labelled as percent sand, increasing toward S, or percent silt increasing toward I. Suppose that percent silt is used on the base; then the other sides must be labelled as increasing percent in the counterclockwise sense started along the base. Hence, the labelling as shown on Fig.(5.1).

Now consider the point P shown on the figure inside the triangle. On the original (rock) scheme, using the bisectors, the reading at E is the percent clay, at F the percent silt and at G the percent sand. This scheme works because, as proved laboriously in Chapter 1, the distances add to the length of a bisector; i.e.,

$$(\mathrm{ME}) + (\mathrm{QG}) + (\mathrm{RF}) = L\sqrt{3}/2, \tag{5.18}$$

where L is the length of each side of the triangle. Hence, each bisector can be marked-off from 0% at the base of the side it meets to 100% at its apex; Eq.(5.18) guarantees that the three abundances add to 100%.

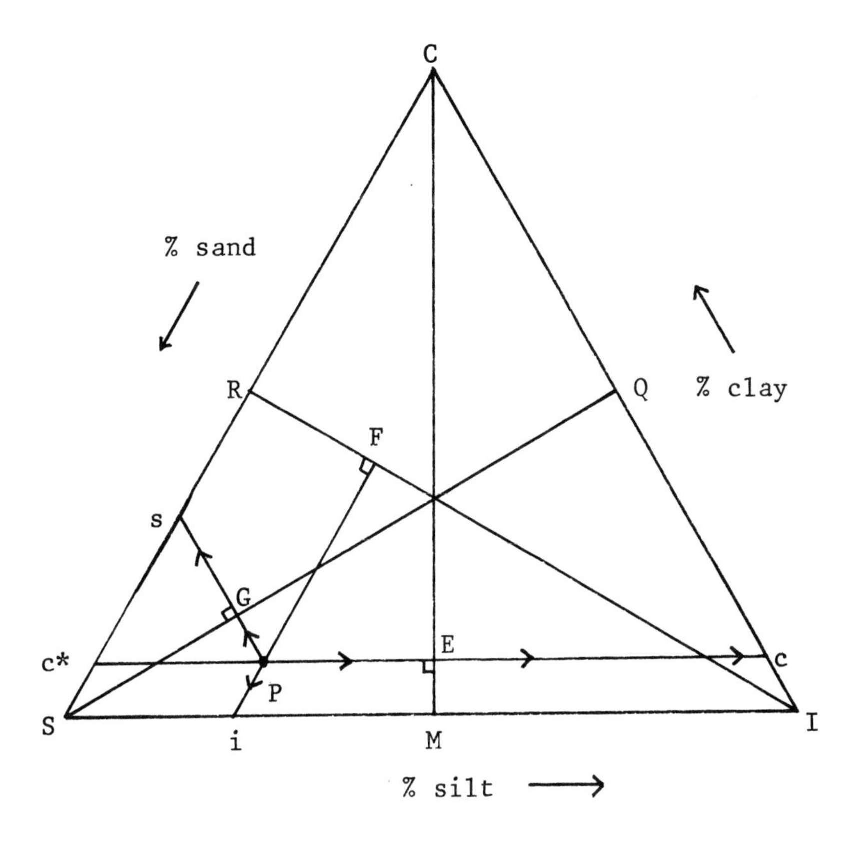

Fig.(5.1) Soil Texture Triangle Chart

Now examine the new scheme, calling for the projections from P to be parallel to the sides of the triangle. These projections will be along lines perpendicular to the bisectors. In doing this, an ambiguity immediately arises, for each projection could go in two directions. For example, from P the line parallel to the base of the triangle could go to the left or right. The rule is to take each projection in the same direction as the percent abundance increases along the side of the triangle that is parallel to the projection. For Fig.(5.1), this means to project in the directions indicated by the arrows.

The projections strike the sides at the points c, s, and i. The assertion is that the number on the side at point c gives the percent abundance of clay, the number at point s the percent of sand, and the number at point i the percent of silt.

For this assertion to hold, an equation similar to Eq.(5.18) must hold; i.e., the three abundances so defined must add up to 100%. For this to be true, the relevant dimensions must add up to L, the length of a side (since the sides are now labelled with percent abundance). That is, it is necessary that

$$(\mathrm{Ic}) + (\mathrm{Cs}) + (\mathrm{Si}) = L. \tag{5.19}$$

To see if Eq.(5.19) holds, note that by projecting from P to the left, parallel to the base, to point c*, two similar triangles are obtained: ΔCSI (the original) and ΔCc*c. By the properties of similar triangles,

$$\frac{(\mathrm{ME})}{(\mathrm{MC})} = \frac{(\mathrm{Ic})}{(\mathrm{IC})}. \tag{5.20}$$

But this means that

$$\frac{(\mathrm{RF})}{(\mathrm{RI})} = \frac{(\mathrm{Si})}{(\mathrm{SI})}, \tag{5.21}$$

and

$$\frac{(\mathrm{QG})}{(\mathrm{QS})} = \frac{(\mathrm{Cs})}{(\mathrm{CS})}, \tag{5.22}$$

since an equilateral triangle presents the same appearance regardless of which side is drawn as the base.

Now, in Eqs.(5.20), (5.21), (5.22), the denominator on the left is the length of a bisector, or $L\sqrt{3}/2$; each denominator on the right is the length of a side, or L. With this in mind, solve Eq.(5.20) for (Ic), Eq.(5.21) for (Si) and Eq.(5.22) for (Cs) and substitute into Eq.(5.19) to get

$$(\mathrm{ME}) + (\mathrm{RF}) + (\mathrm{QG}) = \frac{\sqrt{3}}{2}L. \tag{5.23}$$

But this is just what is proven in the developement of the original scheme in Section (1.8). Therefore, Eq.(5.19) does hold: the three abundances will add up to 100%, as required. Also, Eqs.(5.20), (5.21), (5.22) guarantee that the same abundances will be found whether the "bisector" or "side" scheme is used. The two apparently different ways of reading a triangular composition diagram lead to identical results.

****

EXAMPLE 3

The clay particle abundance in a certain soil sample is 31.9% of the particles smaller than 2 mm, but only 23.6% of the entire sample. What percent of the entire sample are particles smaller than 2 mm?

The size 2 mm is the upper limit to the clay, silt, sand population; larger particles are gravel. If N denotes the total number of particles smaller than 2 mm, N_{C} the number of clay particles, and N_{total} the total number of particles, then the data can be written

$$N_{\mathrm{C}} = 0.319(N),$$

$$N_{\mathrm{C}} = 0.236(N_{\mathrm{total}}).$$

Setting the equations equal yields

$$0.319(N) = 0.236(N_{\mathrm{total}}),$$

$$\frac{N}{N_{\mathrm{total}}} = \frac{0.236}{0.319},$$

$$\frac{N}{N_{\mathrm{total}}} = 0.740,$$

so the desired abundance is 74.0%.

****

5.3 Vertical Soil Stress

In a geologically quiet region, the subsurface vertical stress at a certain depth in soil is, just as in a rock layer, due to the weight of the material situated above. If the soil has a uniform bulk density between the surface and the point at depth D where the stress needs to be known, then, as for rocks,

$$\sigma = \rho g D, \tag{5.24}$$

where σ is the stress and ρ the bulk density of the soil. Needless to say, g is the acceleration due to gravity, $g = 9.8\ \mathrm{m/s^2}$.

If the soil is organized into horizontal layers, with the density uniform in each layer, then, by strict analogy with rock layers,

$$\sigma = \sum_i \rho_i g y_i, \tag{5.25}$$

where y_i are the appropriate thicknesses of the various layers of soil situated between the point at depth D and the surface.

A possible complication in dealing with soils is that the bulk density may vary in a continuous manner between certain depths. This may occur as a result of the water content varying with depth, and/or compaction due to reduction in void ratio at greater depths. It is even possible that the density of the solid portions may increase at depths much less than the depths at which such an effect occurs in rocks.

If, then, the bulk density ρ varies in a continuous manner from the surface down to depth D, the expression for the stress at that depth must be the integral version of Eq.(5.25); that is, replace the thicknesses y_i with the differential thickness dy and the replace the sum with an integral to get

$$\sigma = \int_0^D \rho g\, dy. \tag{5.26}$$

****

EXAMPLE 4

A dry soil has a bulk density ρ_0 and void ratio e_0 at the surface. The void ratio decreases linearly with depth, reaching a value of zero at depth D. The density of the solids remains constant. Find an expression for the vertical stress at depth D.

Since the soil is dry, the fraction of saturation is $f = 0$. Equation (5.9) then yields

$$\rho = \frac{\rho_s}{1+e},$$

where ρ_s is the density of the solids. At the surface $e = e_0$, so the surface bulk density is

$$\rho_0 = \frac{\rho_s}{1+e_0}.$$

The generic linear expression for the variation of the void ratio e with depth y is

$$e = my + b,$$

where m is the slope and b is the e-intercept of the corresponding graph. It is required that $e = e_0$ at $y = 0$, and $e = 0$ at $y = D$. Substituting these two requirements successively into the preceding equation gives

$$e_0 = b,$$

$$0 = mD + b.$$

Hence $b = e_0$ and $m = -e_0/D$. Putting these results into the generic linear expression for void ratio gives the variation of void ratio with depth:

$$e = e_0\left(1 - \frac{y}{D}\right).$$

Now put this expression for e into that for ρ and rearrange, using the formula above for ρ_0, to find that

$$\rho = \rho_0\left(\frac{1}{1-ay}\right),$$

where a is a shorthand:

$$a = \left(\frac{e_0}{1+e_0}\right)\frac{1}{D}.$$

Now apply Eq.(5.26), using this relation for the density, to find the stress at depth D:

$$\sigma = \rho_0 g \int_0^D \frac{dy}{1-ay},$$

$$\sigma = \left(\frac{\rho_0 g}{a}\right)\ln\left(\frac{1}{1-aD}\right).$$

Finally, substitute for the shorthand a to find that

$$\sigma = (\rho_0 g D)\left(\frac{1+e_0}{e_0}\right)\ln(1+e_0).$$

This is an algebraic problem, and therefore a numerical value is not substituted for g. The expression above is in terms of the (algebraically) given quantities ρ_0, e_0 and D, as required in such calculations.

****

5.4 Effective Stress

Often it is of sufficient accuracy to ignore subtle variations in density, and to consider the soil to be stratified in horizontal, plane-parallel layers, within each layer the density being constant. Then the stress at any depth can be calculated from Eq.(5.25), as is done with rock layers in Chapter 2. Distinct soil layers often are called *horizons*.

For example, consider a soil's water profile, that is, the variation of water content with depth. Suppose that the soil is dry at the surface. It is common experience that at a sufficiently great depth the soil will be found completely saturated. Over a certain region, the surface connecting points of minimum depth at which the soil is completely saturated is called the *water table* (diagram symbol $\underline{\nabla}$). Below the water table the soil is completely saturated ($f = 1$). Directly above the water table the soil is not dry: there exists a complex transition zone leading to the actual dry ($f = 0$) soil. But the effects of this transition zone can sometimes be overlooked. In this event, the soil is considered dry above the water table, and completely saturated below it. If there is no compaction, then the soil is thereby divided into two horizontal layers, with the bulk density within each layer constant.

Of course, the water table could, under exceptional circumstances, be at the surface itself. Then there is only one layer. Another special case is that of a soil layer under water: the soil beneath a lake, or reservoir, for example. This situation constitutes two parallel layers, the top one being just water ($e = \infty, f = 1$).

Figure (5.2) is of the first situation described above. The soil is dry down to the water table at depth D, and completely saturated at greater depth. It is presumed that the void ratio has the same value both above and below the water table. (This assumption would break down at great enough depth.) The bulk density of the dry soil is ρ_0 and of the saturated soil is ρ_{sat}. [These bulk densities are related through Eq.(5.9).] How does the vertical stress σ vary with depth?

For points above the water table, $0 \le y \le D$, Eq.(5.25) gives $\sigma = \rho_0 g y$. For points below the water table, $y \ge D$, the same equation gives

$$\sigma = \rho_0 g D + \rho_{\text{sat}} g (y - D). \tag{5.27}$$

The depth y is measured from the surface. The expressions for the stress at points above and at points below the water table must yield the same result at the water table itself, i.e., at $y = D$.

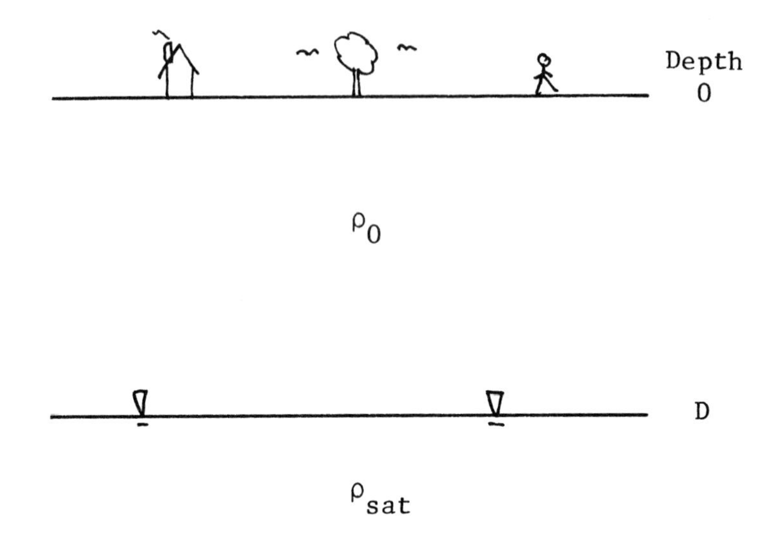

Fig.(5.2) The Water Table and Stress

Now suppose that the voids, which below the water table are completely filled with water, are physically connected, like a randomly-arranged network of tiny pipes. Then part of the stress in Eq.(5.27) can be attributed to the water. This hydrostatic stress of the water, usually called pressure p is given by

$$p = \rho_w g(y - D), \tag{5.28}$$

where ρ_w is the density of water. The quantity $(y - D)$ is the height of the water column. If this contribution be subtracted out of the total stress, Eq.(5.27), then the remaining part is borne by the solids of the soil. This part is called the *effective stress* σ_{eff}. That is,

$$\sigma_{\text{eff}} = \sigma - p. \tag{5.29}$$

Substituting Eqs.(5.27) and (5.28) into Eq.(5.29) and solving for the effective stress gives

$$\sigma_{\text{eff}} = \rho_0 g D + (\rho_{\text{sat}} - \rho_w) g(y - D). \tag{5.30}$$

The effective stress is important for many reasons. Being the stress borne by the solids, if it is too large for the particular soil, then compaction through collapse of the voids can take place, provided that the water to be squeezed out of the voids has a place to go. If the water in the soil is pumped out, then the remaining solids can support only an overburden stress of σ_{eff}, and not of the original "total" stress σ.

****

EXAMPLE 5

In a certain region the soil profile, down to the greatest depth of engineering interest, consists of three layers, or horizons: soils 1, 2, and 3. On Fig.(5.3) are the locations, soil solids densities and void ratios of the three soils. The water table is at a depth of 24.9 m, in soil 2. Find the effective stress at a depth of 49.0 m, in soil 3.

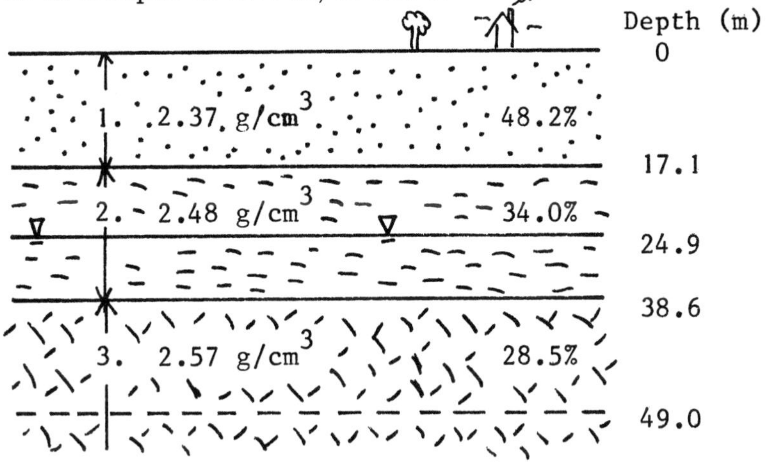

Fig.(5.3) Example 5

In effect, there are four layers, since soil 2 has a dry layer and a saturated layer. From the given values of ρ_s and e, the dry bulk densities can be calculated for soils 1 and 2 from Eq.(5.9) with $f = 0$. Also from Eq.(5.9), the saturated bulk densities of soils 2 and 3 can be found, using $f = 1$ and $\rho_w = 1000\ \mathrm{kg/m^3}$. These results are: $\rho_{1,\mathrm{dry}} = 1599\ \mathrm{kg/m^3}$; $\rho_{2,\mathrm{dry}} = 1851\ \mathrm{kg/m^3}$; $\rho_{2,\mathrm{sat}} = 2104\ \mathrm{kg/m^3}$; $\rho_{3,\mathrm{sat}} = 2222\ \mathrm{kg/m^3}$. Now apply Eq.(5.25) to get the vertical stress at the indicated depth of 49.0 m. Omitting the SI base units,

$$\sigma = \sum_i \rho_i g y_i,$$

$$\sigma = (1599)(9.8)(17.1) + (1851)(9.8)(7.8) + (2104)(9.8)(13.7) + (2222)(9.8)(10.4),$$

$$\sigma = 918.4\ \mathrm{kPa}.$$

The thickness of the zone of water between the water table at depth 24.9 m and the level at depth 49.0 m is 24.1 m. Therefore, the hydrostatic pressure at depth 49.0 m is

$$p = \rho_w g y,$$

$$p = (1000\ \mathrm{kg/m^3})(9.8\ \mathrm{m/s^2})(24.2\ \mathrm{m}),$$

$$p = 236.2\ \mathrm{kPa}.$$

Therefore, the effective stress at the depth of 49.0 m is

$$\sigma_{\mathrm{eff}} = \sigma - p,$$

$$\sigma_{\text{eff}} = 918.4 \text{ kPa} - 236.2 \text{ kPa},$$

$$\sigma_{\text{eff}} = 682 \text{ kPa}.$$

****

5.5 Shear Strength

The failure (rupture) in a layer of soil under an applied load usually occurs as a result of the shear strength of the soil being exceeded on a plane within the soil along which slippage is physically possible. This is true even if the loading appears to be axial, for an axial load sets up shear stresses within the soil.

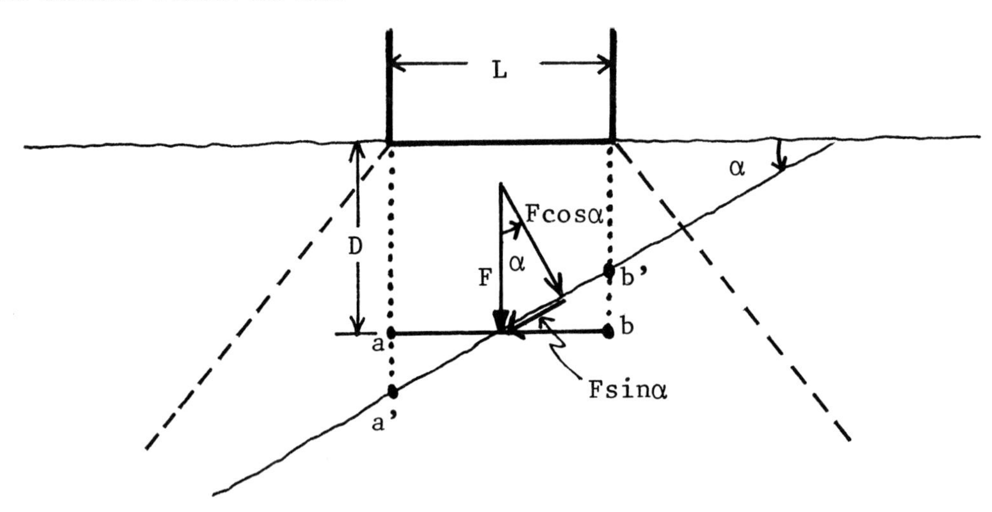

Fig.(5.4) Normal and Shear Stresses

Consider the building of weight W shown on Fig.(5.4), constructed on soil. The weight is actually not transmitted vertically downward through the soil layer: rather, a bulb of stress (dashed lines) emanates from the area of contact. (This is reminiscent of Pascal's principle for a confined liquid: an external load is transmitted to all of the sides of the container.) Let F be that part of the building's weight that is transmitted down (dotted lines). At a depth D, this acts perpendicular to the horizontal plane with edge ab. Suppose that the building has a square base, so the dimension perpendicular to the paper is L (this assumption is not critical to what follows). Then, there is an axial stress σ_{B} on this surface, due to the building, given by

$$\sigma_{\text{B}} = \frac{F}{L^2}. \tag{5.31}$$

Now examine the situation along a plane inclined at a dip angle α to the horizontal [shown on Fig.(5.4) as a potential slip plane]. Relative to this inclined plane, the force F

is not perpendicular; rather, there is a perpendicular component $F\cos\alpha$ and a tangential component $F\sin\alpha$. The perpendicular component gives rise to a compressive stress, which does not tend to cause failure. The tangential component gives rise to a shear stress, which can cause failure if the wedge of soil above (or below) the plane can move.

Consider, then, the tangential force. It acts on a rectangular area, two edges being of length a'b' and the other two of length L. But a'b' $= L/\cos\alpha$. Therefore, the shear stress on this plane is

$$\tau = \frac{F\sin\alpha}{L(L/\cos\alpha)},$$

$$\tau = \frac{1}{2}\sigma_{\mathrm{B}}\sin 2\alpha. \tag{5.32}$$

This result, Eq.(5.32), does not really depend on the building having a square base. Any shape can be constructed from a very large number of very small squares, so that Eq.(5.32) is valid regardles of the shape of the base.

This situation resembles the stability of intact rock on a slope. The shear stress in Eq.(5.32) is a driving stress; for the soil to remain at rest, a resisting stress is needed. The maximum available resisting stress, as given in Section (4.2) but renamed just τ here, is

$$\tau = c + \sigma_{\mathrm{n}}\tan\phi, \tag{5.33}$$

where c is the cohesion (resistance at zero normal stress), σ_{n} is the normal stress, and ϕ is the angle of friction. In short, the concepts of cohesion and angle of friction apply to soils as well as to rock. Equation(5.33) is often called the *Mohr-Coulomb equation*

For example, clay soils have $\phi \approx 0°$ and a cohesion c that varies with water content; a dry clay may have $c \approx 100$ kPa. On the other hand, a dry sand has $c \approx 0$ and angle of friction $\phi \approx 30°$.

The normal stress σ_{n} may, of course, be due just to the weight W of the soil above the plane (i.e., due to the overburden). In this case $\sigma_{\mathrm{n}} = (W\cos\alpha)/A$, where A is the contact area on the plane, $L^2/\cos\alpha$ in the notation of Fig.(5.4), so that

$$\sigma_{\mathrm{n}} = \left(\frac{W}{L^2}\right)\cos^2\alpha.$$

The factor W/L^2 is the vertical stress σ due to the overburden discussed in Section(5.3), and therefore

$$\sigma_{\mathrm{n}} = \sigma\cos^2\alpha.$$

One last complication. If the soil being examined is below the water table, then the pressure p due to the water exerts a buoyant force on the soil normal to the plane of potential failure (unless the water drains away under the applied stress). The normal stress is reduced by this pressure, giving

$$\sigma_{\mathrm{n}} = \sigma\cos^2\alpha - p. \tag{5.34}$$

Substituting Eq.(5.34) in Eq.(5.33) gives the Mohr-Coulomb equation the form

$$\tau = c + (\sigma \cos^2 \alpha - p) \tan \phi. \tag{5.35}$$

The quantity τ calculated by Eq.(5.35) is the maximum resistive stress available under the circumstances described by the values of σ, α and p.

When soil is tested in the laboratory, the load represented by σ is mechanically applied, and slippage along planes perpendicular to the load is examined ($\alpha = 0$). If the soil is dry, $p = 0$. The expression for the shear strength is correspondingly simplified.

****

EXAMPLE 6

The shear strength of a soil along a horizontal layer at a depth of 38.2 m is measured to be 429 kPa. The water table is at a depth of 15.0 m. The dry soil has a unit weight of 20.8 kPa/m and the saturated soil a unit weight of 28.3 kPa/m. The angle of friction of the saturated soil is 25.0°. Find the cohesion of the saturated soil.

Call the depth to the water table D and the depth to the horizontal plane H. The stress on the plane due to the overburden is

$$\sigma = \gamma_{\text{dry}} D + \gamma_{\text{sat}} (H - D),$$

$$\sigma = (20.8 \text{ kPa/m})(15 \text{ m}) + (28.3 \text{ kPa/m})(38.2 \text{ m} - 15 \text{ m}),$$

$$\sigma = 968.6 \text{ kPa}.$$

The pore pressure is

$$p = \gamma_{\text{w}} (H - D),$$

$$p = (9.8 \text{ kPa/m})(38.2 \text{ m} - 15 \text{ m}),$$

$$p = 227.4 \text{ kPa}.$$

Now use Eq.(5.35); since $\alpha = 0$, $\cos^2 \alpha = 1$, so that

$$\tau = c + (\sigma \cos^2 \alpha - p) \tan \phi,$$

$$429 \text{ kPa} = c + (968.6 \text{ kPa} - 227.2 \text{ kPa}) \tan 25°,$$

$$c = 83.4 \text{ kPa}.$$

****

5.6 Bearing Capacity

In Section(5.3) the vertical stress in soil is evaluated. However, it is likely that horizontal stresses also are present. A horizontal stress might arise, for example, from the hydrostatic pressure exerted by water in the voids; this pressure acts in all directions and, therefore, also in the horizontal direction. In general, though, it is not possible to write a simple formula for the horizontal stress, for it could arise from many causes.

In spite of the lack of a general formula for the horizontal stress, suppose that it is not zero in a certain layer of soil, and that a vertical load is going to be applied to this material (perhaps the load is the weight of a building to be constructed). The question arises: How great a vertical load can be applied without triggering soil failure?

The existence of a horizontal stress affects the answer. The soil is "confined"; i.e., under a confining compressive stress on its sides, and this affects the vertical compressive strength. Just as for rock, the confined compressive strength is greater than the unconfined compressive strength; the maximum load that can be applied to a free-standing column of soil is less than the load that can be sustained by a column of soil under a confining lateral stress.

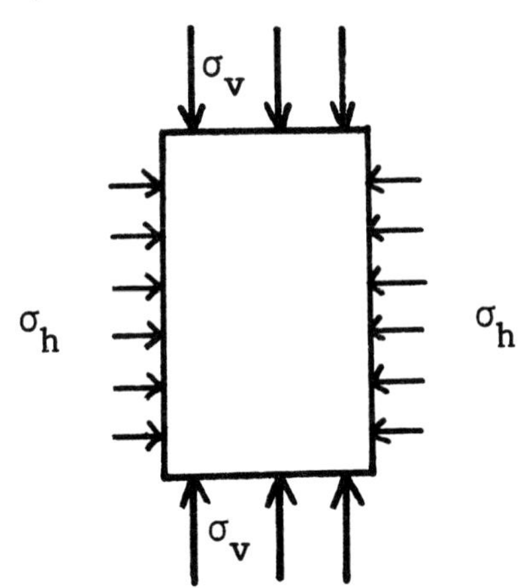

Fig.(5.5) Triaxial Compression Test

Soil can be tested for confined compressive strength, or *bearing capacity* in the laboratory, in a so-called *triaxial compression test.* A cylindrical sample of the soil is placed in a container and a confining hydrostatic pressure is exerted by a surrounding liquid. An additional load is mechanically applied to the ends of the cylinder. These stresses are shown on the sample in Fig.(5.5). The axis of the cylinder is vertical. The confining pressure is σ_h, which acts in the horizontal plane on the sides of the cylinder. The vertically directed axial stress on the ends of the cylinder is σ_v; this represents the external load to be supported. The test is called "triaxial" because stresses exist in three dimensions.

The hydrostatic pressure is exerted against the ends of the cylinder as well as against the sides. This is included in σ_v, so that $\sigma_v \geq \sigma_h$.

The test consists in gradually increasing the value of σ_v until failure. The data consists of the value of σ_v at failure together with the value of the confining stress σ_h. Also recorded is the pore presure due to water in the voids of the soil sample itself. (This is zero if the soil is dry.) The void pressure is subtracted from both σ_v and σ_h, converting them to effective stresses. Henceforth, then, σ_h and σ_v will stand for the effective lateral confining stress and the effective axial loading stress, respectively, at failure of the soil sample.

The test is then repeated, using another sample of the same soil, but at a higher confining stress σ_h. Again, the value of the axial loading stress at failure is noted. Effective stresses are calculated if the soil sample is saturated with water. It is expected that this second axial stress at failure will be greater than the value in the first test, which is conducted at a smaller confining pressure.

After two tests, then, the data consists of the effective stresses at failure, σ_{h1} and σ_{v1} for the first test, and σ_{h2} and σ_{v2} for the second. The next task is to determine the cohesion c and internal angle of friction ϕ for the soil from this data.

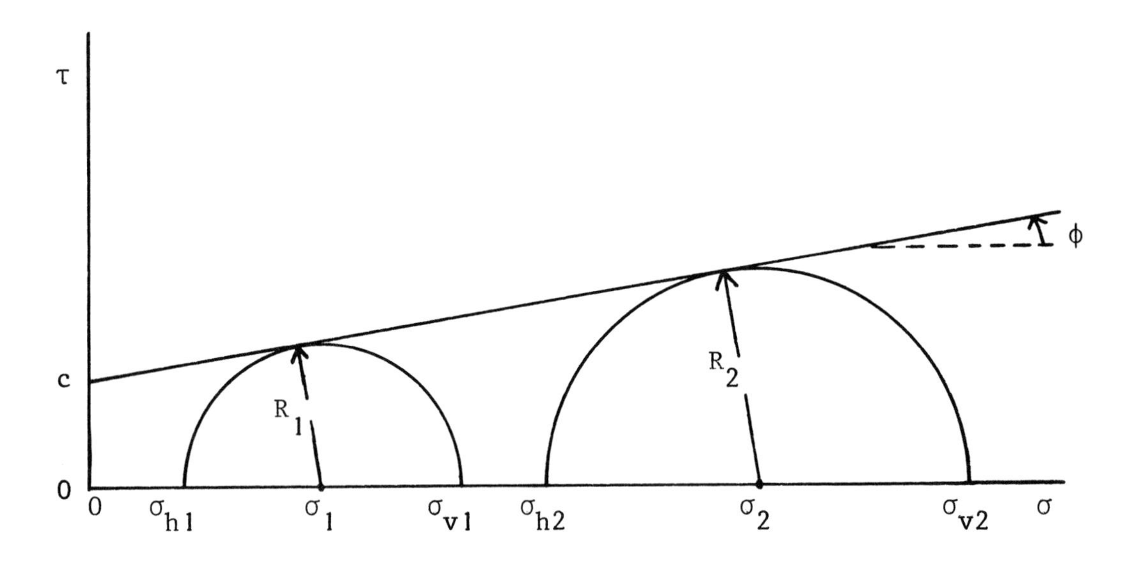

Fig.(5.6) Mohr's Semicircles

A graph is constructed as in Fig.(5.6). The abscissa is labelled with compressive stress σ, and the ordinate with shear stress τ. The two values σ_{h1}, σ_{v1} from test 1 are marked on the σ-axis. Then, from the point labelled σ_1 on the σ-axis, where $\sigma_1 = (\sigma_{v1} + \sigma_{h1})/2$ a semicircle is drawn with radius $R_1 = (\sigma_{v1} - \sigma_{h1})/2$. The same construction is done for the data from test 2. (It may happen that $\sigma_{h2} < \sigma_{v1}$; in this event the semicircles overlap; this does not affect the analysis.) These circles are known as *Mohr's circles.*

The next step is to construct the straight line tangent externally to the two semicircles, as shown in Fig.(5.6). This line has a τ-intercept equal to the soil's cohesion c; the slope of the line is the angle of internal friction of the soil. That is, the equation of the line is

$$\tau = c + \sigma \tan \phi. \tag{5.36}$$

This has the same form as the Mohr-Coulomb equation, Eq.(5.33).

It is not necessary to actually draw the semicircles and tangent line to find the values of c and ϕ from those of σ_{v1}, σ_{h1}, σ_{v2}, σ_{h2}. From the analytic geometry of circles and lines, it can be shown (see Problem 35) that the equation of the straight line forming the upper external common tangent to the two semicircles is

$$\tau = \frac{(R_2 - R_1)\sigma + (R_1\sigma_2 - R_2\sigma_1)}{\sqrt{(\sigma_2 - \sigma_1)^2 - (R_2 - R_1)^2}}, \tag{5.37}$$

where σ_1 and σ_2 are the centers of the "circles" (abbreviation for "semicircles") and R_1 and R_2 are their radii; that is

$$\sigma_1 = \frac{1}{2}(\sigma_{v1} + \sigma_{h1}), \tag{5.38}$$

$$R_1 = \frac{1}{2}(\sigma_{v1} - \sigma_{h1}), \tag{5.39}$$

$$\sigma_2 = \frac{1}{2}(\sigma_{v2} + \sigma_{h2}), \tag{5.40}$$

$$R_2 = \frac{1}{2}(\sigma_{v2} - \sigma_{h2}). \tag{5.41}$$

Comparing Eq.(5.37) with Eq.(5.36) leads to the conclusion that

$$\tan\phi = \frac{R_2 - R_1}{\sqrt{(\sigma_2 - \sigma_1)^2 - (R_2 - R_1)^2}}, \tag{5.42}$$

$$c = \frac{R_1\sigma_2 - R_2\sigma_1}{\sqrt{(\sigma_2 - \sigma_1)^2 - (R_2 - R_1)^2}}. \tag{5.43}$$

With the equation for the Mohr-Coulomb straight line now known, the two circles used to obtain the slope and τ-intercept of that line have served their purpose. An important practical question can be asked and now answered. That is: For a soil with known values of c and ϕ, what is the largest axial stress σ_v that can be applied if the confining pressure is σ_h? This value of σ_v must be that which, when it is plotted with σ_h on a diagram like that of Fig.(5.6), the circle will be tangent to the Mohr-Coulomb line whose equation is Eq.(5.36) with slope and intercept given by Eqs.(5.42) and (5.43).

The configuration is shown on Fig.(5.7). (Again, with c and ϕ determined from tests 1 and 2, those circles are no longer needed.) The circle to be drawn now is that for σ_v and σ_h at failure, as in the question asked above. Let σ_0 be the value of σ at the center of the circle. The radius R drawn from σ_0 to the point of tangency of the circle with the Mohr-Coulomb straight line meets that line at 90°. Also drawn is the perpendicular from the point of tangency to the σ-axis, meeting the axis at a value of σ labelled σ_R. Finally, also drawn is the line from the τ-intercept of the Mohr-Coulomb line running parallel to the σ-axis to the perpendicular drawn just previously.

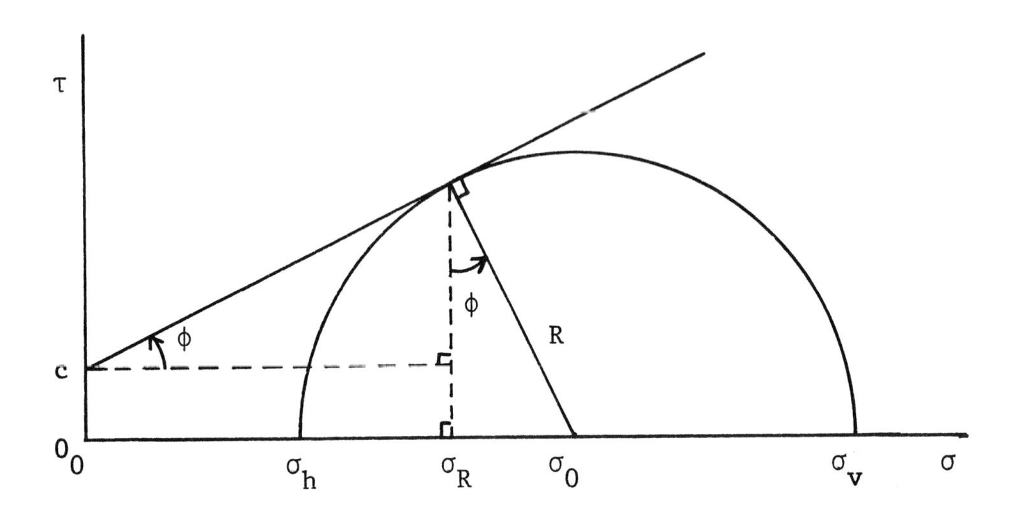

Fig.(5.7) Finding the Axial Stress

In the two right triangles created in Fig.(5.7), note that

$$\tan\phi = \frac{R\cos\phi - c}{\sigma_{\mathrm{R}}},$$

and

$$\sigma_0 - \sigma_{\mathrm{R}} = R\sin\phi.$$

Eliminate σ_{R} between these two equations to get

$$\sigma_0 = \frac{R}{\sin\phi} - \frac{c}{\tan\phi}.$$

But, by equations analogous to Eqs.(5.38) and (5.39),

$$\sigma_0 = \frac{1}{2}(\sigma_{\mathrm{v}} + \sigma_{\mathrm{h}}),$$

$$R = \frac{1}{2}(\sigma_{\mathrm{v}} - \sigma_{\mathrm{h}}).$$

Substituting these equations and rearranging leads to

$$\sigma_{\mathrm{v}} = \left(\frac{1+\sin\phi}{1-\sin\phi}\right)\sigma_{\mathrm{h}} + \left(\frac{2\cos\phi}{1-\sin\phi}\right)c. \tag{5.44}$$

No doubt it has been noticed that the rationale for drawing the circles in the first place is not given. The reason is that an understanding of the theory behind Mohr's circles requires a rather detailed knowledge of the state of stress in a solid, a proper exposition of which requires a level of treatment higher than that at which this text is written.

The procedure outlined in this section can be applied to rock as well as to soil. The values of the compressive strengths in rock usually are much larger than those encountered with soils.

****

EXAMPLE 7

The table shows the values of axial compressive strength at the given confining pressure in tests on a soil sample. Calculate the compressive strength in the third test.

Test	Pressure (Pa)	Axial Strength (Pa)
1	694	3550
2	1570	7420
3	3540	

By Eqs.(5.38) and (5.39), for test 1 it is found by direct substitution that $\sigma_1 = 2122$ Pa and $R_1 = 1428$ Pa. Similarly, by Eqs.(5.40) and (5.41) for test 2, the results are $\sigma_2 = 4495$ Pa and $R_2 = 2925$ Pa. Substitute these values into Eqs.(5.42) and (5.43) to find that $\phi = 39.11°$ and $c = 115.1$ Pa. Put these values of ϕ and c, along with $\sigma_{\text{h}} = 3540$ Pa from test 3 into Eq.(5.44) to find that for this test $\sigma_{\text{v}} = 16{,}100$ Pa.

****

5.7 Problems

1. Calculate the bulk density of an initially dry soil after it has been contaminated with oil from a ruptured freight car. The solids in the soil have density 2.87 g/cm^3 and the oil a specific gravity of 0.620. The soil's void ratio is 39.0% and the fraction of saturation with the oil is 90.0%.

2. Find the void ratio for a soil for which the void ratio is 3.82 times the porosity.

3. A soil sample has a volume of 470 cm^3. The mass of the sample measured immediately after being removed from the ground is 977 g, but after thorough drying the mass is 893 g. The void ratio of the soil is 36.6%. Calculate (*a*) the fraction of saturation of the original sample, and (*b*) the density of the solids.

4. A soil has a bulk density of 1.52 g/cm^3. The density of the solids is 1.86 g/cm^3 and the void ratio equals 34.0%. What fraction of the voids are occupied with water?

5. A soil sample has a diameter of 82.2 cm and length 3.64 m. The void ratio is 37.2% and the fraction of saturation 63.8%. How many liters of water are in the sample?

6. A soil has a bulk density of 1.82 g/cm^3, void ratio 0.566 and density of solids 2.43 g/cm^3. (*a*) Find the water content. (*b*) The total volume of solids in a certain sample of this soil is 422 cm^3. What volume of water does this sample contain?

7. A sample of soil has a water content of 73.6%, a void ratio of 4.28, and solids with a specific gravity equal to 2.10. (*a*) Find the volume of 1.00 kg of this soil. (*b*) Calculate the fraction of saturation.

8. Let ρ_{dry} be the bulk density of a soil sample of bulk density ρ after it has been dried in an oven so that all the water has been driven off. Show that

$$\rho = \rho_{\text{dry}}(1 + wc),$$

where wc is the water content of the soil before drying.

9. A certain clay has a liquid limit of 43 and a plasticity index of 20. What is its plastic limit?

10. Find the volume of water that infiltrates a 15.0 cm thick layer of soil of surface area 3.36 ha as the soil is taken from its plastic limit of 23.0% to its liquid limit of 49.0%. The bulk density of the dry soil is 2.25 g/cm^3.

11. (*a*) Show that, for a rectangular trench, the subsidence ΔH in terms of void ratio is given by

$$\Delta H = H_0 \left(\frac{e_0 - e}{e_0 + 1} \right),$$

where H_0 is the depth when the soil has void ratio e_0, and $H_0 - \Delta H$ is the depth when the void ratio has decreased to e. See Section (1.6). (*b*) A layer of soil in a rectangular trench has an initial depth of 1.720 m. Later, the surface of the soil is found to have lowered by 47.0 cm, at which time the void ratio is measured to be 36.4%. Find the initial void ratio.

12. What size particle has $\phi = 0$?

13. The diameters of particles classified as "coarse or medium sand" have phi values between -1.00 and $+1.32$. Find the diameter range in mm.

14. A particle has a size 3.000 times the size of a smaller particle. Find the difference in their phi values.

15. Find the phi value of a particle with a diameter of 0.00322 mm.

16. In a certain soil sample, 28.2% of the particles are greater than 2 mm in size. Of the remaining particles, 16.4% are sand, 58.2% are silt, and 25.4% are clay. What percent of the whole sample are clay particles?

17. Silt particles comprise 36.2% of the silt, sand, clay components of a soil, but only 24.3% of the entire sample. What percent of the entire sample consists of particles larger than 2 mm?

18. In a particular soil sample, 17.2% of all the particles present are gravel, 21.6% are clay, 38.5% are silt and the rest are sand. What percent of the non-gravel particles are sand?

19. On a soil texture triangular composition diagram, show that if all three projections are made in the wrong sense (direction), the three "abundances" so indicated add up to 200%, and that the correct individual abundances cannot be obtained simply by dividing by two.

20. Show that a triangular composition diagram constructed by labelling the sides of the triangle as shown in Fig.(5.8), and in which a point P is projected along lines perpendicular (not parallel) to the sides does not "work", since the three abundances so derived total 150%, rather than the required 100%. That is, show that, if L is the length of each side,

$$L_{\mathrm{A}} + L_{\mathrm{B}} + L_{\mathrm{C}} = \frac{3}{2} L.$$

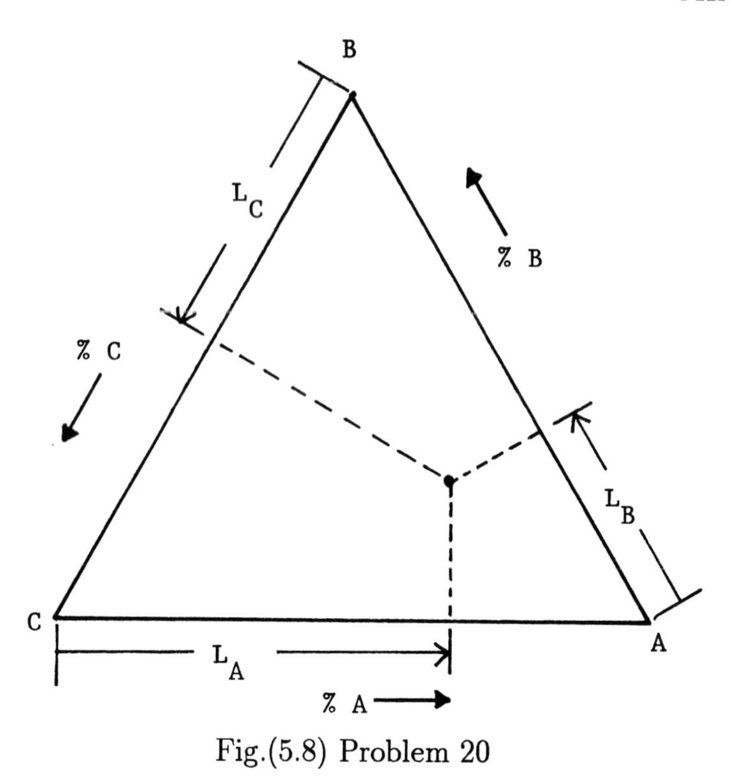

Fig.(5.8) Problem 20

21. What are the percent abundances of silt, sand, clay for a soil that falls at the point shown on Fig.(5.9)?

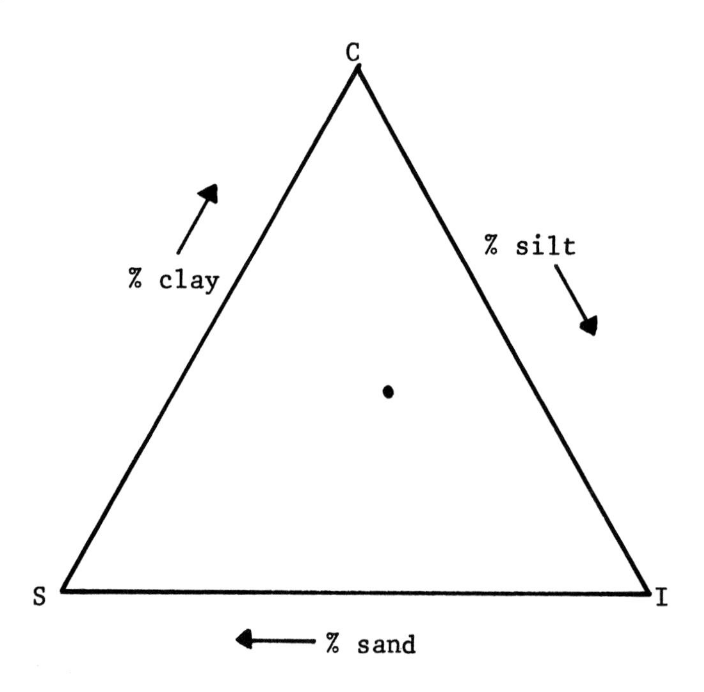

Fig.(5.9) Problem 21

22. Beneath the ground in a certain region, the soil density is found to vary with depth according to the expression

$$\rho = \rho_0 \left(1 + \frac{y^2}{D^2}\right),$$

where ρ_0 is the surface density and y is depth below the surface. The parameter D is the depth at which the soil density would be twice the surface density if the soil density continued to vary according to the given expression down to that depth. However, bedrock is encountered at depth $\frac{3}{4}D$. Find the vertical stress at the soil-bedrock interface.

23. In a particular area of land, the density of the soil varies with depth in a linear manner; i.e., the density at depth y is given by

$$\rho = \rho_0(1 + \frac{y}{D}),$$

where ρ_0 is the surface density and D is the depth at which $\rho = 2\rho_0$. Show that the stress at depth D is

$$\sigma = \frac{3}{2}\rho_0 g D.$$

24. In an isolated region, the density of the soil varies with depth y according to

$$\rho = \rho_0 e^{y/H},$$

where $\rho_0 = 2.71$ g/cm^3 is the density of soil at the surface and $H = 420$ m. This equation is valid down to depth 70.0 m, at which level bedrock is encountered. (*a*) Find the density of the soil resting on the bedrock. (*b*) Find the vertical stress at the soil-bedrock interface.

25. A certain soil has a dry density of 1.87 g/cm^3. When saturated with water its density becomes 2.36 g/cm^3. The effective stress at a depth of 38.2 m in this soil is 622 kPa. Find the depth of the water table.

26. Show that the bulk densities of a soil when dry, ρ_{dry}, and completely saturated, ρ_{sat}, with water, density ρ_{w}, are related by

$$\rho_{\text{sat}} = \rho_{\text{dry}}(1 + \frac{e\rho_{\text{w}}}{\rho_{\text{s}}}),$$

where e is the void ratio of the soil and ρ_{s} the density of its solids.

27. For the soil profile shown in Fig.(5.10), consisting of soils A and B, find the effective stress in the middle of soil B. The densities shown are bulk densities: dry density for soil

above the water table, saturated density for soil below the water table.

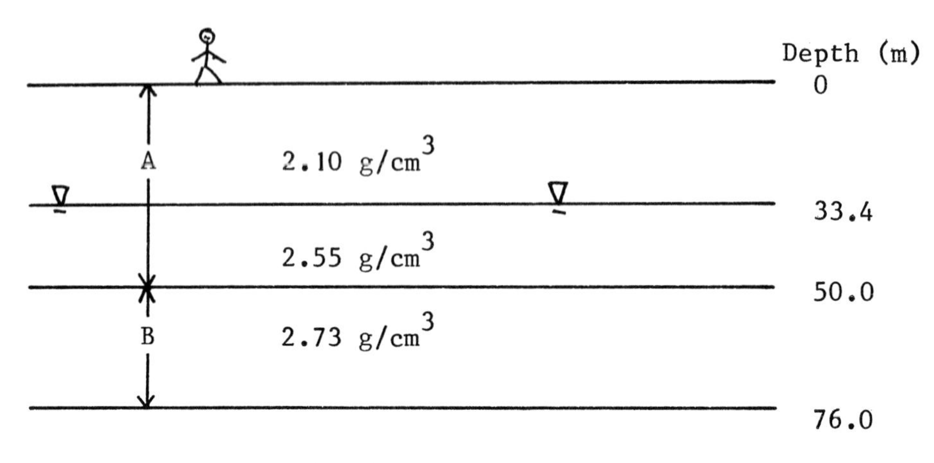

Fig.(5.10) Problem 27

28. The density of the solids in a particular soil is 2.65 g/cm^3. From the surface down to 18.0 m the soil has void ratio 48.0%, but below this depth the void ratio is unknown. The water table is at a depth of 11.0 m. If the total vertical stress at a depth of 23.0 m is 450 kPa, find the void ratio of the soil below 18.0 m, the void ratio presumed to be constant.

29. The solids in a special soil have density 2730 kg/m^3. The void ratio is 53.5%. Find the greatest possible value of the soil's water content.

30. An axial compressive force of 240 kN is applied to a circular cylinder of soil of diameter 4.38 cm. Find the shear stress on a plane with a normal at 32.0° to the axis of the cylinder.

31. In Fig.(5.4), what is the dip angle of the plane on which the shear stress is $\frac{1}{4}\sigma_B$?

32. A soil is being tested for shear strength in the laboratory. The soil is dry. An applied normal load of 142 kPa produces a shear strength of 178 kPa; under a normal load of 205 kPa the shear strength increases to 224 kPa. Find (*a*) the cohesion and (*b*) the angle of friction of the soil.

33. In Example 6, find the dip angle of the plane, with median depth 38.2 m, along which the shear strength of the soil is 375 kPa.

34. At what depth in a dry soil that has a cohesion of 32.0 kPa, a friction angle of 24.0° and density 2.09 g/cm^3 is the shear strength in a horizontal plane equal to 156 kPa?

35. Show that Eq.(5.37) is really the equation of the line tangent externally to the two circles shown on Fig.(5.6).

36. In Example 7, at what confining pressure is the axial compressive strength equal to 12.0 kPa?

37. A particular rock has an axial compressive strength of 118 MPa when the confining pressure is 6.80 MPa. The internal angle of friction is 42.0°. Find (*a*) the cohesion and (*b*) the axial compressive strength when the confining pressure is increased to 14.0 MPa.

38. Certain clay soils have an internal friction angle equal to zero. (*a*) Draw a Mohr circle for such a soil under no confining pressure. (*b*) From the diagram, show that the unconfined axial compressive strength equals twice the soil's cohesion.

39. A certain sandy soil has zero cohesion. At a confining pressure of 875 kPa, its axial compressive strength equals 2330 kPa. Find the angle of friction of the soil.

40. A soil has an axial compressive strength of 398.3 kPa when the confining pressure is 150.0 kPa, and an axial compressive strength of 621.1 kPa when the confining pressure is 272.0 kPa. Find the soil's (*a*) cohesion and (*b*) angle of friction.

Chapter 6

Rivers

6.1 Water

Water is a compound, each molecule of water consisting of two atoms of hydrogen (H) and one atom of oxygen (O); hence, its chemical symbol H_2O.

The Earth is the only planet in the solar system on whose surface liquid water exists. The water is found mainly in the oceans, covering 70.8% of the Earth's surface to an average depth of 3.8 km. Water in the solid phase, ice, covers Greenland and the Antarctic continent, which together constitute about 10% of the land area of the Earth. Water vapor exists in the atmosphere.

Usually the word "water" is used to refer to water in the liquid state; "ice" and "steam" refer to the solid and gaseous phases of water. (Most substances do not have separate names for their different phases.) The density of pure liquid water is about 1.00 g/cm^3 = 1000 kg/m^3. Ocean water contains about 35 g of dissolved salts per kilogram of water, so that the density of sea and ocean water is greater than that of pure water, being about 1025 kg/m^3.

At atmospheric pressure, the freezing point of water is 0°C. Water has the unusual property of expanding as it freezes, leading to many engineering problems. The density of ice is 917 kg/m^3. The boiling point of water is 100°C. However, some molecules escape from any open body of water, so that there is water vapor in the atmosphere even though the temperature of the air near the surface is much less than 100°C.

****

EXAMPLE 1
As seen from above the South Pole, the Antarctic continent is roughly a semicircle in outline, with a radius of 2500 km. The average thickness of the ice cover is 3000 m. If all of this ice melts (perhaps through global warming triggered by a greenhouse effect, in turn induced by atmospheric industrial emissions), to what height would coastal dikes have to be constructed to protect coastal land areas from inundation by ocean water?

Use SI base units in thc calculations. With R_{ice} the radius and h_{ice} the thickness of the ice cover, the volume V_{ice} of ice is

$$V_{\text{ice}} = \frac{1}{2}\pi R_{\text{ice}}^2 h_{\text{ice}},$$

$$V_{\text{ice}} = \frac{1}{2}\pi(2.5 \text{ X } 10^6 \text{ m})^2(3000 \text{ m}),$$

$$V_{\text{ice}} = 2.95 \text{ X } 10^{16} \text{ m}^3.$$

When this ice melts, it does not produce the same volume V_w of water, since the densities of ice and water are different. However, the mass involved is unchanged by the melting, so that

$$M_{\text{ice}} = M_w,$$

$$\rho_{\text{ice}} V_{\text{ice}} = \rho_w V_w,$$

$$(917 \text{ kg/m}^3)(2.95 \text{ X } 10^{16} \text{ m}^3) = (1025 \text{ kg/m}^3)V_w,$$

$$V_w = 2.64 \text{ X } 10^{16} \text{ m}^3.$$

The volume V_o of water in the oceans can be expressed as

$$V_o = A_o h_o,$$

where A_o is the total area of the oceans and h_o is their average depth. If the volume of the ocean water changes by ΔV_o, but the effective area of the ocean basins does not change (due to construction of the dikes), then the average depth of the oceans must change by Δh_o given from

$$\Delta V_o = A_o(\Delta h_o).$$

In the scenario of this example, $\Delta V_o = V_w$, i.e., the change in volume of ocean water equals the volume of ocean water produced from the melted ice from Antarctica. The area of the oceans is $70.8\% = 0.708$ of the surface area of the Earth. Since the Earth is a sphere (very nearly), its area is $4\pi R^2$, where R is the radius of the Earth. The average radius of the Earth is $R = 6370$ km. (Different values can be given for this quantity, since there are different ways to define the average radius of an object that is not precisely a sphere in shape.) Putting all this together gives

$$V_w = [A_o]\Delta h_o,$$

$$V_w = [(0.708)(4\pi R^2)]\Delta h_o,$$

$$2.64 \text{ X } 10^{16} \text{ m}^3 = [(0.708)(4\pi)(6.37 \text{ X } 10^6 \text{ m})^2]\Delta h_o,$$

$$\Delta h_o = 73 \text{ m}.$$

The impact of such a rise in sea level on near coastal communities is easy to imagine (or, perhaps, it is not so easy to imagine).

****

6.2 Rainfall Equation

Water that falls as precipitation entered the atmosphere by evaporation, mainly from the oceans. When the precipitation is over an ocean area, the water returns directly to its source. Water that falls as precipitation on to land areas will have to follow a more circuitous path if it is to return to the ocean.

If the precipitation is snow (ice crystals), then the water remains where it fell until it melts. This may happen almost immediately, or perhaps only after several months. But, once the snow has melted, the possible courses the water may follow are essentially the same as for water that fell as rain.

Sometimes, after a heavy rainfall, water can be seen evaporating back into the atmosphere, especially from road surfaces and parking lots, for example. But evaporation can also take place from water that fell on soil. If the surface on which the precipitation fell is such as to make it difficult for the water to soak into the ground, significant evaporation may take place.

The water that does not evaporate either soaks into the ground (infiltration) or does not do so. The water that does not infiltrate forms surface water runoff. This water runs downhill under gravity. It may thereby flow directly into a stream or river. However, the water may follow a long, time-consuming sequence of paths that ultimately lead to discharge into a river. The whole journey may take several years.

For example, the water may be collected by a storm drain, be stored in a reservoir, used in homes or industrial facilities, be consumed and subsequently evacuated by people, etc. Relatively little water is actually destroyed in chemical reactions. To a high degree of accuracy, it can be said that the total quantity of water is conserved.

The water that does infiltrate into the ground, the ground water runoff, may be absorbed by the roots of plants, small and large. The plants later exhale the water as vapor back into the atmosphere. This botanical process is called transpiration.

The remainder of the infiltration, after perhaps years of moving through soil and rock, has much the same ultimate fate as surface water runoff. It may be withdrawn from the ground in wells, to be subjected to the various tasks as surface water runoff, and eventually be discharged into a river or directly into the ocean. Or, it may reach the ocean by moving entirely through underground rocks.

The quantities of water involved in evaporation and transpiration are added together and called *evapotranspiration*, symbol ET. Similarly, the surface water runoff and the ground water runoff that is not transpired are added together and called simply *runoff*, symbol RO. Hence, the water from precipitation, symbol PT, forms either evapotranspiration or runoff, where these terms refer to the actual quantities of water that undergo the respective processes.

The quantities PT, ET, and RO are expressed as the heights that the water involved would reach if the water were allowed to pile-up on the area on which the precipitation from which it originated fell. To obtain the volume of water involved, multiply the height by the area A of the land surface over which the precipitation took place. That is,

$$V_{\text{prec}} = (PT)A, \tag{6.1}$$

$$V_{\text{evap}} = (ET)A, \tag{6.2}$$

$$V_{\text{runo}} = (RO)A. \tag{6.3}$$

The SI base unit of area is the square meter, m^2. But this is a very small area when dealing with land. A larger metric unit is the square kilometer: 1 km^2 = 1 X 10^6 m^2. Often, a unit intermediate between these two is desired. In metric units, this is provided by the *hectare* , ha, where 1 ha = 1 X 10^4 m^2. The hectare can be considered as a sort of metric acre, although actually 1 ha = 2.471 acre, so the hectare is larger than an acre.

****

EXAMPLE 2

During a storm, 3.72 cm of rain fell over an area of 426 km^2. Find the volume of water that fell (a) in m^3 and (b) in ha·cm.

(a) The data are: PT = 3.72 cm and A = 426 km^2. Converting to SI base units gives PT = 0.0372 m and A = 4.26 X 10^8 m^2. Therefore, the volume V_{prec} that fell is, by Eq.(6.1),

$$V_{\text{prec}} = (PT)(A),$$

$$V_{\text{prec}} = (0.0372 \text{ m})(4.26 \text{ X } 10^8 \text{ m}^2),$$

$$V_{\text{prec}} = 1.58 \text{ X } 10^7 \text{ m}^3.$$

(b) Since 1 ha = 10,000 m^2, the area in ha is

$$A = (4.26 \text{ X } 10^8 \text{ m}^2)(1 \text{ X } 10^{-4} \text{ ha/m}^2),$$

$$A = 42{,}600 \text{ ha}.$$

The volume of water in ha·cm must be

$$V_{\text{prec}} = A(PT),$$

$$V_{\text{prec}} = (4.26 \text{ X } 10^4 \text{ ha})(3.72 \text{ cm}),$$

$$V_{\text{prec}} = 1.58 \text{ X } 10^5 \text{ ha} \cdot \text{cm}.$$

By comparing the results of (a) and (b), note that, in effect, it has been proven that 1 ha·cm = 100 m^3.

****

The discussion above on the disposition of precipitated rain water can be summarized in the *rainfall equation*:

$$PT = ET + RO. \tag{6.4}$$

This relation holds only over periods of time long enough so that the two processes on the right hand side of the equation have had a chance to take place. For example, if the precipitation is all in the form of snow and the temperature remains below 0°C for several weeks, say, then for a time period that begins when the precipitation begins and ends one minute after the precipitation ends, $PT \neq 0$, but $ET = 0$ and $RO = 0$. In analyzing water supplies in a certain region, annual (one year) time intervals often are used, this being a time period over which the processes usually will have taken place and the rainfall equation will therefore be satisfied.

****

EXAMPLE 3
Over a 15,800 km^2 land area the annual precipitation is 82.5 cm. Of this, 37.0% returns to the atmosphere via evapotranspiration. What volume of water, each year, enters the rivers that drain this land area?

The water that enters the rivers is the runoff, so the desired volume V_{runo} is given from Eq.(6.3), $V_{\text{runo}} = (RO)A$, where RO is the runoff and A is the land area. By the rainfall equation, Eq.(6.4), $RO = PT - ET$. Since $ET = (0.370)PT$, the rainfall equation implies that

$$RO = PT - (0.370)PT,$$

$$RO = (0.630)PT.$$

Therefore, the volume of water entering the rivers each year is

$$V_{\text{runo}} = (0.630)(PT)A,$$

$$V_{\text{runo}} = (0.630)(0.825 \text{ m})(1.58 \text{ X } 10^{10} \text{ m}^2),$$

$$V_{\text{runo}} = 8.21 \text{ X } 10^9 \text{ m}^3.$$

****

6.3 Discharge

Examine now the flow of water in rivers, the water arising from runoff. For this discussion, the terms river, stream, creek, brook, etc., can be used interchangeably. The results of this section hold also for a canal, a "human-built" channel, and for a pipe.

Figure (6.1) is a sketch of an idealized river. The depth of the river is d, its width w, and the water is flowing at speed v. This river is an idealization (or simplified model) of actual rivers in many ways, amongst them the fact that: (i) Rivers do not have rectangular cross sections; their actual cross sections are irregular in shape and change with location along the river and with time; (ii) Rivers seldom run in straight line paths, but meander; (iii) The speed v of the water has different values within a cross section. Therefore, d should be considered as the average depth, w the average width and v the average speed over a given cross section of the river. It also is assumed here that the river flow is not turbulent (no water is moving backward).

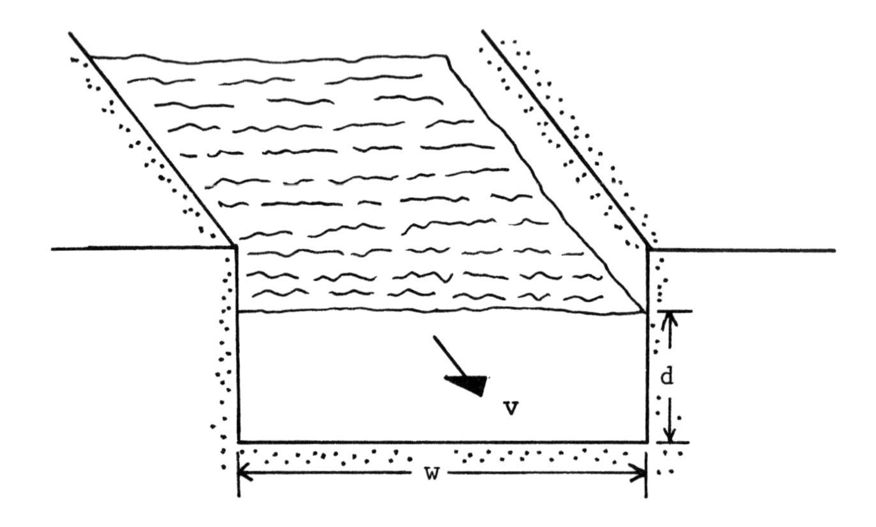

Fig.(6.1) Idealized River

For any situation in which water (or any liquid) is flowing in a channel, either open (e.g., a river) or closed (e.g., a pipe), the *discharge* Q is defined by

$$Q = \frac{V}{t}, \tag{6.5}$$

where V is the volume of water passing through a perpendicular cross section of the channel in the time interval t. The SI base units of discharge are m^3/s, since cubic meter is the SI base unit of volume and second the SI base unit of time. The discharge is also known as the "volume flow rate", "volume flux", or simply "flow", although the term flow is sometimes used to refer to the speed v of the water.

Strictly speaking, Eq.(6.5) applies only if the discharge Q does not change in value over the time t. If the discharge does vary with the time, so that $Q = Q(t)$, then the volume of water that passes in time t is given by

$$V = \int_0^t Q(t)\, dt. \tag{6.6}$$

The volume V so found and then divided by the time t, as in Eq.(6.5), yields the *average* discharge.

A graph of discharge Q versus the time t is called a *hydrograph.* On a hydrograph, the volume of water is represented as the area between the discharge line and the t-axis over the time interval of interest. (This does not apply if the logarithm of the discharge is plotted.)

Sometimes the *mass flow rate* Q_{m} is important; this is defined by

$$Q_{\mathrm{m}} = \frac{m}{t}, \tag{6.7}$$

where m is the mass of water passing in time t. Since $m = \rho V$, where ρ is the density of water, it follows that, using Eq.(6.5),

$$Q_{\mathrm{m}} = \frac{\rho V}{t},$$

$$Q_{\mathrm{m}} = \rho Q,$$

$$m = \rho Q t. \tag{6.8}$$

The SI base units of mass flow rate Q_{m} are kg/s.

Now, the discharge of a river must be determined by its size (cross-sectional area) and the speed at which the water is flowing. To express this in definite form, consider Idealized River again, this time referring to Fig.(6.2).

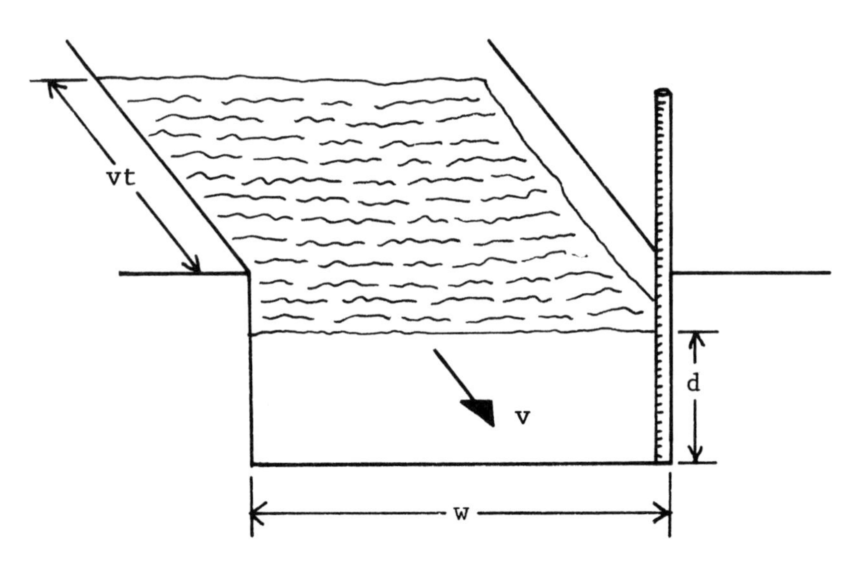

Fig.(6.2) Idealized River Discharge

As shown, a depth pole gauge stands at the "front" cross section. All of the water that will flow past the pole in the immediately forthcoming time interval t is upstream from the pole at distances ranging from zero to vt, since vt is the distance that water, flowing at constant speed v, can travel in time t. (Water at greater distances upstream than vt will not reach the pole in time t.) The water that will reach the pole forms a rectangular cylinder with cross-sectional area wd and length vt. The volume of the cylinder of water is

$$V = (wd)(vt).$$

Therefore, the discharge is

$$Q = \frac{V}{t},$$

$$Q = \frac{wdvt}{t},$$

$$Q = wdv. \tag{6.9}$$

In the form of Eq.(6.9), the units of Q are (m)(m)(m/s) = m^3/s, in agreement with the units from Eq.(6.5).

If the cross section of the river can be better approximated with a non-rectangular shape for which there is a formula for the area, then replace wd in Eq.(6.9) with that cross-sectional area A to get

$$Q = Av. \tag{6.10}$$

****

EXAMPLE 4
Circle River flows in a channel with a semicircular cross section of width 34.2 m. After a storm upstream, the river is found to be just within its banks and flowing at 13.5 m/s; see Fig.(6.3). (*a*) Calculate the discharge. (*b*) Find the volume of water that flows through a cross section in 2.00 h.

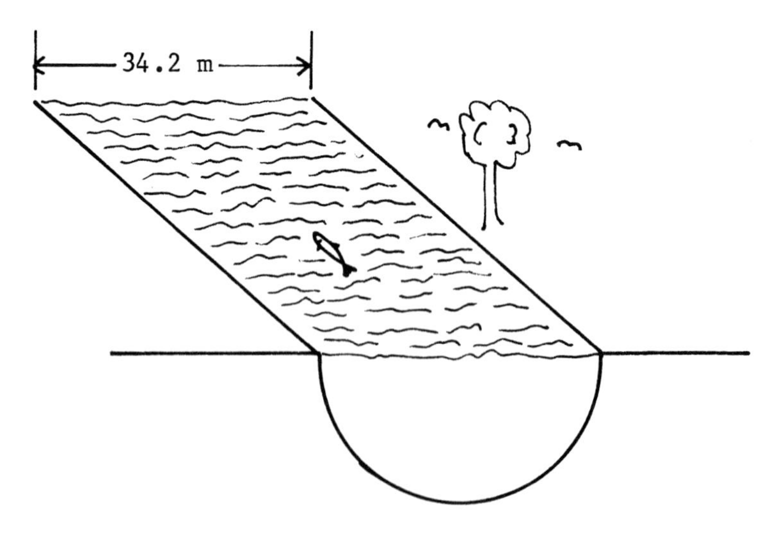

Fig.(6.3) Example 4

(*a*) With the cross section being a semicircle, it is appropriate to use Eq.(6.10), with the formula for the area of a semicircle, i.e., one-half the area of a circle. Since the width of the river is the diameter D of the semicircle, Eq.(6.10) gives for the discharge

$$Q = \frac{1}{2}(\frac{1}{4}\pi D^2)v,$$

$$Q = \frac{1}{8}\pi(34.2 \text{ m})^2(13.5 \text{ m/s}),$$

$$Q = 6200 \text{ m}^3/\text{s}.$$

(*b*) With Q evaluated, the volume of water that passes can be found from Eq.(6.5). But Q is in m^3/s and $t = 2.00$ h, so the time units must be reconciled. Since 1 h = 3600 s,

$$V = Qt,$$

$$V = (6200 \text{ m}^3/\text{s})[(2)(3600 \text{ s})],$$

$$V = 4.46 \text{ X } 10^7 \text{ m}^3.$$

****

Since the source of the water that flows in rivers is the runoff from precipitation, the discharge of a river draining a certain land region must be related to the *intensity* (amount as a function of time) of the precipitation over that region.

Examine the land area shown in Fig.(6.4). Its boundary is drawn so that all the runoff in this region flows into the streams shown, which themselves are tributaries of the region's principal river, called here Drain River. The land region shown, whose runoff drains into Drain River, is called Drain River's *drainage basin* (also known as its *catchment area*). The boundary of a drainage basin is a *drainage divide*. At its mouth M, Drain River enters the ocean.

Now consider point P, shown on one of Drain River's tributaries. The region inside the dotted line can be considered as the drainage basin of that part of the Drain River system that sits upstream of point P. The runoff from this region determines the discharge at P, just as the runoff from the entire drainage basin determines the discharge at the mouth of Drain River (tidal variations excluded).

The volume V of water associated with runoff RO due to precipitation over drainage area A is $V_{\text{runo}} = (RO)A$. If this water takes time t to pass that point on the river where it exits the drainage region [e.g., point P for the catchment area delineated with the dotted line in Fig.(6.4)], then the discharge of the river at that location due to this runoff is given from Eqs.(6.3) and (6.5) by

$$Q = \frac{V}{t},$$

$$Q = \frac{(RO)A}{t},$$

$$(RO)A = Qt. \tag{6.11}$$

This discharge is in addition to the discharge, if any, present in the river before the precipitation considered above took place (the river's *base flow*). If t is one year, then it is likely, with all four seasons passed through, that all sources of the river's water are accounted for. In this event, Q can be considered to be the average discharge during the year. If Eq.(6.11) is combined with Eq.(6.9), then it is found that

$$A(RO) = wdvt. \tag{6.12}$$

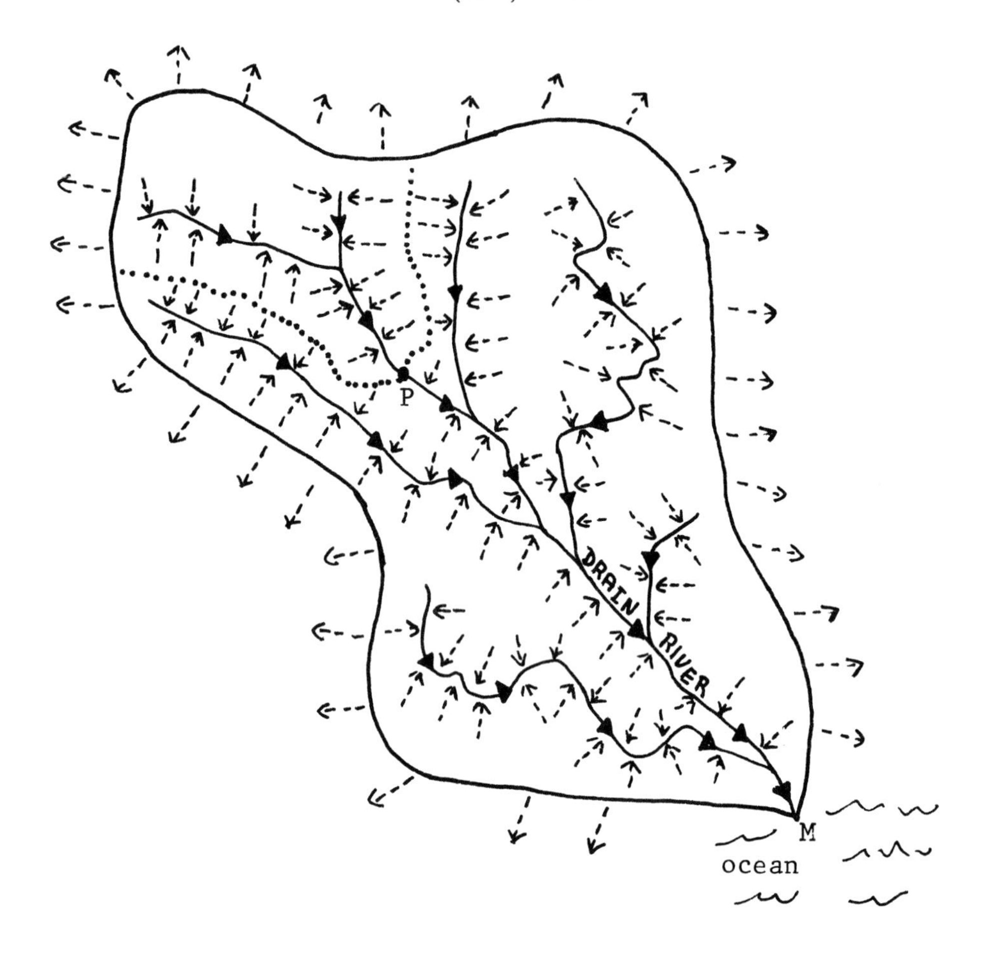

Fig.(6.4) Drain River Catchment Area

****

EXAMPLE 5

The annual precipitation over a 7200 km^2 drainage basin is 82.2 cm. Of this, 38.0 cm returns to the atmosphere. The runoff drains into a river 10.5 m wide and 3.20 m deep on the average. Find the average current speed of the river.

Converting all length units to the meter, the SI base unit, Eq.(6.4), the rainfall equation, yields for the runoff

$$RO = PT - ET,$$

$$RO = 0.822 \text{ m} - 0.380 \text{ m},$$

$$RO = 0.442 \text{ m}.$$

Now use Eq.(6.12). There are 3.16 X 10^7 seconds in one year, and therefore

$$(RO)A = wdvt,$$

$$(0.442 \text{ m})(7.2 \text{ X } 10^9 \text{ m}^2) = (10.5 \text{ m})(3.20 \text{ m})v(3.16 \text{ X } 10^7 \text{ s}),$$

$$v = 3.00 \text{ m/s}.$$

This is the water speed averaged over one year at the location where the river exits the catchment area. At times the speed will be greater, and at times less.

****

6.4 Hydraulic Geometry

Suppose that a stream monitoring, or gauging, station is placed at a certain point on the banks of, or in, a river. If it measures the current speed v, depth d, and width w of the river, the discharge Q can be calculated by Eq.(6.9): $Q = wdv$. But suppose, instead, that the instrument measures only the discharge Q. Can the values of w, d, and v be calculated therefrom?

This question is not merely academic. The discharge of a river at any fixed location is usually not constant with time. The discharge can vary significantly, sometimes over a time of only a few hours. For example, if there has been heavy precipitation, or melting of snow, upstream, it is to be expected that much of the runoff generated will, after some delay, show up as increased discharge of the river. It can be quite important to know how the river will respond, through changes in the values of depth, width, and current speed.

Of course, these changes cannot be predicted on the basis of Eq.(6.9) alone, for one equation cannot generally be solved for three unknowns. More equations are needed.

Now, many rivers have been "under observation" for a long time. Measurements of depth, width, speed and discharge have been, and still are being, made at different locations on a number of rivers and streams. To get a clear picture of the river's behavior under different seasonal conditions, data is taken frequently (sometimes more than once per hour).

After data has been taken for many years, the task is to discover if there are any relations between the values of w, d, v, Q other than the one represented by Eq.(6.9). Equation (6.9) is a *theoretical formula*, in that it is derived from theory or previously known relations (specifically, the geometric formula for the volume of a cylinder and the definition of average speed). From the accumulated data, however, an *empirical formula* or formulas are sought;

that is, formulas not derived fom other equations, but which seem to "fit" the data, for reasons that may not, at least yet, be understood.

One commonly used method to find an empirical formula, if one exists, is to display the data on a graph. Since the discharge is the driving agent behind changes in width, depth, and speed of the water, graphs of w vs. Q, d vs. Q, and v vs. Q are made. On any particular graph, if the data points fall everywhere, generating a "scatter diagram", the implication is that there is no correlation between the quantities being plotted. But, if the points tend to fall along a line, straight or curved, then a relation between the properties is indicated.

It may be no surprise to learn that data taken as described above does tend to indicate relations between the measured quantities when displayed on graphs. Specifically, measured values of width w and discharge Q, obtained from a fixed location on a particular river, tend to fall about a straight line when plotted on log-log paper. The same is true for depth d and discharge Q, and for speed v and discharge Q.

Note that log-log paper is indicated. This means, for example, that it is log$w$ vs. log$Q$ that are really being plotted, even though, for convenience, the axes are labelled with values of w and Q. The use of log-log paper makes it easy to display data with a very wide range of numerical values (over several powers of ten). However, it also tends to make the scatter in the data look less severe than it actually is, since the log is a slowly-varying function of its argument. (For example, if the data varies between 10 and 10,000, the log only varies between 1 and 4.)

The algebraic relation indicated by a straight line on a log-log plot can be found by recalling that the equation $y = mx + b$ is a straight line on a linear (ordinary) graph of y vs. x. On the graph considered, log$w$ is on the y-axis and log$Q$ is on the x-axis. A straight line therefore means that

$$y = mx + b,$$

$$\log w = m(\log Q) + b,$$

$$w = (10^b)Q^m.$$

Writing $a = 10^b$ gives

$$w = aQ^m. \tag{6.13}$$

A straight line on a log-log plot indicates a power law relation between the variables.

Equation(6.13) is a relation between a geometric quantity, the width, and a hydraulic quantity, the discharge. Hence, such relations are said to constitute the *hydraulic geometry* of the river.

****

EXAMPLE 6

Values for the width and discharge for Damp River, measured at the town of Sharp Bend, cluster about the line shown on Fig.(6.5). (The individual data points from which the line is obtained are not shown.) Using the two indicated points on the line, find the values of a and m in Eq.(6.13).

The values of w and Q are needed at two points on the line because there are two quantities to be calculated, for which two equations are required. The points labelled 1 and 2 are chosen since their associated values of w and Q can easily be read from the graph.

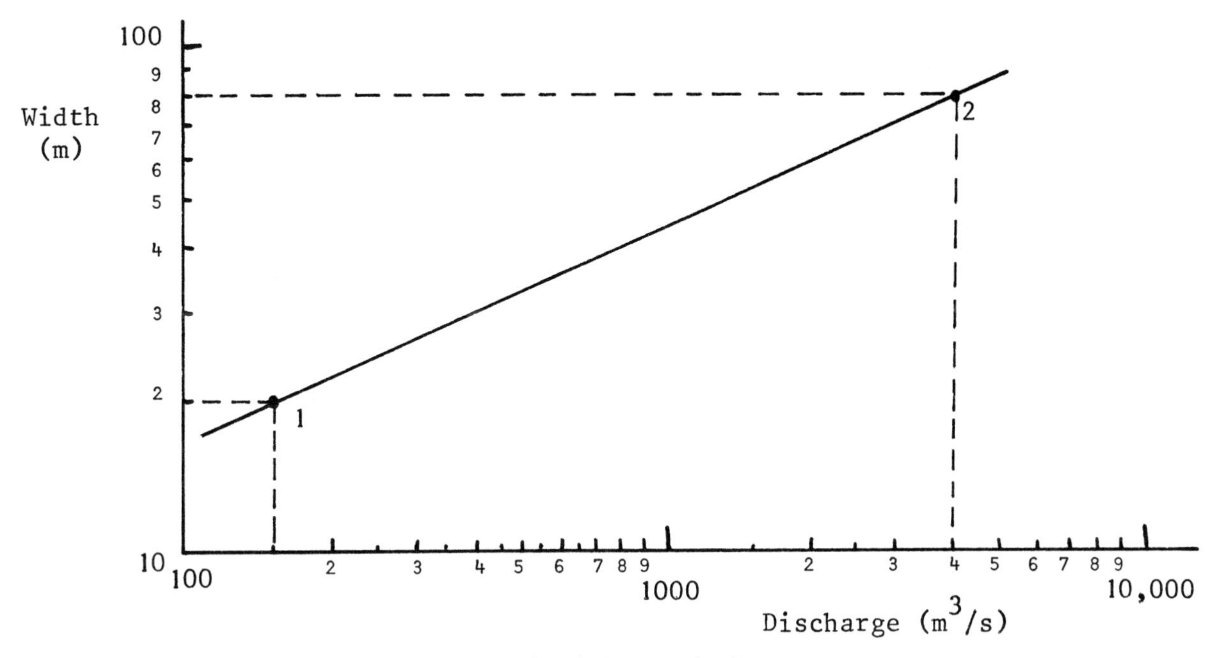

Fig.(6.5) Example 6

The exponent m in Eq.(6.13) is the slope of the line; thus,

$$m = \frac{\log w_2 - \log w_1}{\log Q_2 - \log Q_1},$$

$$m = \frac{\log(w_2/w_1)}{\log(Q_2/Q_1)}.$$

The indicated values must be carefully read from Fig.(6.5) and then substituted into this equation. Note that the ratio of widths and the ratio of discharges will cancel out all units, so that

$$m = \frac{\log(80/20)}{\log(4000/150)},$$

$$m = 0.422.$$

(Natural logarithms, ln's, could have been used in place of log's.) To find the value of the coefficient a in Eq.(6.13) now that the value of m has been obtained, it is only necessary to apply Eq.(6.13) to any one point on the line. Selecting point 1 gives

$$w_1 = aQ_1^m,$$

$$20\ \mathrm{m} = a(150\ \mathrm{m}^3/\mathrm{s})^{0.422},$$
$$a = 2.41\ \mathrm{m}^{-0.266} \cdot \mathrm{s}^{0.422}.$$

Notice that the units get complicated, due to the non-integral value of the exponent m. This is common in empirical equations. In such cases, the units often are simply omitted, with it being understood that the equation only applies for the specified set of units. In the example here, it should be written that

$$w = 2.41Q^{0.422},$$

only when w and Q are expressed in SI base units (m and m^3/s, respectively).

****

As mentioned above, straight lines also appear as the best fit to data for depth vs. discharge and speed vs. discharge. Hence, expressions similar to Eq.(6.13) can be written for each of these pairs of variables also. That is,

$$w = aQ^b, \tag{6.14}$$

$$d = cQ^e, \tag{6.15}$$

$$v = fQ^g. \tag{6.16}$$

[Note that the exponent m in Eq.(6.13) has been renamed b in the corresponding Eq.(6.14); this avoids confusion with mass or meter.] The values of the exponents b, e, and g (not gravity) and of the coefficients a, c, and f must be determined at the location of interest on the given river; there are no "global" values.

These three empirical equations must be consistent with Eq.(6.9), the theoretical relation between w, d, v and Q. To see what this requirement implies, substitute Eqs.(6.14), (6.15) and (6.16) for w, d, and v into Eq.(6.9) to obtain

$$Q = wdv,$$

$$Q = (aQ^b)(cQ^e)(fQ^g),$$
$$Q = (acf)Q^{b+e+g},$$
$$1Q^1 = (acf)Q^{b+e+g}.$$

Now the last equation above must hold for all possible values of Q. This means that the coefficients of Q on the two sides of the equation must be equal, and that the exponents on the two sides of the equation must be equal. That is,

$$acf = 1, \tag{6.17}$$

$$b + e + g = 1. \tag{6.18}$$

The product of the coefficients in Eqs.(6.14), (6.15), (6.16) must be 1, and the sum of the exponents must be 1. Equations (6.17) and (6.18) can be used in several ways. First, if all of the coefficients have been determined from measured data, then the closeness of their product to 1 gives an indication of the accuracy of their values; similarly for the sum of the exponents. Alternatively, if two of the coefficients have been obtained from data, Eq.(6.17) can be used to find the third; similarly, Eq.(6.18) can be used to find the third exponent if the other two are known.

****

EXAMPLE 7
At Ghost Town on Wild River, measured values of width in m, depth in m, water speed in m/s, and discharge in m^3/s are found to obey the relations

$$w = 2.62Q^{0.382},$$

$$d = 0.773Q^{a},$$

$$v = bQ^{0.426}.$$

When the discharge is 470 m^3/s, what are the values of (*a*) the depth, (*b*) the speed, and (*c*) the width of the river?

(*a*) To find the depth d, the exponent a must be calculated from Eq.(6.18). [Note that the notation for the coefficients and exponents in this example does not match that used in Eqs.(6.14), (6.15), (6.16) and therefore in Eqs.(6.17), (6.18): there is no industry standard for the symbols.] The exponents must add to 1, and therefore

$$0.382 + a + 0.426 = 1,$$

$$a = 0.192.$$

There is now a complete operating equation for the depth into which only the value of the discharge needs to be substituted. In keeping with remarks about units in empirical equations, the units are left out during the calculation. Proper units must always be attached to results of calculations. The operating equation for the depth, then, is written

$$d = 0.773Q^{0.192},$$

for SI base units. When $Q = 470$ m^3/s, the depth is

$$d = 0.773(470)^{0.192},$$

$$d = 2.52 \text{ m}.$$

(*b*) To find the current speed v, the coefficient b must first be evaluated. By Eq.(6.17), the product of the coefficients is 1, and thus

$$(2.62)(0.773)b = 1,$$

$$b = 0.4938.$$

Do not write out the units of b. The speed can now be found for the discharge of 470 m^3/s:

$$v = 0.4938Q^{0.426},$$

$$v = 0.4938(470)^{0.426},$$

$$v = 6.79 \text{ m/s}.$$

(c) The width w can be found immediately from the given empirical equation, since both the coefficient and exponent are given; doing this gives

$$w = 2.62Q^{0.382},$$

$$w = 2.62(470)^{0.382},$$

$$w = 27.5 \text{ m}.$$

Alternatively, w could have been found from Eq.(6.9), since the depth and speed have already been calculated. If this route is followed, the result is

$$Q = wdv,$$

$$470 \text{ m}^3/\text{s} = w(2.52 \text{ m})(6.79 \text{ m/s}),$$

$$w = 27.5 \text{ m}.$$

This value agrees with the first determination since a sufficient number of significant figures are carried in the calculation.

****

6.5 Sediment Load

Rivers transport small solid particles of matter, called *sediment*, suspended in the water. These particles are small enough not to settle out; rather, they are kept in suspension by the irregular, sometimes turbulent, nature of the flow of the water, even if the density of the particles is greater than the density of water. However, they will settle to the bottom, or bed, of the river if the flow speed becomes small enough. Many of these are particles of soil eroded from the land surface by the runoff as it drains toward the river.

The amount of sediment carried by a river (the *sediment load*) can be expressed as the mass of sediment carried either per unit mass of water, or per unit volume of water. The present work uses the mass of sediment per unit mass of water. A typical value for this sediment load is several grams of sediment per kilogram of water.

The motion of the water carries sediment past any point through a cross section of the river. It is often important to know the mass of sediment carried through a cross section of the river over a certain time. (For example, if there is a flood downstream, the sediment may be deposited in the floodplain; sediment may be deposited, or help cause erosion, around the foundations of bridges built across the river.)

In a time interval t, the mass m of water carried through a cross section is given by Eq.(6.8):

$$m = \rho Q t,$$

where ρ is the density of water and Q is the discharge of the river. If this mass of water carries a mass m_{sed} of sediment, the *sediment load* SL (single symbol) is defined by

$$SL = \frac{m_{\mathrm{sed}}}{m}, \tag{6.19}$$

so that

$$m_{\mathrm{sed}} = (SL)m,$$

$$m_{\mathrm{sed}} = (SL)\rho Q t. \tag{6.20}$$

The associated rate of sediment transport is called the *sediment yield* of the river, SY; that is

$$SY = \frac{m_{\mathrm{sed}}}{t}. \tag{6.21}$$

Usually, in evaluating the sediment yield, the time interval t is taken to be one year. The sediment yield observed over shorter periods of time may be expected to show seasonal variations.

****

EXAMPLE 8

A river 28.0 m wide, 8.30 m deep and flowing at 3.50 m/s carries a sediment load of 9.30 g per kilogram of water. How many metric tons of sediment passes under a bridge over the river in one day?

Calculate the discharge from Eq.(6.9): $Q = wdv$. Using the data given it is found that $Q = 813.4\ \mathrm{m^3/s}$. For one day, $t = 86{,}400$ s. By Eq.(6.20),

$$m_{\mathrm{sed}} = (SL)\rho Q t,$$

$$m_{\mathrm{sed}} = (9.30\ \mathrm{g/kg})(1000\ \mathrm{kg/m^3})(813.4\ \mathrm{m^3/s})(86{,}400\ \mathrm{s}),$$

$$m_{\mathrm{sed}} = 6.54\ \mathrm{X}\ 10^{11}\ \mathrm{g},$$

$$m_{\mathrm{sed}} = 6.54\ \mathrm{X}\ 10^{5}\ \mathrm{t},$$

since the metric ton (t) is defined, in grams, by 1 t = 1 Mg.

****

6.6 Chezy Formula

Water in a river flows downhill under gravity. The source of a river is at a greater elevation, relative to sea level, than its termination point (which may not actually be the sea, but another river into which it flows). The river bed is, therefore, a surface inclined slightly relative to the horizontal. It might seem that it should be possible to derive a formula for the speed of the water at any location along the river, much as the speed of a block sliding under gravity down an inclined plane can be derived from Newton's second law. However, the situation of water flowing under gravity down an inclined channel is very much more complex than a block sliding down a plane. Only the simplest case of uniform flow is discussed here, and even this analysis is an approximation. But, in practical terms, an exact analysis is hardly necessary, because of the inherent uncertainties, or gaps, in knowledge of the changing geometry of a river. (For example, it is not known precisely how the cross section of a river varies along its length at any instant of time.)

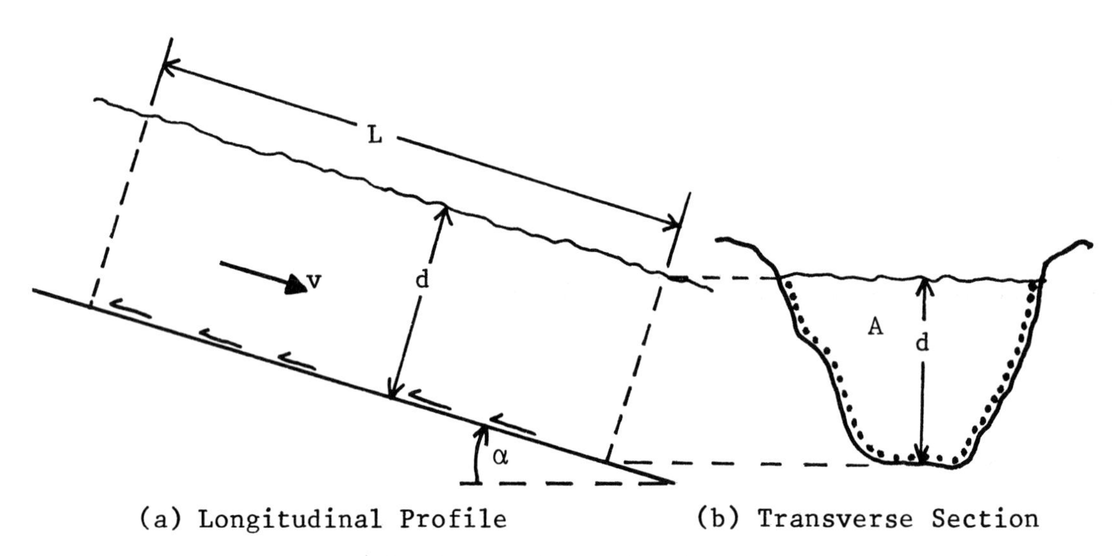

Fig.(6.6) Two Views of a River

Consider the river geometry shown in longitudinal profile and transverse cross section in Fig.(6.6). Over a distance L the river flows at a constant speed v. It is assumed that the cross section does not change in size or shape, and that no water is added to, or removed from, the river over this distance.

Let m be the mass of water contained in the channel over the distance L. Newton's second law, applied to the water, requires that

$$\Sigma F = ma, \tag{6.22}$$

where a is the acceleration of the water and ΣF the net force on the water. If $a \neq 0$, then the speed changes; if $a = 0$, the speed remains constant.

The forces acting are a driving force due to gravity and a resisting force originating in friction between the water and the sides of the channel. The driving force is

$$DF = mg \sin \alpha, \tag{6.23}$$

where α is the inclination angle of the channel. This is the same equation for the driving force as applies to a block of rock on an inclined rock slope; see Chapter 4.

In contrast to the driving force, there is no simple, theoretically convincing, formula for the resisting force RF of the water against the side of the channel. Apart from other complications, the force depends on the speed of flow, the presence or absence of turbulence, and the relative roughness or smoothness of the channel walls. However, experiments suggest that, if the flow is uniform and non-turbulent, then the resisting force RF is given, to a good approximation, by

$$RF = \frac{1}{2} C_{\text{ch}} A_{\text{ch}} \rho v^2. \tag{6.24}$$

In Eq.(6.24), ρ is the density of water, A_{ch} the area of the sides of the channel with which the water is in contact (the wetted area), and C_{ch} is a constant, the *channel coefficient*, the numerical value of which depends on the characteristics of the sides of the channel as to being smooth or rough. Note from the equation that C_{ch} is dimensionless. This force opposing the motion of the water acts parallel to the sides of the channel and therefore is a shear force, as indicated on Fig.(6.6).

Substitute Eqs.(6.23) and (6.24) into Eq.(6.22); express the mass of the water as $m = \rho AL$, where A is the cross-sectional area of the channel, to get

$$\rho ALg \sin \alpha - \frac{1}{2} C_{\text{ch}} A_{\text{ch}} \rho v^2 = ma.$$

Near its source the speed of the river is small. As the water moves away from the source, the speed v may be expected to increase. Eventually, the speed reaches a value such that the resisting force, the second term on the left in the equation above, comes to equal the driving force, the first term on the left. At that speed, the left-hand side of the equation is zero, and therefore $a = 0$. This value of the speed is the constant speed at which the water flows in the section shown in Fig.(6.6). Setting $a = 0$ in the preceding equation yields for this constant speed

$$ALg \sin \alpha - \frac{1}{2} C_{\text{ch}} A_{\text{ch}} v^2 = 0,$$

$$v = \sqrt{\frac{2ALg \sin \alpha}{A_{\text{ch}} C_{\text{ch}}}}. \tag{6.25}$$

Now the wetted area A_{ch} of the channel can be written as

$$A_{\text{ch}} = PL, \tag{6.26}$$

where P is the *wetted perimeter* of the channel cross section. This is shown on Fig.(6.6); P is the length of the dotted line marking water in contact with the channel on the transverse cross section, the dots being the "tails" of the shear stress harpoon-arrows on the longitudinal profile. Substituting Eq.(6.26) into Eq.(6.25) eliminates the length L to give

$$v = \sqrt{2(\frac{A}{P})\frac{g \sin \alpha}{C_{\text{ch}}}}. \tag{6.27}$$

The quantity A/P has units of length, and is called the *hydraulic radius* R_{H} of the channel; that is

$$R_{\text{H}} = \frac{A}{P}. \tag{6.28}$$

Also, the *Chezy coefficient* C is defined by

$$C = \sqrt{\frac{2g}{C_{\text{ch}}}}. \tag{6.29}$$

Since C_{ch} is dimensionless, C has units $\sqrt{\text{m}}/\text{s}$. Putting all this together results in

$$v = C\sqrt{R_{\text{H}} \sin \alpha}. \tag{6.30}$$

Equation (6.30) is known as the *Chezy formula* for the steady speed of water flowing in an open, slightly-inclined channel (α small).

****

EXAMPLE 9
A river with a rectangular cross section has an angle of inclination of 0.21°. Find the speed of the water over a particular stretch of the river where its width is 28.4 m and its depth 9.30 m. The channel coefficient equals 0.018.

The Chezy coefficient and the hydraulic radius must be calculated. Taking the Chezy coefficient first, by Eq.(6.29)

$$C = \sqrt{\frac{2g}{C_{\text{ch}}}},$$

$$C = \sqrt{\frac{2(9.8 \text{ m/s}^2)}{0.018}},$$

$$C = 33.0 \sqrt{\text{m}}/\text{s}.$$

Calculate the hydraulic radius by Eq.(6.28). For a rectangular channel, the wetted perimeter P is twice the depth plus the width, so that

$$R_{\text{H}} = \frac{A}{P},$$

$$R_{\text{H}} = \frac{(28.4 \text{ m})(9.3 \text{ m})}{2(9.3 \text{ m}) + 28.4 \text{ m}},$$

$$R_{\mathrm{H}} = 5.62 \text{ m}.$$

The speed of the water can now be found by Eq.(6.30):

$$v = C\sqrt{R_{\mathrm{H}} \sin \alpha},$$

$$v = (33.0 \sqrt{\text{m}}/\text{s})\sqrt{(5.62 \text{ m}) \sin 0.21^\circ},$$

$$v = 4.7 \text{ m/s}.$$

****

6.7 Problems

1. In one year the Amazon River discharged $5.49 \text{ X } 10^{13}$ m^3 of water to the South Atlantic Ocean. What thickness of ice covering Greenland must melt to provide this same volume of water? The area of Greenland is $2.15 \text{ X } 10^6$ km^2; assume that even the coastal regions of Greenland are covered with ice.

2. A stream has a width of 5.20 m, depth 3.70 m, and current speed 63 cm/s. (*a*) Calculate the discharge. (*b*) How much water flows through a cross section in 1.00 h? Give the volume and weight of the water.

3. Figure (6.7) shows a section of a braided river; the river breaks up into many (in this case, three) twisted, but quasi-parallel, channels that coalesce downstream. The discharge at point A on the trunk stream is 490 m^3/s. The discharge in channel 1 is 163 m^3/s and in channel 2 is 77 m^3/s. At point B in channel 3 the width is 18.2 m and the water speed 3.10 m/s. Find the depth at point B in channel 3.

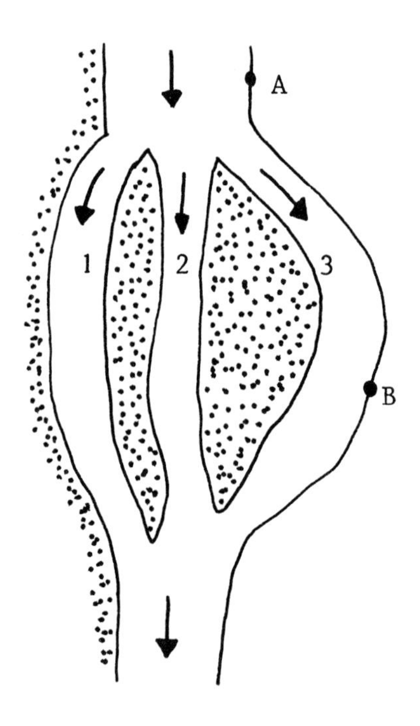

Fig.(6.7) Problem 3

4. Tidal currents in narrow channels connecting coastal bays with the ocean can be very swift. Water flows through the channel into the bay as the tide rises, and back out to the ocean as the tide falls. Consider Rectangle Bay, shown in Fig.(6.8). The bay is connected to the ocean by a channel 38 m wide and 12.8 m deep at mean (i.e., average) water level in the

bay. The graph shows the diurnal (i.e., daily) variation in water level in the bay. Calculate the average speed, in km/h, of the tidal current in the channel.

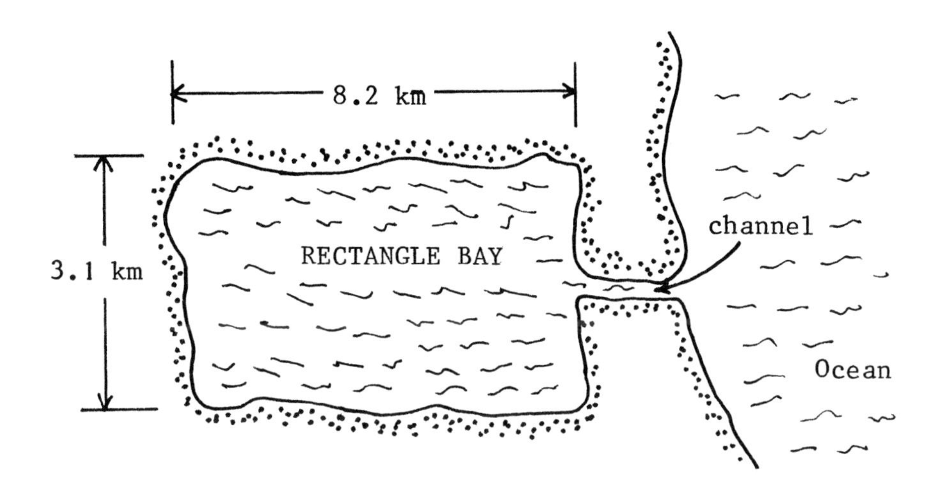

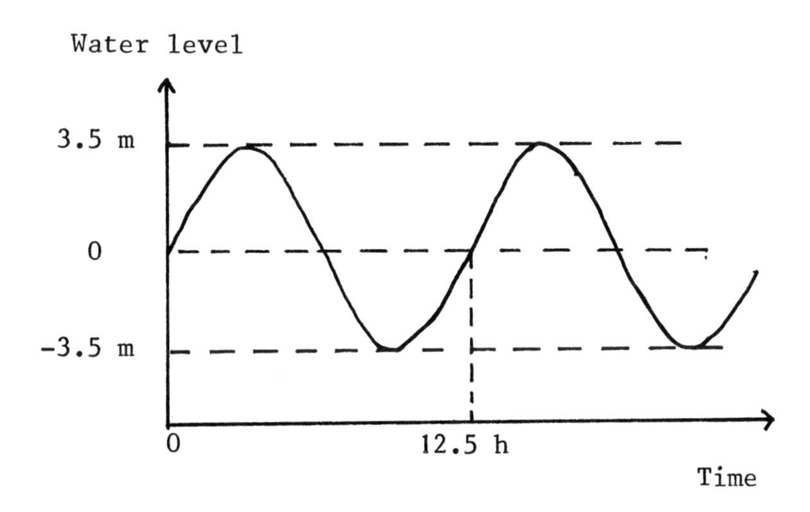

Fig.(6.8) Problem 4

5. The cross section of Isosceles River, as shown in Fig.(6.9) with levees built for flood prevention, is, not unexpectedly, an isosceles triangle. The width of the channel is 23.0 m and the depth at midchannel is 9.92 m. The levees are 4.50 m high. The river is rising, due to rapid melting of snow in the mountains upstream. (*a*) When the depth of the water at midstream is 6.30 m, the current speed is 4.36 m/s. Find the discharge. (*b*) As the rising river reaches the top of the levees, the water speed is clocked at 7.14 m/s. Find the discharge

now, at flood stage.

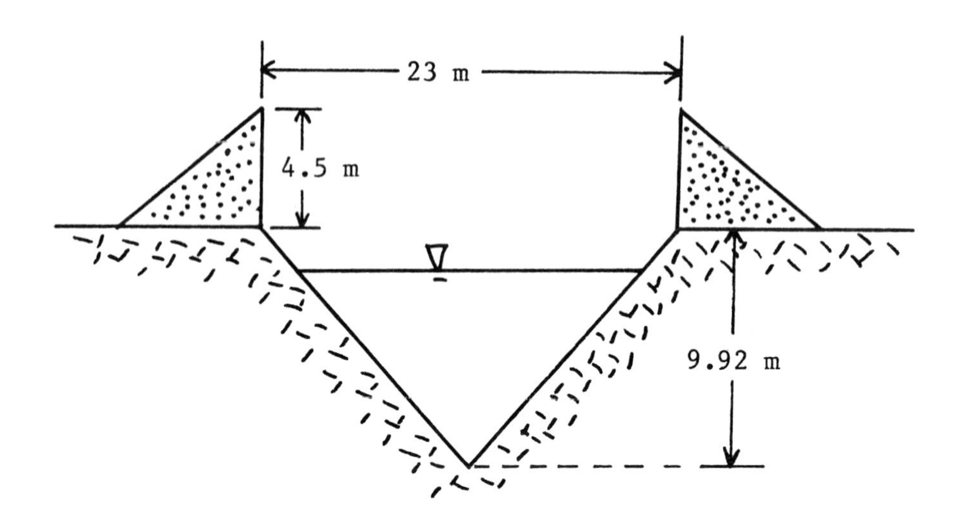

Fig.(6.9) Problem 5

6. The river shown in cross section in Fig.(6.10) is flowing at 3.20 m/s. How long does it take for 15,000 metric tons of water to pass through the cross section? (Note that one bank is steeper than the other.)

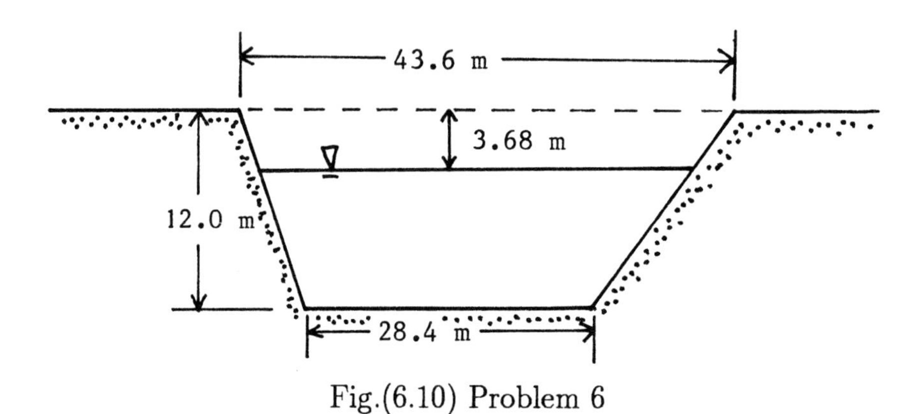

Fig.(6.10) Problem 6

7. At Circle Bay, 3.00 km in diameter, the high tides occur 12 h 25 min apart. The circular bay is connected to the ocean by a tidal channel of width 26.0 m and depth 9.20 m. The average speed of the current in the channel is 8.40 km/h. By how much does the water level in the bay fall from one high tide to the next low tide?

8. The Trans-Alaska pipeline is 1273 km long, 1.22 m in diameter, and carries 3.0 X 10^5 m^3 of oil per day. How long does it take for the oil to travel the length of the pipeline?

9. In the 1992 flood event at San Buenaventura, the discharge in a channel of the Ventura River increased from 25 m^3/s to 1322 m^3/s over 4.0 h. Assuming that the discharge increased at a uniform rate, find the volume of water that flowed through the channel during this time.

10. The hydrograph of Flood River over a time interval of two weeks is shown on Fig.(6.11). Find the volume of water that flowed though a cross section of the river between days 2 and 12.

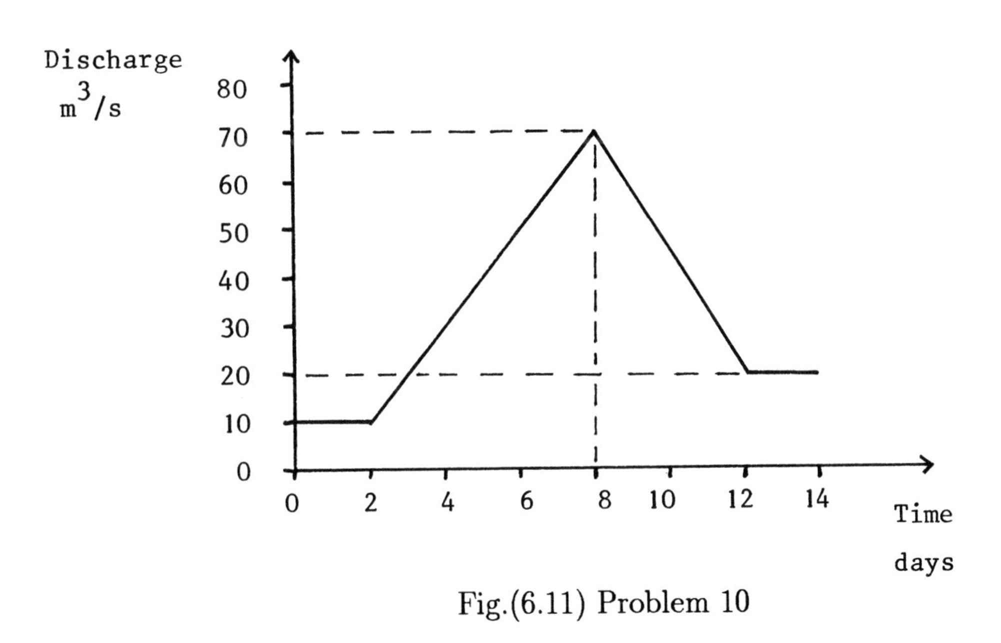

Fig.(6.11) Problem 10

11. Over a certain land area the annual precipitation is 88.2 cm. Of this 36.0 cm returns to the atmosphere by evapotranspiration. The remainder forms runoff that drains into a river 16.3 m wide, 2.84 m deep and flowing at 1.43 m/s. Find the area of the land in km^2.

12. The drainage basin of a river 16 m wide has an area of 2.7 Mha. The annual precipitation is 76 cm and the runoff 53 cm. What volume of water is returned to the atmosphere each year from this basin?

13. Over a 15,800 km^2 catchment area, 72.0% of the precipitation forms runoff. The discharge of the draining river is 234 m^3/s. Find the annual precipitation.

14. A plain in Spain on which it rains has an area of 3.11 Mha. The annual precipitation is 49.0 cm, of which 17.0 cm undergoes evapotranspiration. The draining river has width 21.2 m and current speed 1.40 m/s. Find (*a*) the depth of the river, and (*b*) the volume of water returned to the atmosphere each year.

15. The average speed of the water is 2.10 m/s in a river 37.2 m wide and 6.30 m deep. The catchment region has an area of 28,400 km^2, over which 62.0% of the precipitation forms runoff. Find the annual precipitation.

16. How high H must the levees be built so that when the river is flowing at 8.35 m/s and the discharge is 4220 m^3/s, the river crests just at the top of the levees, as shown on Fig.(6.12)?

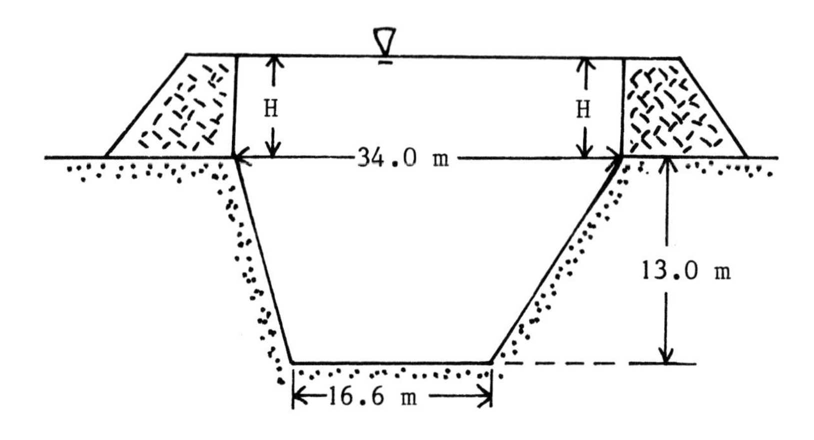

Fig.(6.12) Problem 16

17. At a particular place on a particular river, measurements of width w in m, speed v in m/s, depth d in m, and discharge Q in m^3/s yield the following equations:

$$w = aQ^{0.47},$$

$$d = 0.72Q^b,$$

$$v = 0.88Q^{0.31}.$$

At a time when the discharge is 195 m^3/s, what are the values of (a) the width, and (b) the depth of the river?

18. At Certain Village on Certain Stream, the width w in m, depth d in m, and current speed v in m/s are related to the discharge Q in m^3/s by the equations

$$w = 3.0Q^{0.23},$$

$$d = 0.65Q^{0.42},$$

$$v = aQ^b.$$

(a) Find the current speed when the discharge is 200 m^3/s. (b) After a storm the discharge doubles. By how much does the width of the stream increase?

19. The width, depth, water speed, and discharge of a river, at a particular point on the river, obey the relations

$$w = aQ^{0.461},$$

$$d = 1.36Q^{0.250},$$

$$v = 0.892Q^{b}.$$

The variables must be expressed in SI base units. When the depth is 8.68 m, what are (*a*) the width, and (*b*) the speed of the water?

20. At Special Town on Special River, the width w in m, depth d in m, speed v in m/s, and discharge Q in m^3/s are related by

$$w = 0.398Q^{a},$$

$$d = 1.22Q^{0.182},$$

$$v = bQ^{0.273}.$$

Find (*a*) the depth, and (*b*) the speed of the river when the width is 43.9 m.

21. A river 31 m wide and 6.4 m deep flows with a speed of 5.5 m/s. It carries a sediment load of 12 g per kg of water. What mass of sediment passes any point in 4.0 h?

22. A river has a width of 15.2 m, depth 3.61 m, and a speed of 2.47 m/s. It carries a sediment load of 8.20 g per kg of water. How much sediment flows under a bridge over the river in 2.00 days?

23. A river has a discharge of 722 m^3/s. It carries a sediment load of 1.40 g per kg of water. How long will it take for 1250 metric tons of sediment to pass through a cross section of the river?

24. A river with a discharge of 430 m^3/s carries 3.04 X 10^6 metric tons of sediment past a bridge each week. Calculate the sediment load of the river, in grams of sediment per kilogram of water.

25. Measured values of the depth d and discharge Q at a certain location on a particular river yield the straight line shown on Fig.(6.13) as the average of all the data points. Using

the two points indicated, find the values of a and b in the empirical equation $d = aQ^b$.

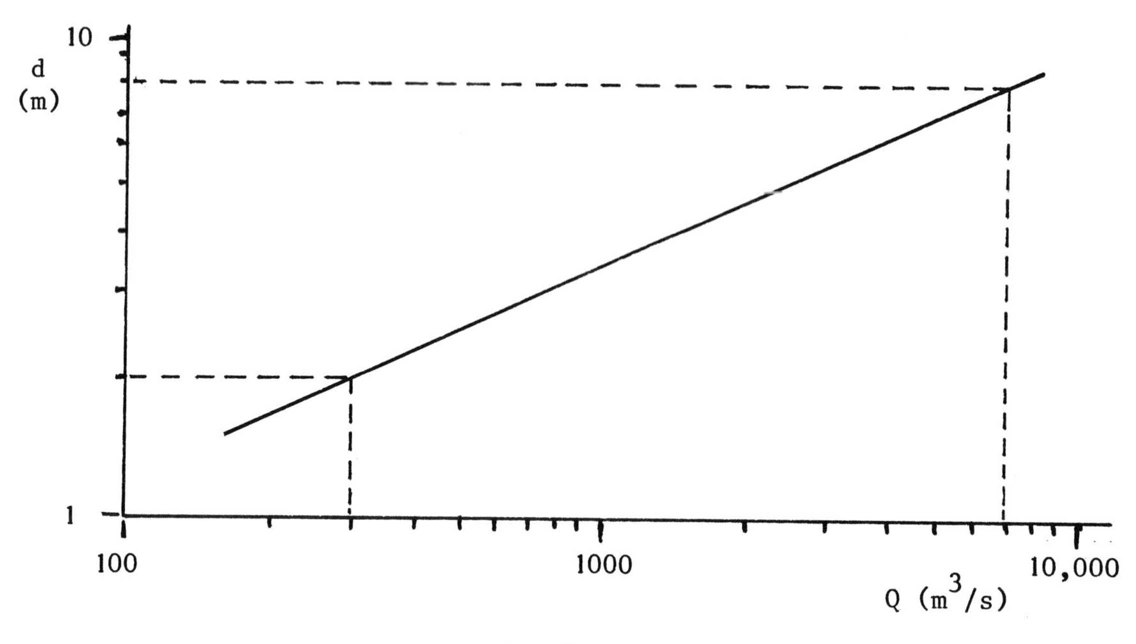

Fig.(6.13) Problem 25

26. A river draining a land area of 910 km^2 has an average width of 15 m and an average depth of 2.3 m. Annual precipitation in the region is 76 cm, of which 17 cm is dispersed by evapotranspiration. The sediment load of the river is 8.0 g per kg of water. The density of the soil is 3.1 g/cm^3. How long, in years, will it take for the land surface to be worn down by 5.0 cm? (Assume that the sediment load of the river is entirely eroded soil and that no new soil is being created or deposited.)

27. Each day the Mississippi River discharges about 2.0 X 10^6 metric tons of sediment into the Gulf of Mexico. To account for this, sediment must be eroded from the land surface at about 6.0 mm/century. The river's drainage basin has an area of 3.0 X 10^6 km^2. Calculate the average density of the surficial layers of soil in the drainage basin.

28. Calculate the hydraulic radius of a circular pipe of radius R that is completely filled with flowing water.

29. Find the hydraulic radius of a river with an equilateral triangle for the channel cross section, edge length 14.5 m, when the water is (a) level with the river's banks, and (b) has subsided to one-half of this overflow depth.

30. What would have to be the inclination of the river in Example (6.9) in order that its current speed be twice that in the example, all other quantities being equal?

31. Water is flowing in a rectangular channel 39.4 m wide and inclined at 0.180°. The Chezy coefficient equals 51.2 $\sqrt{m}/s$. The water is flowing at 6.27 m/s. Find the depth of the water in the channel.

32. A canal with concrete sides has been constructed, through monumental error, with an isosceles triangle cross section, symmetry axis vertical. The apex angle of the triangle is 50.0°. The channel coefficient for water on concrete equals 0.00246. The canal is inclined at 0.320° to the horizontal. Find the speed of flow of the water when the depth of the water, measured at midchannel, is 18.0 m.

33. A river has a cross section closely resembling a trapezoid. The bed width is 17.7 m, and the two sides slope upwards and outwards from the bed, each at an angle of 32.0° to the horizontal. The Chezy coefficient equals 35.0 $\sqrt{m}/s$ and the river's inclination is 0.650°. Find water's speed when the depth of the water in the middle of the stream is 5.22 m.

34. Calculate the missing quantities in the table below, which refers to some of the world's longest rivers.

River	Q (m^3/s)	SY (t/y)	SL (g/kg)
Amazon		3.64 X 10^8	0.0542
Yangtze	2.18 X 10^4		0.726
Brahmaputra	1.98 X 10^4	7.26 X 10^8	
Ganges		1.45 X 10^9	2.45

Chapter 7

Dams and Reservoirs

7.1 Forces on Dams

Dams are built for a variety of reasons, the principal ones being for the production of electricity by the controlled fall of water through electrical generators, to control the release of water for irrigation, or for downstream flood control. But dams also impact upstream regions through the creation of a reservoir.

Construction methods vary. Dams can be made of concrete, or earthen material (soil and rock). Earth dams are not considered here, one reason being that they have front surfaces that slope at significant angles. The analysis of forces on such a dam presents mathematical complexities out of proportion to the additional physical principles necessarily invoked. Concrete dams generally have steeper faces. In this chapter, the upstream face will be taken to be vertical. A concrete dam may be "arched" into valley cliffs which help in supporting the force exerted by the water on the dam. Alternatively, the dam may simply sit on the ground, not being supported at its ends at all. It is this so-called *gravity dam* that is analyzed here. (Many of the ideas outlined in this chapter apply also to a retaining wall or levee, designed to hold back much shallower bodies of water than the reservoir typically leaning on a dam.)

Assume, then, that the overall structure is a concrete gravity dam, that the upstream face in contact with the reservoir water is vertical, and also that the ground on which the dam is built (the foundation) is horizontal and remains dry (so that possible buoyancy effects can be ignored). The first task is to find an expression for the force exerted by the water on the wetted upstream face of the dam.

The hydrostatic (gauge) pressure p of the water is given by

$$p = \rho g y, \tag{7.1}$$

where y is the depth below the surface of the water at which the gauge pressure is p, and ρ is the density of water. In this chapter, the water is assumed fresh, so that $\rho = 1000\ \mathrm{kg/m^3}$. The use of gauge pressure is appropriate since the pressure of the atmosphere acts on both the front and back faces of the dam and therefore cancels out.

The value of the pressure is $p = 0$ at the surface ($y = 0$) of the reservoir, and increases linearly to $p = \rho g D$ at the bottom, where D is the depth of the water in the reservoir (or river, perhaps, if the structure is a levee). See Fig.(7.1). Since the pressure increases uniformly with depth, the average pressure $\overline{p}$ is given as the average of the numerical values of the pressures at the top and bottom of the reservoir; i.e., by

$$\overline{p} = \frac{1}{2}(0 + \rho g D),$$

$$\overline{p} = \frac{1}{2}\rho g D. \tag{7.2}$$

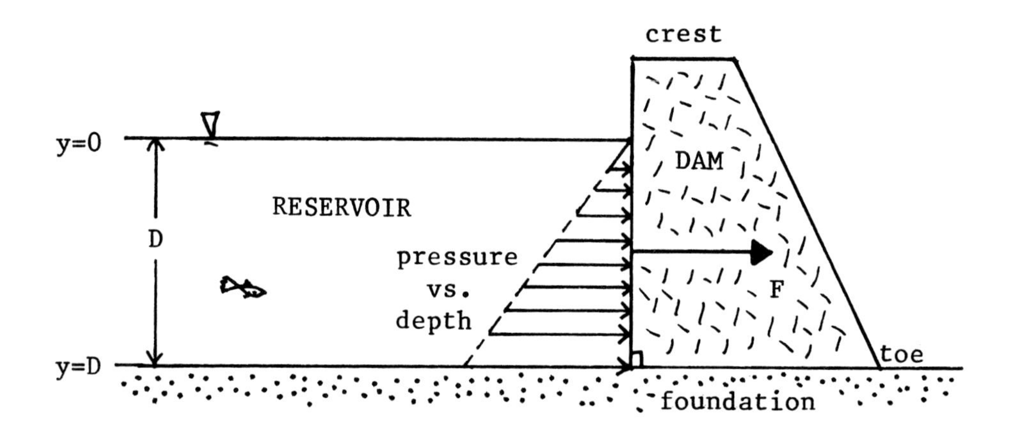

Fig.(7.1) Pressure at Dam Face

Since pressure is force/area, the force F exerted by the water on the dam is

$$F = \overline{p} A_w, \tag{7.3}$$

where A_w is the wetted area of the upstream face. Now, Fig.(7.1) is a cross section of a three-dimensional structure: the face of the dam and the water are assumed to extend for a distance L (length of the dam) perpendicular to the cross section. Hence, $A_w = DL$. Therefore, by Eqs.(7.2) and (7.3) the force of the water (the *hydrostatic force*) on the dam is

$$F = (\frac{1}{2}\rho g D)(DL),$$

$$F = \frac{1}{2}\rho g L D^2. \tag{7.4}$$

To determine the direction of the force, recall that the hydrostatic pressure exerted by an ideal liquid is directed perpendicular to the sides of the liquid's container. Therefore the pressure and force are directed normal to the dam face, to the right in Fig.(7.1). Since the dam face is vertical, the force is horizontal. It can be drawn as acting anywhere on the wetted surface. The set of open arrows shows the variation of gauge pressure with depth against the face. The pressure starts from zero (no arrow) at the water surface and increases linearly (dashed straight line) to the maximum value at the base of the face. Such graphs, sometimes without labelling or units, are often included on engineering drawings.

Engineers often express forces as "force per unit length" of the object to which the force is applied, in this case the dam of length L, but omit the length in writing out the formula. That is, Eq.(7.4) would be written in such "engineeringeese" as

$$F = \frac{1}{2}\rho g D^2.$$

Apparently, the reader is expected to supply the missing "L". Of course, the equation above is dimensionally incorrect unless it is realized that F does not stand for force, but for force per unit length. In this book, equations always are properly written, dimensionally correct and with all terms shown explicitly.

The hydrostatic force F tends to push the dam off its foundation. In the language of Chapter 4, this force is a driving force DF: it is the force tending to move the dam from its desired equilibrium position, to make it slide, and therefore write

$$DF = \frac{1}{2}\rho g L D^2. \tag{7.5}$$

For the dam to remain at rest, a resisting force, equal in magnitude and opposite in direction to the driving force, must act on the dam. For a gravity dam that simply rests on the surface upon which it is built, the resisting force can only be the force of static friction f_{s} that acts at the dam-foundation interface. For equilibrium, $f_{\mathrm{s}} = DF$ in magnitude.

As discussed in Chapter 4, the force f_{s} of static friction cannot exceed $f_{\mathrm{s,max}}$, where $f_{\mathrm{s,max}} = \mu_{\mathrm{s}} R$, μ_{s} being the coefficient of static friction between dam and ground, and R the normal force exerted by the ground on the dam; see Fig.(7.2). For the situation of a dam resting on a horizontal foundation, the normal force is given by $R = W$, W the weight of the dam, since $R = W \cos\alpha$, given in Chapter 4, reduces to $R = W$ for a horizontal surface ($\alpha = 0°$). Hence,

$$f_{\mathrm{s}} \leq \mu_{\mathrm{s}} W.$$

The maximum possible value of the resisting force of static friction is, again in the spirit of Chapter 4, given the symbol RF, so that

$$RF = \mu_{\mathrm{s}} W. \tag{7.6}$$

The weight W of the dam is given by

$$W = \rho_{\mathrm{d}} V_{\mathrm{d}} g,$$

where ρ_{d} is the density of the dam (assumed uniform throughout), perhaps the density of concrete. The volume V_{d} of the dam can be expressed as

$$V_{\mathrm{d}} = LA, \tag{7.7}$$

where A is the cross-sectional area of the dam (i.e., of the cross section perpendicular to the length L of the dam). Therefore,

$$RF = \mu_s \rho_d ALg. \tag{7.8}$$

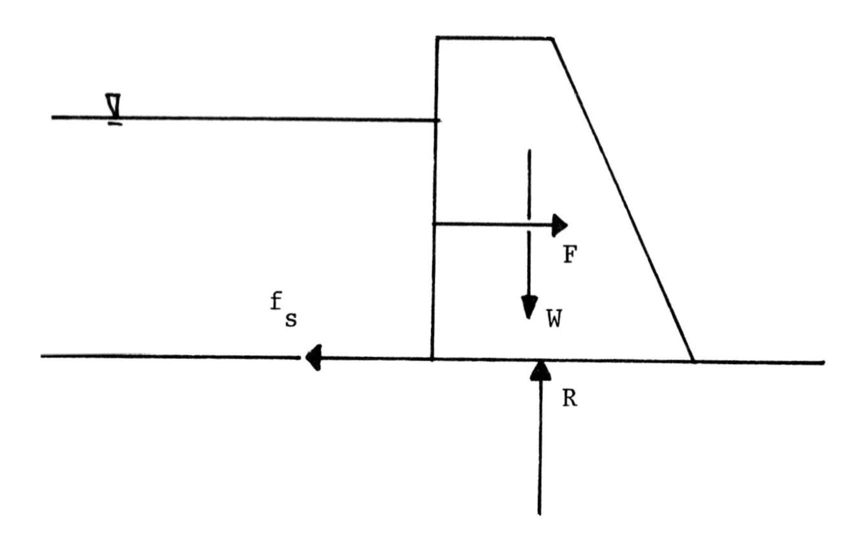

Fig.(7.2) Forces on Dam

7.2 Factor of Safety Against Sliding

The ratio of the maximum available resisting force RF to the driving force DF of the water is called the *factor of safety against sliding* FS; that is

$$FS = \frac{RF}{DF}. \tag{7.9}$$

Combining Eqs.(7.4) and (7.8) with Eq.(7.9) yields

$$FS = \frac{\mu_s \rho_d LAg}{\frac{1}{2}\rho g L D^2},$$

$$FS = 2\mu_s \left(\frac{\rho_d}{\rho}\right)\left(\frac{A}{D^2}\right). \tag{7.10}$$

For the dam to remain at rest, the value of the factor of safety against sliding must be greater than unity: $FS > 1$. If $FS = 1$, the dam will slide, in either the up or down stream direction, on the least disturbance. If $FS < 1$, the dam is pushed off its foundation by the water in the reservoir, and starts sliding in the downstream direction.

The cross-sectional area A of the dam has a value that depends on the size of the dam and the shape of the cross section. Typically, dams are thicker at the base than at the top, not least because the water pressure is greater at the base. For example, for the cross section of Fig.(7.3), a trapezoid which can be considered as a rectangle plus a right triangle,

$$A = aH + \frac{1}{2}H(b-a),$$

$$A = \frac{1}{2}H(a+b). \tag{7.11}$$

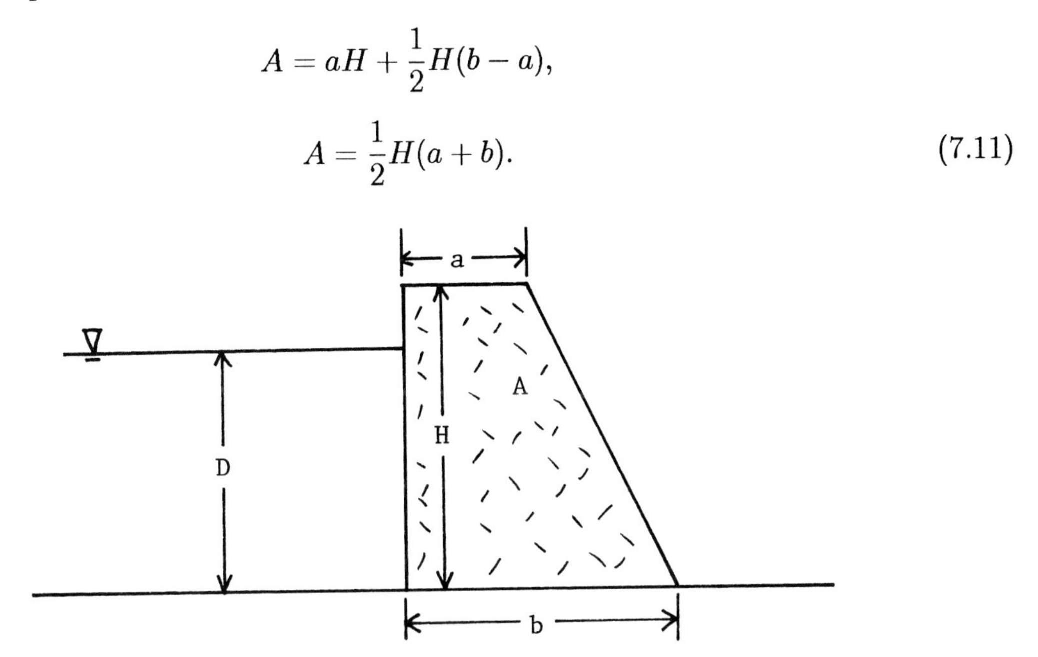

Fig.(7.3) Dimensions of Dam

****

EXAMPLE 1

For the dam shown in Fig.(7.4), which is 220 m long, how far d is the water below the top of the dam when the factor of safety against sliding equals 1.80? The coefficient of static friction between dam and foundation is 0.730.

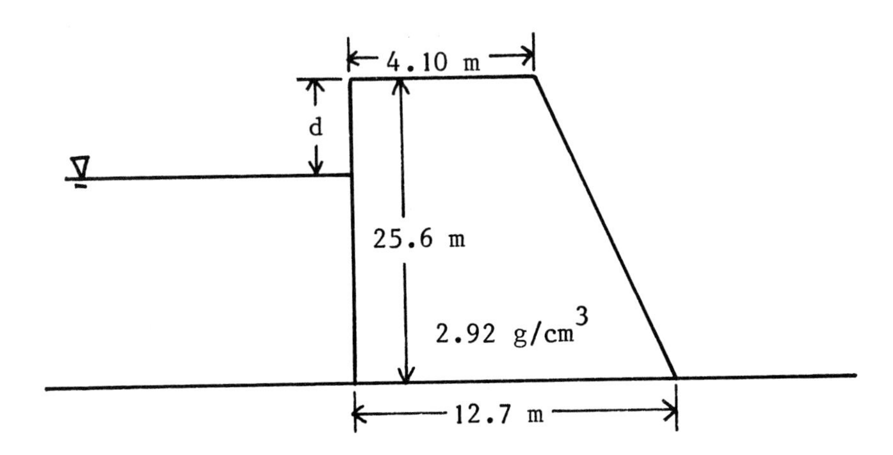

Fig.(7.4) Example 1

Clearly, the distance sought is $d = 25.6 - D$, where D is the depth of the water. Hence, solve for D and then calculate d. From Eq.(7.11),

$$A = \frac{1}{2}H(a+b),$$

$$A = \frac{1}{2}(25.6 \text{ m})(4.10 \text{ m} + 12.7 \text{ m}),$$

$$A = 215.0 \text{ m}^2.$$

Since $\rho_d/\rho = (2.92 \text{ g/cm}^3)/(1.00 \text{ g/cm}^3)$, $\rho_d/\rho = 2.92$ (which amounts to the specific gravity of the building material), Eq.(7.10) gives

$$FS = 2\mu_s \left(\frac{\rho_d}{\rho}\right)\left(\frac{A}{D^2}\right),$$

$$1.80 = 2(0.730)(2.92)\left(\frac{215.0 \text{ m}^2}{D^2}\right),$$

$$D = 22.6 \text{ m}.$$

Therefore, $d = 25.6 - 22.6$, $d = 3.0$ m.

****

It can be seen from Eq.(7.10) that the factor of safety is directly proportional to μ_s, the coefficient of friction between the dam and the ground. It should be recalled from the discussion of the factor of safety of a slab of rock on a slope, that the value of the coefficient of friction depends on the physical conditions at the time. For example, when the ground is dry, the coefficient of friction will have a certain value. If water seeps into the pore spaces of the foundation rock or soil (a very likely occurrence with a reservoir only meters away), the value of the coefficient of friction will decrease, and so therefore will the value of the factor of safety. In cases where the factor of safety could be reduced to values close to unity, one or more possible design safeguards may be incorporated. One of these is to build the dam into the foundation, "locking in" the dam; the additional segment is called a *keyway*; see Fig.(7.5).

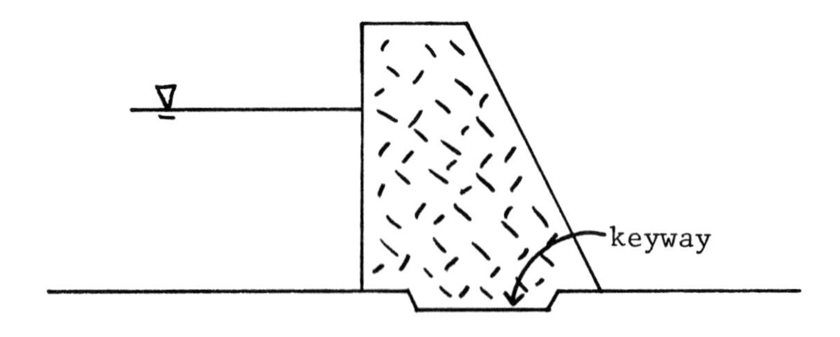

Fig.(7.5) Keyway

7.3 Torques on Dam

There are two requirements to be fulfilled for a structure to remain in equilibrium: one is that the net force on the structure be zero, so that it is in translational equilibrium; the other is that the net torque on the structure be zero, so that it is in rotational equilibrium. Therefore, to obtain a full picture of the safety of a dam, its rotational tendencies must be examined, via an analysis of the torques acting.

The situation of the dam has some features in common with the treatment of rock topples (the tipping over of a block of rock on a slope), as described in Section (4.5). Tendencies toward rotation are analyzed by examining the torques acting on the object. In Eq.(4.35), torque τ is defined by $\tau = rF$, where r is the moment arm of the force F. Torques that tend to tip the object over, and those tending to restore the object once tipped even slightly to its original position, must be considered.

Turn first to the torque exerted by the water on the face of the dam. To compute the torque, expressions must be written for both the force and the moment arm. The hydrostatic force is related to the pressure exerted on specific areas. The moment arm is the perpendicular distance from the point of application of the force to the axis of likely rotation.

Begin by identifying the axis of rotation. The water exerts a force at all points on the wet part of the face of the dam. The direction of the force is perpendicular to the face. Therefore, wherever on the face of the dam the force is drawn, the tendency of the force is to tip the dam over: in effect, to try to rotate the dam about an axis through O at the base of the dam on the downstream side (the *toe*). In Fig.(7.6), this tendency is to tip the dam clockwise about the toe at point O (really a line through O). The shape of the cross section of the dam is not important here.

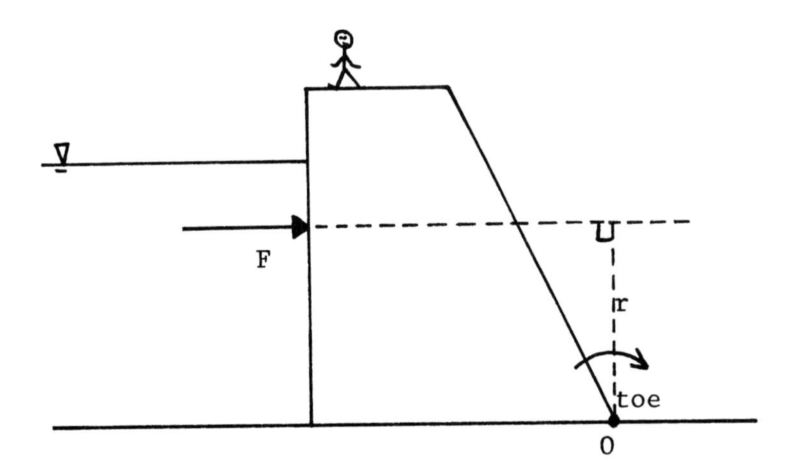

Fig.(7.6) Rotation of Dam by Water

The moment arm for the force is the line drawn from O to meet the line of action of F at a 90° angle. In Fig.(7.6), the moment arm is the vertical line marked with its length r. The numerical value of r depends on just where the force F is drawn as acting. Unlike treatment of translation, when dealing with rotation the forces must be shown at the actual point of application.

Force equals pressure times area, and the pressure varies with depth. Therefore, for equal areas, the force varies with depth. The length of the moment arm is also different for areas at different depths, because they are at different distances from the axis through O at the toe of the dam. It follows that the torque cannot be calculated from the net force F, but the actual distribution of the hydrostatic force over the face of the dam must be considered.

To manage these complexities, recourse must be had to calculus. To relate pressure to force, the proper element of area must be chosen. The pressure varies with depth, that is vertically, but does not vary in the horizontal direction at any fixed depth. The rules of calculus thereby indicate that the area chosen must be very short (length dy) in the vertical direction (so that the pressure over the element of area is constant), but can extend the entire length L of the dam in the horizontal direction, because no change in pressure is encountered in the horizontal direction. Figure (7.7) shows such a rectangular area dA at a depth y. The force acting on this area is dF. (Differentials must be used since the quantities dA and dF are very small.)

The differential form of the force-pressure relation is

$$dF = p(dA). \tag{7.12}$$

Since the area is rectangular, $dA = L\,dy$. The pressure p is given by Eq.(7.1). Thus,

$$dF = (\rho g y)(L\,dy). \tag{7.13}$$

The moment arm r is most easily visualized in the cross-sectional view of Fig.(7.7), from which it is seen that

$$r = D - y. \tag{7.14}$$

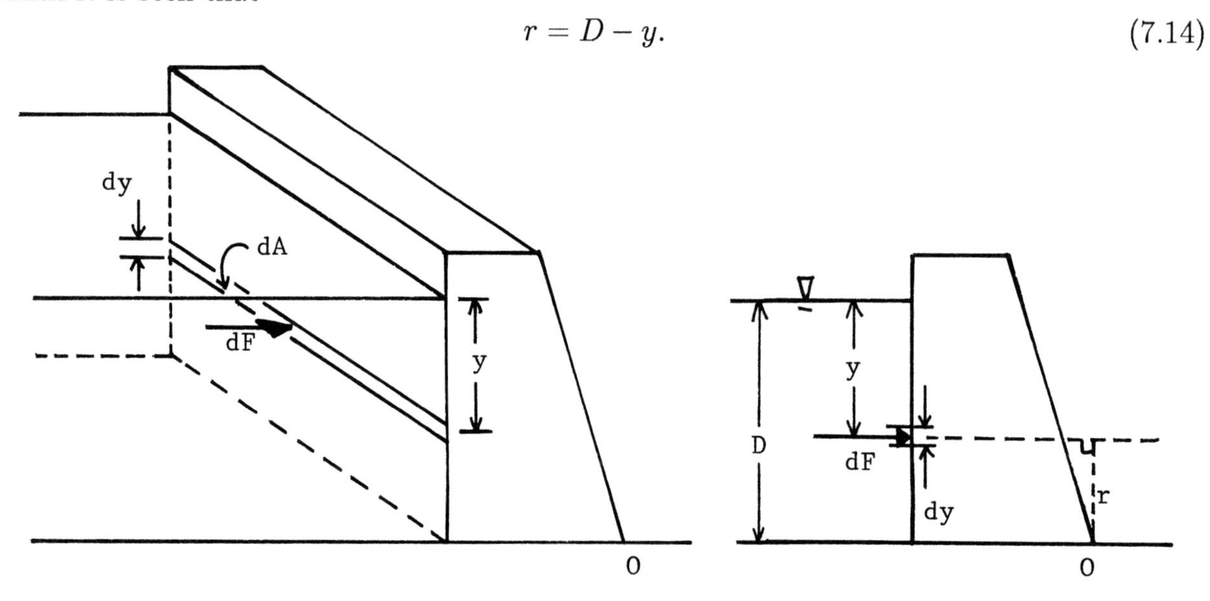

Fig.(7.7) Torque Due to Water

Even though the force dF is very small, the value of the associated moment arm r is not. Hence, the differential form of Eq.(4.35) is

$$d\tau = r(dF).$$

Substituting Eq.(7.13) for dF and Eq.(7.14) for r gives the associated torque $d\tau$ as

$$d\tau = (D - y)\rho g y L \, dy. \tag{7.15}$$

To find the total torque τ exerted by the water on the dam, integrate Eq.(7.15) over the wetted portion of the dam face. The water extends from depth $y = 0$ to $y = D$, so that

$$\tau = \int d\tau,$$

$$\tau = \int_0^D (D - y)\rho g y L \, dy,$$

$$\tau = \rho g L \int_0^D (D - y) y \, dy,$$

$$\tau = \frac{1}{6}\rho g L D^3. \tag{7.16}$$

This torque, tending to tip the dam over, is caused by the forces exerted by the water. In the discussion of the factor of safety of the dam against sliding, see Eq.(7.9), the disturbing force F exerted by the water is called the driving force DF. By analogy, call the torque exerted by the water the driving torque $D\tau$; that is,

$$D\tau = \frac{1}{6}\rho g L D^3. \tag{7.17}$$

****

EXAMPLE 2

A dam 95.0 m long is made of concrete with a density of 2.88 g/cm^3. When the water surface is 2.80 m below the top of the dam, the torque tending to tip the dam over has the value 444 MN·m. How high is the dam?

Find the depth of the water first, from Eq.(7.17); noting the SI prefix,

$$D\tau = \frac{1}{6}\rho g L D^3,$$

$$444 \text{ X } 10^6 \text{ N} \cdot \text{m} = \frac{1}{6}(1000 \text{ kg/m}^3)(9.8 \text{ m/s}^2)(95 \text{ m})D^3,$$

$$D = 14.2 \text{ m}.$$

Hence, the height of the dam must be 14.2 m + 2.8 m = 17.0 m. (The density of the concrete of which the dam is constructed plays no part in this calculation.)

****

7.4 Resisting Torque

In order that the dam not tip over under the action of the water, there must be present a torque acting on the dam that resists the effect of the torque due to the water. To calculate this resisting torque, the associated force must be identified.

This force can be discovered by imagining that the water actually has succeeded is slightly tipping the dam. Then, suddenly remove the force and torque associated with the water, and see what will happen to the dam. As Fig.(7.8) suggests, if the dam is released from its slightly tipped position, with all the water removed, the dam will simply drop back into its original position under the action of gravity. The weight W of the dam provides the resisting torque R_T; that is

$$R_T = W r_w, \tag{7.18}$$

where r_w is the moment arm of the weight W about the axis through the toe of the dam at O.

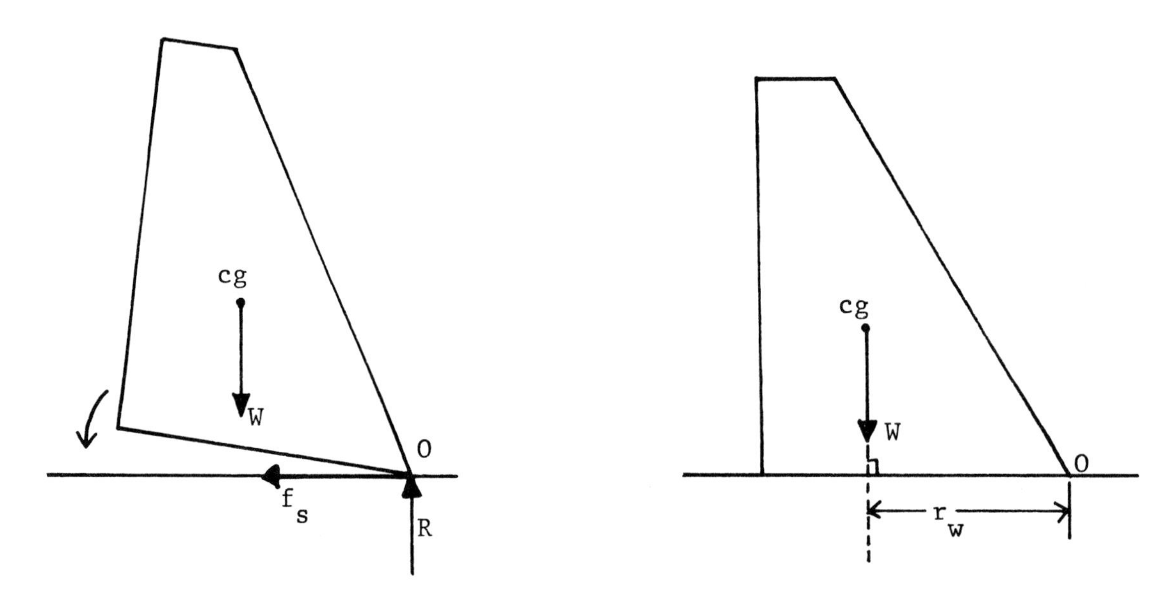

Fig.(7.8) The Resisting Torque

Notice in Fig.(7.8) that the force that resists sliding, the friction force f_s, acts through O and therefore its moment arm about O is zero, yielding zero torque for this force. (The normal force R also acts through O and exerts no torque.) The force responsible for the resisting torque is not the force that resists the driving force of the water.

To complete the calculation of the resisting torque R_T, the length of the moment arm r_w of the weight about the axis through O is needed. The weight W acts at the center of gravity cg of the dam. The moment arm is the line from O meeting the line of action of the weight at 90°. Since the line of action of the weight is vertical, this means that the moment arm is horizontal. To find its length, assume that the dam is tilted by so small an angle that, to all intents and purposes, it is in its undisturbed horizontal position. The value of

r_w will equal the horizontal distance of the center of gravity of the dam cross section from O. (The height of the center of gravity does not matter.)

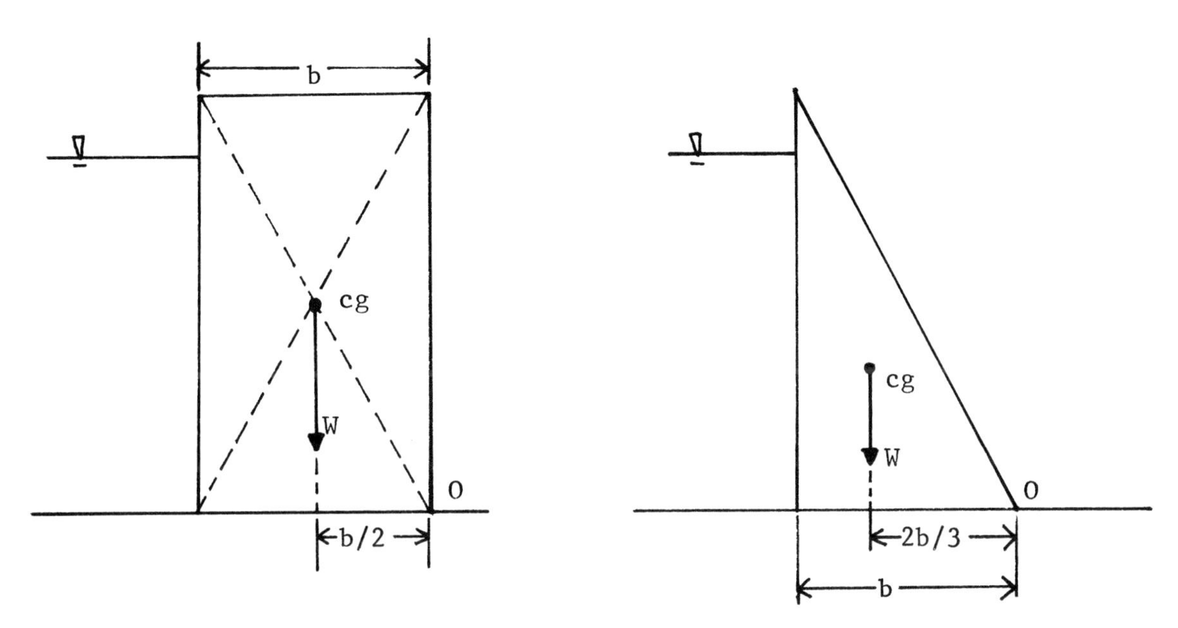

Fig.(7.9) Location of Center of Gravity

The value of r_w will vary with the shape of the dam's cross section. For a rectangular cross section of thickness b, Fig.(7.9) shows that $r_w = \frac{1}{2}b$. For a right triangle cross section, $r_w = \frac{2}{3}b$, where b is the base thickness of the cross section.

It would be difficult to walk across the top of the triangular dam of Fig.(7.9). A more realistic cross section is shown in Fig.(7.10). To calculate r_w for this trapezoidal cross section, use the definition for the location of the center of gravity of an object made of two pieces. The trapezoid can be thought of as a rectangle joined to a right triangle. Therefore

$$r_w = \frac{W_1 r_1 + W_2 r_2}{W_1 + W_2}. \tag{7.19}$$

In this equation, W_1 is the weight of the rectangular piece, r_1 the horizontal distance of its center of gravity from O; W_2 and r_2 are the corresponding quantities for the triangular section. From Fig.(7.10) it is seen that

$$r_1 = \frac{1}{2}a + (b - a),$$

$$r_1 = b - \frac{1}{2}a, \tag{7.20}$$

$$r_2 = \frac{2}{3}(b - a). \tag{7.21}$$

The dimension a is the thickness of the rectangular section, and b is the base thickness of the entire cross section; see Fig.(7.10).

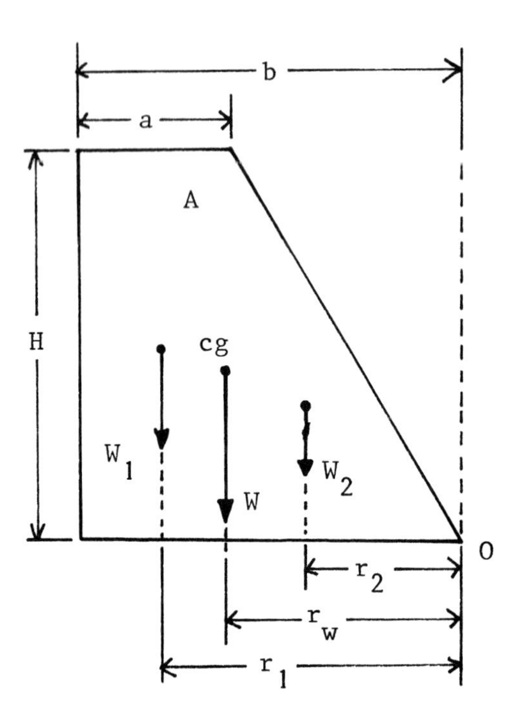

Fig.(7.10) Trapezoidal Dam

Since the total weight W of the dam is $W = W_1 + W_2$, the resisting torque R_T given by Eq.(7.18) can also be written, by virtue of Eq.(7.19), as

$$R_T = W_1 r_1 + W_2 r_2. \tag{7.22}$$

As before, let ρ_d be the density of the building material of which the dam is made. Then

$$W_1 = \rho_d g a H L, \tag{7.23}$$

$$W_2 = \frac{1}{2}\rho_d g H(b-a)L, \tag{7.24}$$

where H is the height of the dam. Now substitute Eqs.(7.20), (7.21), (7.23), (7.24) into Eq.(7.22). In arranging the resulting formula, it will be convenient to express the result in terms of A, the cross-sectional area of the dam. Bearing this in mind, the result is found to be

$$R_T = \frac{1}{3}\rho_d g A L\left[\frac{2b^2 + 2ab - a^2}{a+b}\right]. \tag{7.25}$$

Since the weight of the dam is $W = \rho_d g A L$, this means that

$$r_w = \frac{1}{3}\left[\frac{2b^2 + 2ab - a^2}{a+b}\right]. \tag{7.26}$$

****

EXAMPLE 3
A dam with a right triangle cross section has a height of 56.0 m and a base thickness equal to 21.0 m. The dam is composed of concrete with a density of 2.72 g/cm^3. The torque resisting tipping has the value 26.3 GN·m. Find the length of the dam.

For a right triangle cross section, set $a = 0$ as implied from Fig.(7.10) and $A = \frac{1}{2}bH$. With these substitutions, Eq.(7.25) becomes

$$R_T = \frac{1}{3}\rho_d g b^2 H L.$$

"Plugging in" the numbers, all in SI base units of course,

$$26.3 \text{ X } 10^9 \text{ N} \cdot \text{m} = \frac{1}{3}(2720 \text{ kg/m}^3)(9.8 \text{ m/s}^2)(21.0 \text{ m})^2(56.0 \text{ m})L,$$

$$L = 120 \text{ m}.$$

****

7.5 Factor of Safety Against Rotation

With expressions for the driving and resisting torques now available, the factor of safety against rotation can be evaluated. The factor of safety against rotation is defined in strict analogy with the factor of safety against sliding, as displayed in Eq.(7.9), except that rotation is governed by torques rather than forces. Hence, the factor of safety against rotation is defined by

$$FS = \frac{R_T}{D_T}. \tag{7.27}$$

Here D_T is the driving torque due to the water, given by Eq.(7.17), and R_T is the maximum available resisting torque of Eq.(7.25). Equation (7.25) represents the maximum available resisting torque since it is calculated with the dam slightly tipped: in this position, the normal force R exerted by the ground can provide no tipping torque about O.

To avoid cumbersome notation, the same symbol FS is used for the factor of safety against rotation as for the factor of safety against sliding. The context within which the symbol FS is used will reveal whether it pertains to sliding or tipping.

Substituting Eq.(7.17) for D_T and Eq.(7.25) for R_T into Eq.(7.27) yields for the factor of safety against rotation

$$FS = 6\left(\frac{\rho_d}{\rho}\right)\left(\frac{A r_w}{D^3}\right). \tag{7.28}$$

The density of water is ρ and its depth is D. Now r_w, the moment arm, is given by Eq.(7.26) and A, the cross-sectional area of the dam, by Eq.(7.11). Replacing r_w and A with these expressions gives the formula for the factor of safety in terms of the dam dimensions:

$$FS = \left(\frac{\rho_d}{\rho}\right) \frac{H(2b^2 + 2ab - a^2)}{D^3}. \tag{7.29}$$

The factor of safety against rotation has the same significance for tipping as the factor of safety against sliding has for sliding. The more that FS exceeds unity (1), the safer. If $FS = 1$, the dam is on the verge of tipping over.

****

EXAMPLE 4

The dam shown on Fig.(7.11) is made of material with density 2.62 g/cm^3. With the water level as shown, (*a*) find the value of x so that the factor of safety against rotation equals 1.40, and (*b*) find the minimum value of the coefficient of static friction to prevent sliding.

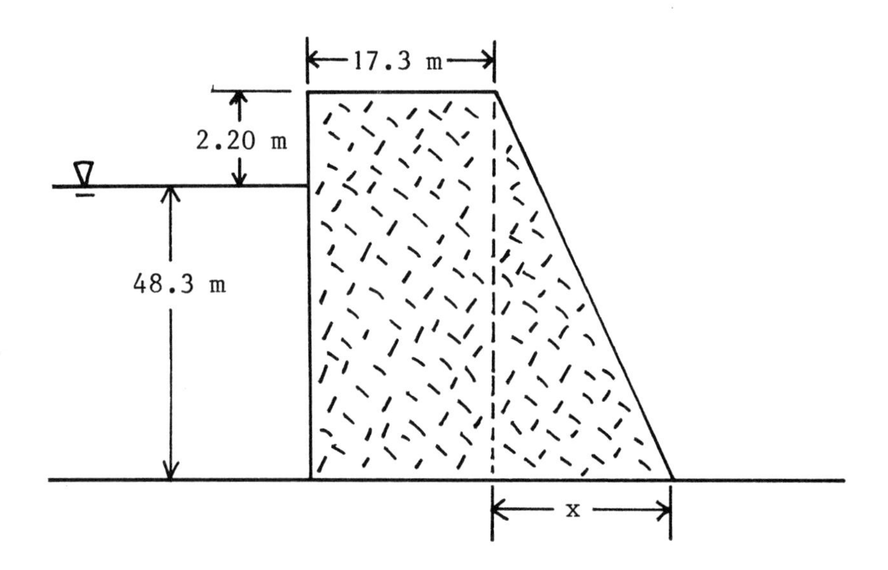

Fig.(7.11) Example 4

(*a*) Taking dimensions from Fig.(7.11) the data are: $\rho_d = 2.62$ g/cm^3, $a = 17.3$ m, $D = 48.3$ m, $H = 50.5$ m (by addition), and $FS = 1.40$ against rotation. Now $x = b - a$ and a is given, so find b, and then solve for x. By Eq.(7.29),

$$2b^2 + 2ab - a^2 = (FS)\left(\frac{D^3}{H}\right)\left(\frac{\rho}{\rho_d}\right).$$

Since the densities enter as a ratio, they may be left in g/cm^3 (for water, $\rho = 1.00$ g/cm^3). Entering the numbers into the equation above, and evaluating the right-hand side, gives

$$2b^2 + 34.6b - 299.3 = 1192.3,$$

$$2b^2 + 34.6b - 1492 = 0,$$

$$b = 20.0 \text{ m}, -37.3 \text{ m}.$$

The negative solution to the quadratic equation has no practical significance. Hence $b = 20.0$ m. Therefore, $x = 20.0 \text{ m} - 17.3 \text{ m} = 2.7$ m.
(b) For the minimum value of μ_s, set the factor of safety against sliding equal to unity (1). By Eq.(7.11), the cross-sectional area of the dam is

$$A = \frac{1}{2}H(a + b),$$

$$A = \frac{1}{2}(50.5 \text{ m})(17.3 \text{ m} + 20 \text{ m}),$$

$$A = 942 \text{ m}^2.$$

By Eq.(7.10),

$$FS = 2\mu_s \left(\frac{\rho_d}{\rho}\right)\left(\frac{A}{D^2}\right),$$

$$1 = 2\mu_s \left(\frac{2.62 \text{ g/cm}^3}{1.00 \text{ g/cm}^3}\right)\left(\frac{942 \text{ m}^2}{(48.3 \text{ m})^2}\right),$$

$$\mu_s = 0.473.$$

****

7.6 Reservoirs

The construction of a dam across a river inevitably leads to the formation of an artificial lake or reservoir behind the dam.

Water enters the reservoir from the river upstream of the dam. But the cross-sectional area of the reservoir is large compared with the cross-sectional area of the river. From the discharge equation, $v = Q/A$, it is evident that this area increase indicates that the water will slow down considerably as it enters the much wider reservoir.

As discussed in Chapter 6, rivers carry a suspended sediment load. The particles are kept in suspension by the motion of the water. If the speed of the water becomes so low that the water is hardly moving, the sediment tends to settle out. This is just the situation in reservoirs, and therefore sediment deposition should be expected.

Some water moves downstream, past the dam. Water may seep under, or even through the dam (especially if it is an earthen dam). Water may be discharged deliberately: for the hydroelectric generation of electricity, to provide water for irrigation in dry seasons, to reduce flooding upstream after heavy rains, etc. Indeed, to keep the reservoir level constant, on average, the rate at which water enters the reservoir must equal the rate that it leaves through evaporation plus discharge past the dam. Of course, there are fluctuations in level, but over a long period of time the inflow rate must equal the outflow rate to maintain a constant desired reservoir level.

Now consider the sediment "budget" of the reservoir. Sediment is carried into the reservoir by the river on which the dam is built. Some sediment flows out when water is discharged past the dam. However, due to settlement of some sediment in the reservoir, the sediment load of the water flowing out is less than that flowing in. The deposition of sediment in the reservoir means that the capacity of the reservoir to store water gradually diminishes. The capacity is measured with the reservoir water level assumed to be at the maximum design height, which will be some distance below the crest of the dam.

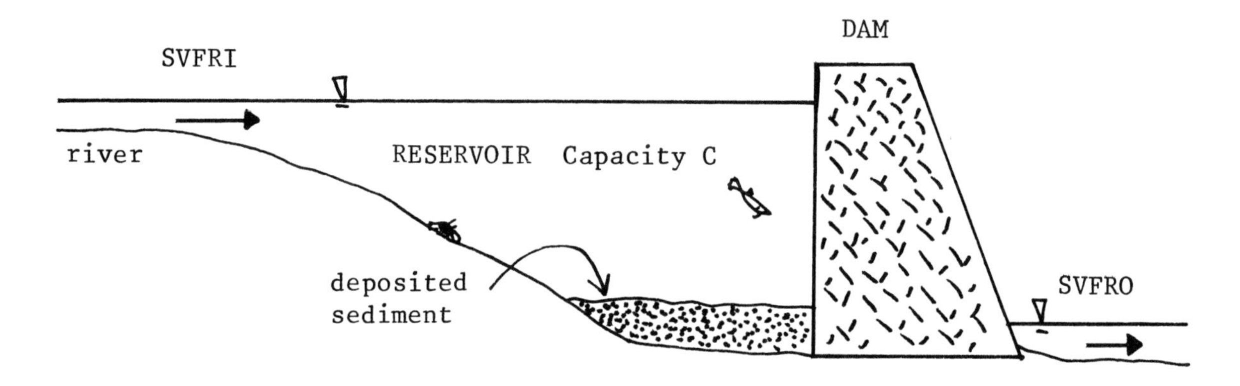

Fig.(7.12) Dam and Reservoir

Use the following notation. The capacity of the reservoir at any time is C. It is assumed that the reservoir is never cleaned, so that C will decrease as sediment accumulates. Write T for the *reservoir lifetime* , the time needed for the reservoir to become completely filled with sediment from the time its capacity is C. The sediment volume flow rate into the reservoir is $SVFRI$ (one symbol) and the sediment volume flow rate out is $SVFRO$. (These symbols avoid subscripted notation.) See Fig.(7.12).

The volume rate at which sediment is deposited in the reservoir is the difference between the inflow rate and the outflow rate of sediment. This difference is the rate at which the reservoir is filling up with sediment. If this rate is constant, then

$$T = \frac{C}{SVFRI - SVFRO}, \tag{7.30}$$

that is, the time needed to fill the reservoir is the volume to be filled divided by the rate of filling.

Dam reservoirs often are rated as to their efficiency in trapping sediment. The *trap efficiency* TE (one symbol) is defined by

$$TE = \frac{SVFRI - SVFRO}{SVFRI}. \tag{7.31}$$

Assume that the trap efficiency remains constant. Multiply numerator and denominator of the right-hand side of Eq.(7.31) by a time interval t. This replaces volume flow rates by actual volumes. Hence, the trap effciency can also be expressed as the ratio of the volume of sediment that becomes trapped in time t to the volume of sediment that flowed into the reservoir in this same time t.

Sometimes the trap efficiency is expressed as a percent. For example, if the water discharged past the dam is "clear" (sediment load equals zero), the $SVFRO = 0$ and $TE = 100\%$. Note that a high trap efficiency means a short reservoir lifetime. Once the reservoir has filled with sediment, the dam serves no useful purpose. The reason a dam is built is to control the flow of water, not to trap sediment.

If Eqs.(7.30) and (7.31) are combined, an expression for the reservoir lifetime in terms of the trap efficiency is obtained; this is

$$T = \frac{C}{(TE)(SVFRI)}. \tag{7.32}$$

The trap efficiency can be put in terms of the mass flow rates of sediment. Use the same symbols for mass flow rates as for volume flow rates except, in them, change the V to M. If ρ_{sed} is the density of the sediments and M_{sed} the mass of sediment that flows in (out) in time t, with V_{sed} the volume of sediment that flows in (out) in the same time, then

$$M_{\text{sed}} = \rho_{\text{sed}} V_{\text{sed}},$$

$$M_{\text{sed}}/t = \rho_{\text{sed}}(V_{\text{sed}}/t),$$

$$SMFRI = \rho_{\text{sed}}(SVFRI), \tag{7.33}$$

$$SMFRO = \rho_{\text{sed}}(SVFRO). \tag{7.34}$$

Substitute Eqs.(7.33) and (7.34) into Eq.(7.31) for the trap efficiency; the density of sediment cancels leaving

$$TE = \frac{SMFRI - SMFRO}{SMFRI}. \tag{7.35}$$

****

EXAMPLE 5

A dam opened on July 1, 1900. The reservoir capacity at opening is 1.53 X 10^6 m^3. The sediment yield of the river flowing into the reservoir equals 73,000 t/y. It is observed that 40,000 t of sediment is discharged past the dam each year. The sediment density is 1.42 g/cm^3. In what year will (did) the reservoir fill up with sediment?

Use Eq.(7.30) to find the reservoir lifetime from opening, when $C = 1.53$ X 10^6 m^3. Because 1 t = 1000 kg, the $SMFRI = 7.3$ X 10^7 kg/y. But

$$SMFRI = \rho_{\text{sed}}(SVFRI),$$

$$7.3 \text{ X } 10^7 \text{ kg/y} = (1420 \text{ kg/m}^3)(SVFRI),$$

$$SVFRI = 51,410 \text{ m}^3/\text{y}.$$

Similarly, it is found that $SVFRO$ = 28,170 m^3/y. By Eq.(7.30),

$$T = \frac{C}{SVFRI - SVFRO},$$

$$T = \frac{1.53 \text{ X } 10^6 \text{ m}^3}{51,410 \text{ m}^3/\text{y} - 28,170 \text{ m}^3/\text{y}},$$

$$T = 65.8 \text{ y}.$$

Therefore, the reservoir became filled in 1900.5 + 65.8 = 1966.3, or March-April, 1966.

****

7.7 Problems

1. Consider the dam shown in Fig.(7.13). (*a*) Calculate the tipping torque exerted by the water on the dam. (*b*) What must be the value of the dimension labelled x in order that the factor of safety against rotation equals 2.30?

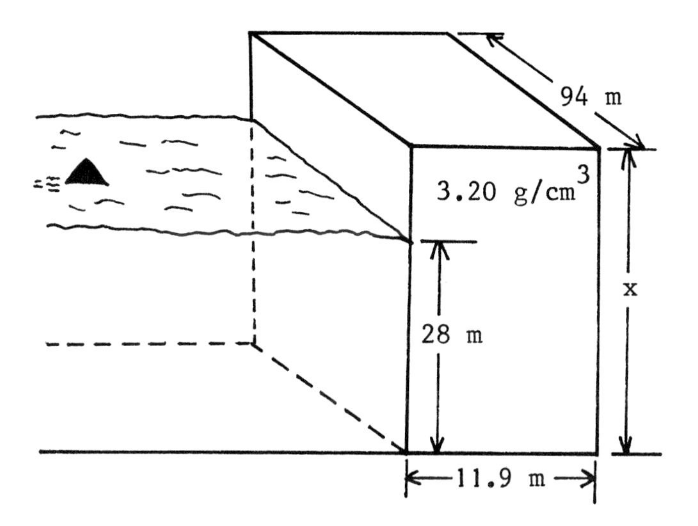

Fig.(7.13) Problem 1

2. For the rectangular dam of Fig.(7.14), find the base dimension b so that the factor of safety against rotation equals 2.73.

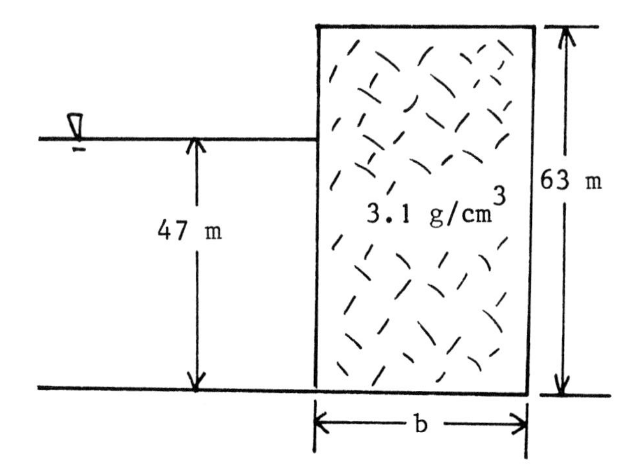

Fig.(7.14) Problem 2

3. What must be the depth of water behind a dam 170 m long and 52.0 m high in order that the force of the water on the dam face be 1.40 GN?

4. A dam is 124 m long and 62.0 m high. It is made of concrete with a unit weight of 32.2 kPa/m. The torque exerted by the water about the toe of the dam is 20.6 GN·m. How far is the water surface below the crest of the dam?

5. A dam 240 m long has a right-triangle cross section 72.0 m high and 53.0 m thick at the base. The concrete has density 2.80 g/cm^3. The coefficient of static friction between dam and foundation is 0.380. (*a*) What is the depth of the water if the factor of safety against sliding is 1.62? (*b*) Find the factor of safety against tipping when the water is at the depth found in (*a*).

6. A dam has the cross section of a right triangle of height 71.0 m. The dam is 214 m long. The density of the construction material is 3400 kg/m^3. What should be the value of the base thickness in order that, when the water surface is 5.20 m below the crest of the dam, the factor of safety against rotation is 2.78?

7. A gravity dam 115 m long has a right triangle cross section of height 42.0 m and base thickness 18.0 m. The concrete has density 2.96 g/cm^3. The dam is seen to begin to slide when the water reaches the crest (top) of the dam. (*a*) Find the coefficient of static friction between the dam and the ground on which it rests when sliding begins. (*b*) Calculate the torque that resists the tipping of the dam.

8. A dam 160 m long has a rectangular cross section of thickness 28.0 m and height 70.0 m. The unit weight of the concrete is 33.3 kN/m^3. The factor of safety against sliding equals 2.20. The reservoir is 55.0 m deep. (*a*) Find the coefficient of static friction between the dam and foundation. (*b*) Calculate the torque that tends to tip the dam over.

9. A gravity dam 130 m long has a right triangle cross section 43.0 m high and 19.0 m thick at the base. It is made of concrete with unit weight 29.4 kPa/m. During a bad storm, the dam was observed to begin to slide when the surface of the reservoir reached 7.00 m below the crest of the dam. (*a*) Calculate the force tending to push the dam from its foundation. (*b*) Find the coefficient of static friction between the dam base and the ground.

10. A reservoir of area 1250 ha and depth 36.0 m is contained by a dam 120 m long with a vertical upstream face. Find the force exerted by the reservoir on the dam.

11. Show that, for the purpose of analyzing possible rotation of a dam, the hydrostatic force can be considered to be acting at a depth of $\frac{2}{3}D$ below the water surface, where D is the depth of the reservoir.

12. In Example 1, suppose it is expected that, in the near future, changing soil conditions will reduce the coefficient of static friction to 0.320. To guard against sliding, a keyway is to be added to the dam. What force must the keyway be designed to provide if the dam will be on the verge of sliding when the reservoir level reaches the top of the dam? (Ignore the weight of the keyway.)

13. Find an expression for the factor of safety against rotation for a dam with a square cross section and with the water just slopping over the top of the dam.

14. The dam shown in Fig.(7.15), 145 m long, is made of material with a density of 2840 kg/m^3. What must be the weight of the dam so that, with the reservoir depth 46.0 m, the factor of safety against rotation is 1.70?

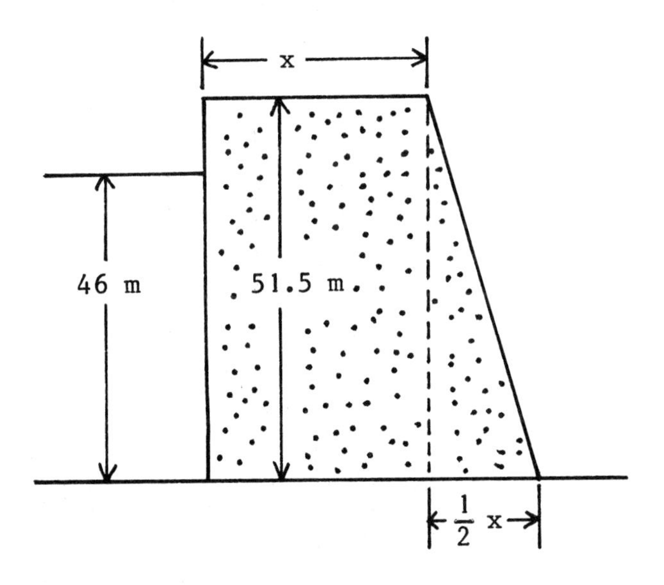

Fig.(7.15) Problem 14

15. Show that the driving force on a rectangular dam of length L, when the reservoir depth is D and the depth of the tailwater (water behind the dam) is d, where $d < D$, is given by

$$DF = \frac{1}{2}\rho g L(D^2 - d^2).$$

16. For a rectangular dam, show that if the factor of safety against sliding equals the factor of safety against rotation, the coefficient of static friction is given by

$$\mu_s = \frac{3b}{2D},$$

where b is the thickness of the dam and D the depth of the reservoir.

17. A service pipe 6.30 cm in diameter is constructed through a dam as shown in Fig.(7.16). The pipe valve at the dam face is removed by terrorists, and water enters the pipe. As an emergency repair, workers push a spare rubber plug into the pipe at the back face of the dam. Find the frictional force the plug must exert against the sides of the pipe to hold back the water.

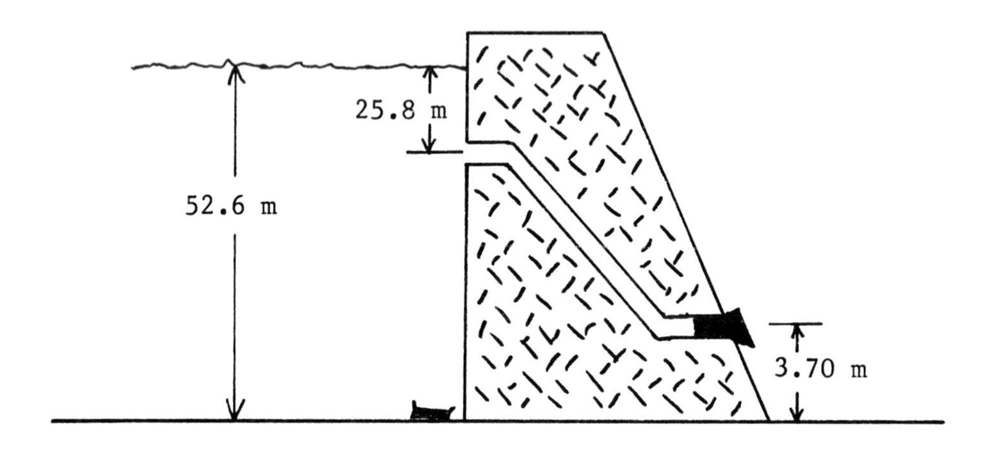

Fig.(7.16) Problem 17.

18. A dam has a trapezoidal cross section 40.0 m thick at the base and 15.0 m thick at the top. The coefficient of static friction between dam and ground equals 0.587. The reservoir is 46.0 m deep. Find the factor of safety against sliding when the factor of safety against rotation equals 4.22.

19. The reservoir in front of a dam has an expected lifetime of 180 years. The capacity of the reservoir at opening is $1.78 \text{ X } 10^6 \text{ m}^3$. Each year, 38,000 m^3 of sediment enters the reservoir. (*a*) Calculate the reservoir trap efficiency. (*b*) What volume of sediment is discharged past the dam in 5.00 years?

20. A dam opened on March 18, 1923 with a reservoir capacity of $1.85 \text{ X } 10^8 \text{ m}^3$. On March 18, 1995 the reservoir became completely filled with sediment. During the reservoir's lifetime, the sediment yield of the river flowing into the reservoir was 8.40 Mt/y. The sediment density was 1.20 g/cm^3. (*a*) What mass of sediment did the reservoir contain on March 18, 1960? (*b*) What volume of sediment was carried past the dam each year?

21. A pipe with a cross-sectional area of 1.66 m^2 discharges into a tank that has a volume of 34,200 m^3. The pipe carries water with a sediment load of 7.72 g per kg of water; the

sediment density is 1.38 g/cm^3. The water flows through the pipe at 6.83 m/s. Filters on the sides of the tank allow all of the water to escape, but they trap 97.5% of the sediment. How long will it take for the initially empty tank to fill with sediment?

22. Show that Eq.(7.26) reduces to expected results for (*a*) a rectangular dam, and (*b*) a right triangular dam.

23. Consider a dam with an upstream face sloping at angle θ to the vertical, as shown in Fig.(7.17). The pressure p acts at 90° to the face of the dam. (*a*) Show that the resulting hydrostatic force has a horizontal component identical with Eq.(7.4), and a vertical component equal to the weight of water above the sloping face; i.e., the water contained between the face and the dashed line in Fig.(7.17). (*b*) Show that the factor of safety against sliding for such a dam exceeds the value given from Eq.(7.10) by $\tan\theta$.

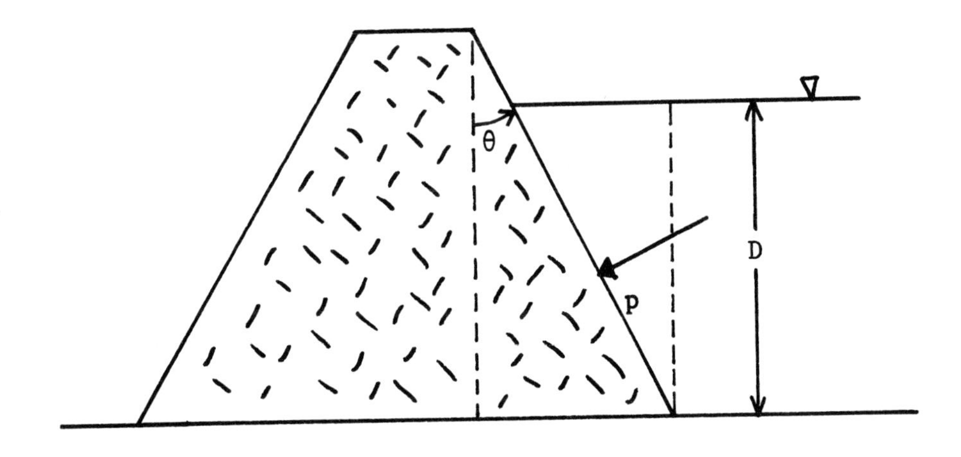

Fig.(7.17) Problem 23

Chapter 8

Groundwater Flow

8.1 Aquifers and Porosity

In order that water (or any other liquid, such as oil) can flow through rock, it is necessary that the rock not only be porous, but also *permeable.* That is, the pores (or at least some of them) must form, in effect, a connected network of tiny pipes through which water can move. Figure (8.1a) shows porous but impermeable rock, and Fig.(8.1b) shows permeable rock.

(a)

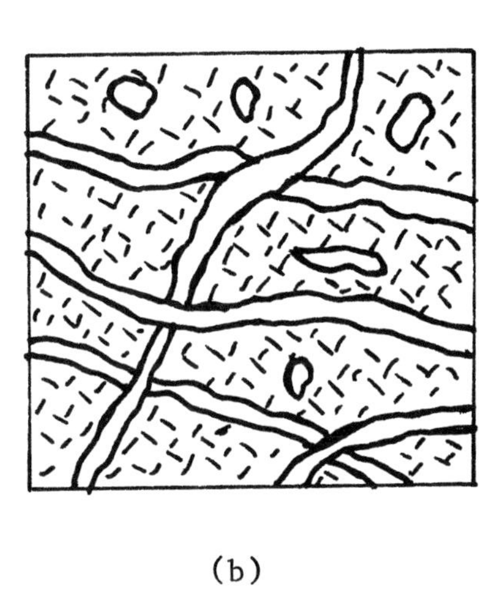

(b)

Fig.(8.1) Impermeable (a) and Permeable (b) Rock

Note that some of the pores in Fig.(8.1b) do not form part of the connected pore system, but are isolated from it. Any water contained in these pores is cut off from the system of connected pores. (Evidently, this water found itself trapped in the isolated pores that formed as the rock solidified.) The water in these isolated pores cannot flow to a well, say, or be pumped out of the rock. Rather, this water is retained by the rock.

Let V_{retain} be the total volume of the isolated pores (and hence of the retained water if the pores are filled), and V_{yield} the volume of all the connected pores (and hence the volume of the water the rock can yield if these pores are filled) in a certain block of rock. As in Chapter 1, the total pore volume is V_{pores}. The volume of the block of rock, as given from its external dimensions, is V.

The total pore volume V_{pores} is given by

$$V_{\text{pores}} = V_{\text{retain}} + V_{\text{yield}}. \tag{8.1}$$

Dividing by V gives

$$\frac{V_{\text{pores}}}{V} = \frac{V_{\text{retain}}}{V} + \frac{V_{\text{yield}}}{V}. \tag{8.2}$$

Now $V_{\text{pores}}/V = n$, the porosity of the rock; see Eq.(1.7). In a similar manner, define the *yield porosity* n_{y} by

$$n_{\text{y}} = \frac{V_{\text{yield}}}{V}, \tag{8.3}$$

and the *retention porosity* n_{r} by

$$n_{\text{r}} = \frac{V_{\text{retain}}}{V}. \tag{8.4}$$

Therefore, by Eq.(8.2),

$$n = n_{\text{r}} + n_{\text{y}}. \tag{8.5}$$

The porosity that "counts" in groundwater flow is the yield porosity n_{y}, sometimes called the *effective porosity*. The isolated pores, and any water they contain, are "de facto" part of the rock matrix.

A formation, or layer, of permeable rock is called an *aquifer*, a layer of impermeable rock an *aquiclude*.

An aquifer bounded only from below by an aquiclude is termed an *unconfined aquifer*; an aquifer bounded both from below and above by an aquiclude is a *confined aquifer*; see Fig.(8.2). Only unconfined aquifers are considered in the present work.

As indicated in Chapter 5, the water table , symbol $\underline{\nabla}$, is the surface below which the pores are completely filled with water (saturated rock or soil) and above which the pores are empty (dry rock or soil). Usually there is a finite transition region between the saturated and dry layers, but it will be assumed that this transition region is so thin that it can be considered a discontinuity.

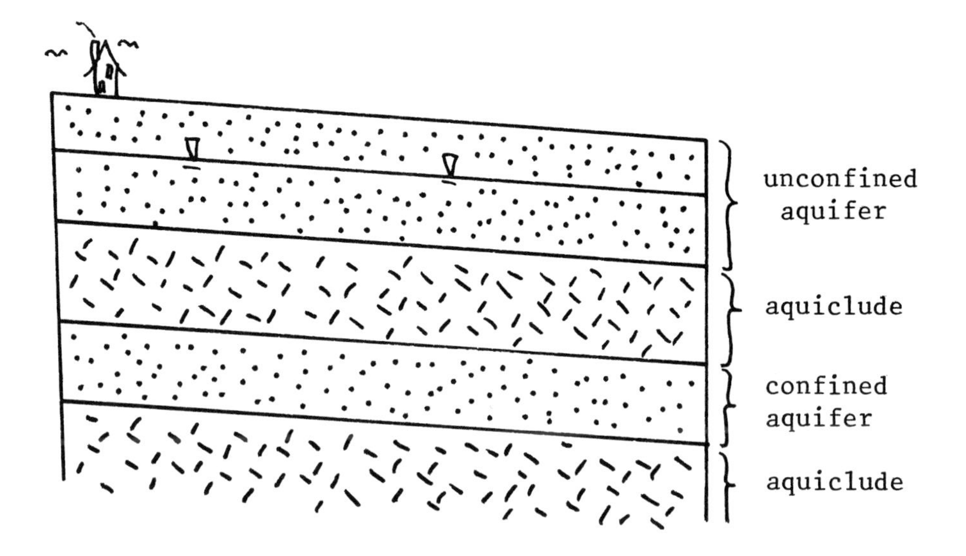

Fig.(8.2) Aquifers

8.2 Aquifer Discharge

Consider a rectangular block of aquifer rock below the water table. The block has horizontal width w, vertical thickness h and horizontal (or near horizontal) length L. Suppose that water is moving through the rock in the direction parallel to the edges of length L and perpendicular to the cross section with edge lengths w and h. Since the water can move only through the connected pores, the discharge Q of the water through the cross section is

$$Q = A_{\mathrm{pores}} v, \tag{8.6}$$

where A_{pores} is the total cross-sectional area of all the connected pores in the aquifer cross section, and v is the speed of the water in the pores as the water passes the cross section. Note that Eq.(8.6) implies the same definition of discharge as for the discharge of a river: $Q = V/t$, where V is the volume of water passing through a cross section in time t.

The quantity A_{pores} is hardly ever measurable directly in the field. Rather, it is the cross-sectional area A of the saturated part of the aquifer itself that can be more easily determined, where $A = wh$. It is convenient to express Eq.(8.6) in terms of A rather than A_{pores}.

To relate A to A_{pores}, imagine that the connected pores through which the water moves constitute tiny pipes extending the length L of the aquifer in the direction of flow. By definition of the yield porosity n_{y}, Eq.(8.3),

$$V_{\mathrm{yield}} = n_{\mathrm{y}}(Lwh). \tag{8.7}$$

The total volume of all the connected pores can now be expressed as $V_{\text{yield}} = A_{\text{pores}}L$. The saturated aquifer cross-sectional area is $A = wh$. Using these two equations, Eq.(8.7) becomes

$$A_{\text{pores}}L = n_{\text{y}}AL,$$

$$A_{\text{pores}} = n_{\text{y}}A, \tag{8.8}$$

where A_{pores} is the total cross-sectional area of all the connected pores cutting the wetted aquifer cross section of area A.

Substituting the last equation into Eq.(8.6) for the discharge gives

$$Q = (n_{\text{y}}A)v. \tag{8.9}$$

Rearrange this to read

$$Q = A(n_{\text{y}}v),$$

and then define v_{D}, by

$$v_{\text{D}} = n_{\text{y}}v, \tag{8.10}$$

to get for the discharge the expression

$$Q = Av_{\text{D}}. \tag{8.11}$$

Equation (8.11) for the discharge now looks like the equation $Q = Av$ for the discharge of a river, except that the real speed v of the water has been replaced with the quantity v_{D} defined by Eq.(8.10). Now v_{D} has units of speed, since n_{y} in Eq.(8.10) has no units, and it is called various names, prominent among them the *Darcy speed* , named after the nineteenth century engineer Henry Darcy. It is not the speed at which the water, or liquid, actually moves, but is only a mathematical construction with units of speed. Nevertheless, v_{D} is sometimes misleadingly called the "percolation speed" or even "flow velocity". The actual velocity at which the liquid moves is v, not v_{D}.

****

EXAMPLE 1

An aquifer has a cross section with a horizontal width of 265 m, and a vertical thickness below the water table of 42.0 m. The water table is 36.0 m below the ground surface. Each day, 3340 m^3 of water is discharged through the cross section. The aquifer rock has an effective porosity of 27.1%. Find (*a*) the Darcy speed, and (*b*) the actual speed of the water as it passes through the aquifer.

(*a*) The discharge Q can be calculated from

$$Q = \frac{V}{t},$$

$$Q = \frac{3340 \text{ m}^3}{86,400 \text{ s}},$$

$$Q = 0.03866 \text{ m}^3/\text{s}.$$

The cross sectional area A of the aquifer through which the water is discharged is

$$A = wh,$$

$$A = (265 \text{ m})(42 \text{ m}),$$

$$A = 1.113 \text{ X } 10^4 \text{ m}^2.$$

For h, use the vertical thickness of that part of the aquifer cross section that is below the water table; i.e., $h = 42$ m. There is no water above the water table. Now use Eq.(8.11):

$$Q = Av_\text{D},$$

$$0.03866 \text{ m}^3/\text{s} = (1.113 \text{ X } 10^4 \text{ m}^2)v_\text{D},$$

$$v_\text{D} = 3.47 \text{ X } 10^{-6} \text{ m/s}.$$

(b) The actual speed v of the water at discharge is given from Eq.(8.10). Convert the porosity to a decimal value, and then find that

$$v_\text{D} = n_\text{y} v,$$

$$3.47 \text{ X } 10^{-6} \text{ m/s} = (0.271)v,$$

$$v = 1.28 \text{ X } 10^{-5} \text{ m/s},$$

$$v = 0.0128 \text{ mm/s}.$$

Groundwater flow can be very slow.

****

8.3 Darcy's Law

So far, the only parameter that has entered the analysis of groundwater flow is the yield porosity of the aquifer rock. No properties of the fluid itself have entered. But different fluids (water and oil, for example) have different densities and viscosities (internal friction), for example. Hence, different fluids could be expected to flow at different speeds through the same aquifer. Nothing in the equations so far presented would allow for such a difference; hence, the set of equations developed to this point must be incomplete.

Additional equations can be sought by asking why should water (or other liquid) flow through the aquifer at all, or why should it not simply sit in place in the pores of the rock (as, indeed, sometimes it does).

Consider an aquifer in which the water table is horizontal everywhere. In this configuration, the water will just sit in the connected pores; there is no tendency for the water to move in any horizontal direction. The reason is that gravity acts vertically down and has no horizontal component. Since the water table is horizontal, there is no component of gravity parallel to the water table.

In the absence of applied external pressure, water will flow through an unconfined aquifer for the same reason that water flows in a river. Water flows downhill under gravity.

In an aquifer in which the water table is inclined at the angle i with the horizontal, gravity g has the component $g \sin i$ parallel to the water table. Water will move in the direction of this component of gravity in an attempt to bring the water table into a horizontal position.

Focus now on the speed of the water. If an object is released from rest on a frictionless table inclined at the angle i with the horizontal, the object's speed v after time t will be $v = (g \sin i)t$. That is, the speed continues to increase with time. But water, trying to move through tiny pores in rock is definitely not in a frictionless environment. In fact, there is so much friction, within the liquid and between the liquid and the inner surface of the pores, that the water moves at constant speed, the driving force of gravity being balanced by the resisting friction forces.

It is extremely difficult, even more difficult than the case of water flowing in an open channel, to derive a formula for this constant speed v from theory alone, so reliance is put on an empirical equation suggested by experiments. The engineer Henry Darcy, who is mentioned above, conducted experiments on the flow of liquids through porous materials. He obtained an empirical formula (now called Darcy's law, naturally enough) that describes his observations.

The water table is presumed to be inclined at the angle i with the horizontal, and it is presumed that i is not too large. (If the angle is too large, the speed of the water may reach values so that turbulence sets in, and Darcy's law is not valid under such circumstances.) As a result of being inclined, the water table drops a vertical distance H over a horizontal distance L. Darcy's law states that the Darcy speed v_{D} is given by

$$v_{\mathrm{D}} = K(\frac{H}{L}). \tag{8.12}$$

The quantity K is called the *hydraulic conductivity.* Since the units of H and L in Eq.(8.12) cancel, K is seen to have the same units as v_{D}; i.e., m/s in SI base units. The numerical value of K depends on the properties of the rock and on properties of the liquid. Specifically, for the rock it is the degree to which the pores are connected, and their size, that enters into K. For the liquid, the viscosity (internal friction) and density affect the value of K. This means that for any one given kind of rock, there are different values of K for water and for oil, for example.

Now, the movement of water through rock is usually a very slow process. The pores are very small, their geometry complex, and friction ever present. This means that the real flow speed v is usually very small. For this reason, hydraulic engineers often use the day (abbreviation d) rather than the second (s) as the unit of time. That is, values of v, v_{D}, and

K are generally expressed in m/d rather than the base units of m/s. For water, values of K for various rocks range between near zero and several hundreds of meters per day.

Equation (8.12) can be combined with Eq.(8.11) to yield this equation for the discharge:

$$Q = KA(\frac{H}{L}). \tag{8.13}$$

Sometimes this equation, rather than Eq.(8.12), is referred to as Darcy's law.

Darcy's law is sometimes expressed in terms of the angle i of inclination of the water table. Since i is small (for Darcy's law to be valid), $i \approx \tan i$ when i is expressed in radians. From Fig.(8.3) notice that $\tan i = H/L$. Hence, for small i, $i = H/L$. Equation (8.13) can now be written as

$$Q = KAi. \tag{8.14}$$

Sometimes, this is called Darcy's law.

In whichever version it is written, Darcy's law provides the additional equation needed to describe the flow of water (or another liquid) through permeable rock. The quantity i (or H/L) is often called the *hydraulic gradient.* Also, the parameter K is sometimes called the *coefficient of permeability.* As mentioned in several places in the present work, engineering nomenclature is not standardized.

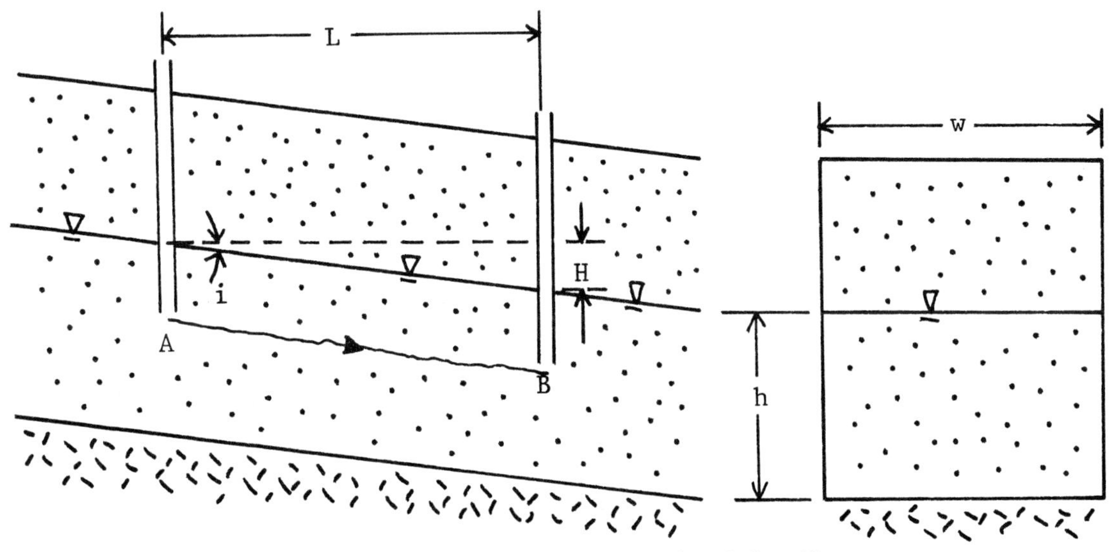

Fig.(8.3) Setting of an Unconfined Aquifer

Figure (8.3) shows an idealized setting of a typical unconfined aquifer. The uphill source of the groundwater, precipitation over a certain land area, is not shown. The ground surface slopes down "gently" (i.e., at a small angle i) from left to right. Usually, the water table is roughly parallel to the ground surface, falling a vertical distance H over a horizontal distance L. Here, the additional assumption is made that the layer of impermeable rock beneath the aquifer also slopes down at the angle i. This means that the vertical thickness

h of the water-bearing part of the aquifer has everywhere the same value. Two wells have been driven into the aquifer, piercing the water table at points A and B. The time t needed for the groundwater to move from A to B can be measured by dumping marked water into the well at A and noting when it shows up at well B. (Water can be marked with dye, or with a small quantity of radioactive heavy water.) Strictly speaking, the travel time from A to B is given by

$$t = \frac{\text{AB}}{v}.$$

Now

$$\text{AB} = \sqrt{L^2 + H^2}.$$

Since i is small, $H^2 << L^2$, so that AB $\approx L$. This approximation is appropriate not only because the hydraulic gradient is small, but because the values of the hydraulic conductivity and effective porosity are not likely to be known with great precision throughout the aquifer. Therefore, it is sufficiently accurate to use

$$t = \frac{L}{v}. \tag{8.15}$$

****

EXAMPLE 2

The longitudinal and transverse cross sections of an unconfined aquifer are shown in Fig.(8.4). Water takes 1.91 y to move from well A to well B. The hydraulic conductivity of the aquifer rock is 135 m/d. (a) Find the yield porosity of the aquifer rock. (b) It is found that 8.42 X 10^5 m^3 of water passes though any cross section of the aquifer in 2.00 weeks. Find the width w of the aquifer.

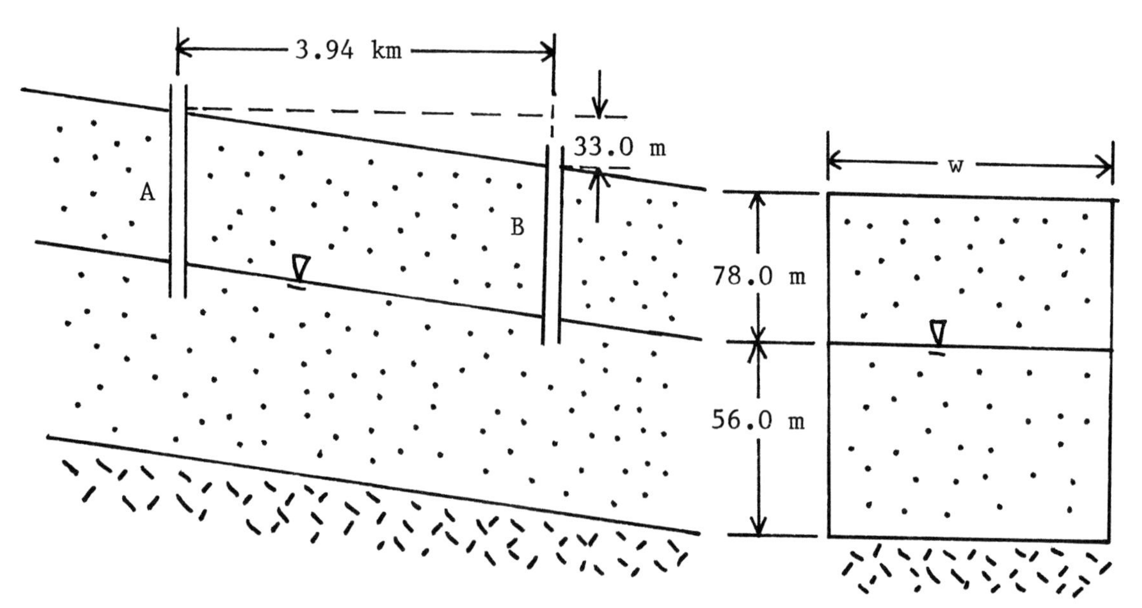

Fig.(8.4) Example 2

(a) Reading distances from Fig.(8.4), the data are: $L = 3940$ m, $H = 33$ m (not 78 m: the water table is parallel to the ground surface, and therefore drops by the same distance as does the ground surface); $h = 56$ m (thickness of aquifer below the water table). Also, $t = (1.91\text{ y})(365.25\text{ d/y})$, $t = 697.6$ d; $K = 135$ m/d. By Eq.(8.12), the Darcy speed is

$$v_{\rm D} = K(\frac{H}{L}),$$

$$v_{\rm D} = (135\text{ m/d})\left(\frac{33\text{ m}}{3940\text{ m}}\right),$$

$$v_{\rm D} = 1.131\text{ m/d}.$$

The actual speed of the water follows from Eq.(8.15):

$$v = \frac{L}{t},$$

$$v = \frac{3940\text{ m}}{697.6\text{ d}},$$

$$v = 5.648\text{ m/d}.$$

By Eq.(8.10),

$$v_{\rm D} = n_{\rm y}v,$$

$$1.131\text{ m/d} = n_{\rm y}(5.648\text{ m/d}),$$

$$n_{\rm y} = 0.200\ (20.0\%).$$

(b) Evidently, the discharge is

$$Q = \frac{V}{t},$$

$$Q = \frac{8.42\text{ X }10^5\text{ m}^3}{14.0\text{ d}},$$

$$Q = 6.014\text{ X }10^4\text{ m}^3/\text{d}.$$

Note that the symbol t for time in this part is not the time for the water to move from A to B. It would be tedious to use two different, or subscripted, symbols for various time intervals. As usual in such situations, the context of the discussion reveals the appropriate identification for t. Now apply Eq.(8.11) with $A = wh$, so that

$$Q = whv_{\rm D},$$

$$6.014\text{ X }10^4\text{ m}^3/\text{d} = w(56\text{ m})(1.131\text{ m/d}),$$

$$w = 950\text{ m}.$$

****

8.4 Problems

1. A rock has a porosity of 43.6%, but 35.0% of the pores are isolated. Calculate the yield porosity of the rock.

2. Write an equation equivalent to Eq.(8.3), but in which the yield porosity is obtained as a percent.

3. The aquifer shown in Fig.(8.5) has a width of 7.10 km. Industrial waste BXC is illegally dumped into well A, near the factory. This material moves with the ground water, and appears at well B after 6.40 y. The effective porosity of the aquifer rock is 0.350. Calculate (*a*) the hydraulic conductivity of the aquifer rock, and (*b*) the aquifer discharge.

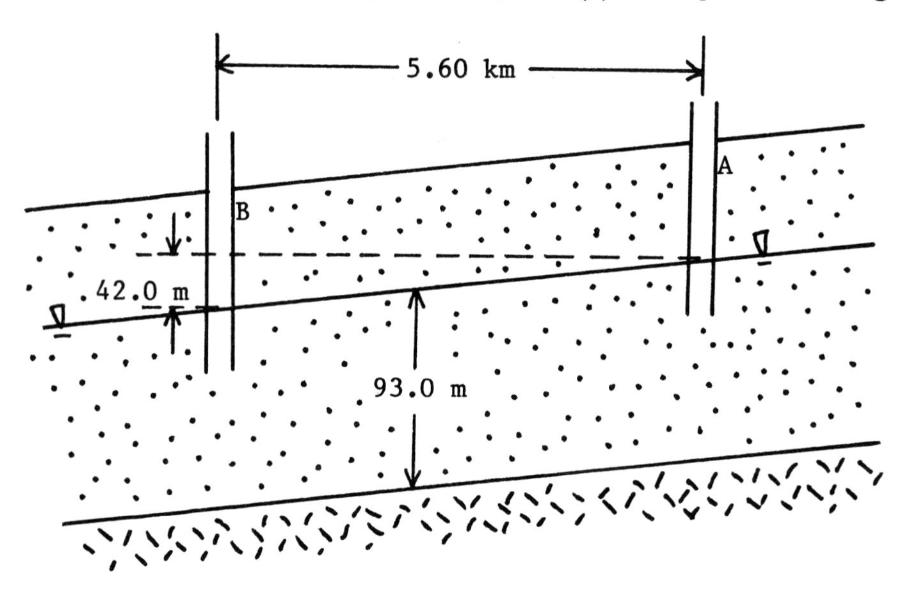

Fig.(8.5) Problem 3

4. An aquifer in a certain region consists of coarse sand with a yield porosity of 39.0% and a hydraulic conductivity of 45.0 m/d. The width of the aquifer is 2.90 km and its vertical thickness below the water table is 73.0 m. The aquifer discharge equals 8380 m^3/d. Calculate the vertical drop of the water table over a horizontal distance of 6.60 km.

5. Show that the hydraulic conductivity K of the rocks of an unconfined aquifer like that shown on Fig.(8.4) can be expressed as

$$K = \frac{v v_D}{v_i},$$

where v_D is the Darcy speed, v the true speed of the groundwater, and v_i the vertical speed of the water induced by the slope of the water table.

6. Find the porosity of an aquifer rock with a retention porosity of 23.3%, and through which the speed of the water is 3.40 times the Darcy speed.

7. The aquifer shown on Fig.(8.6) has a width of 1.70 km. Water takes 2.50 y to move from the upper well to the lower. The hydraulic conductivity of the aquifer is 31.0 m/d. Find (*a*) the yield porosity of the aquifer, (*b*) the volume of water that passes any cross section in 3.00 weeks, and (*c*) the total volume of connected pore space in the aquifer between the wells and above the water table.

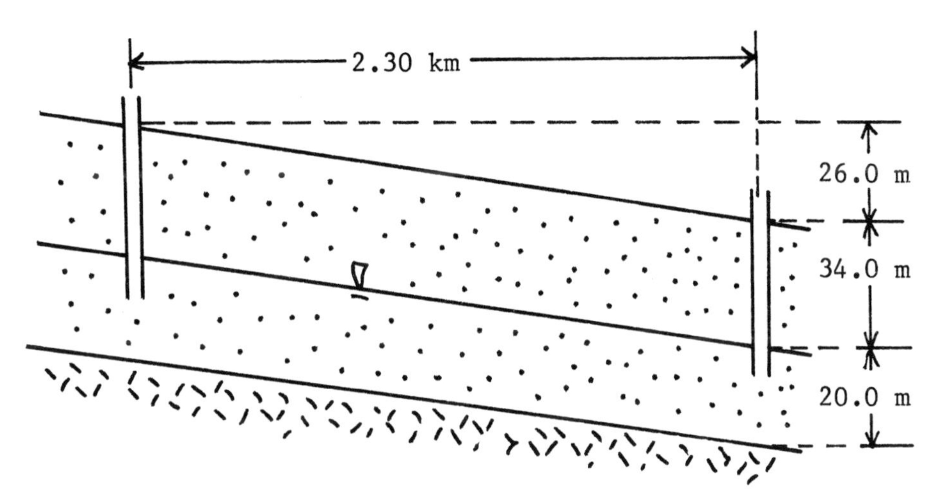

Fig.(8.6) Problem 7

8. Consider the aquifer shown in Fig.(8.7); the aquifer is 830 m wide. The yield porosity of the medium is 47.3% and its hydraulic conductivity equals 143 m/d. (*a*) How long does it take the groundwater to move from well A to well B? (*b*) How long does it take for 150,000 metric tons of water to pass through a cross section of the aquifer?

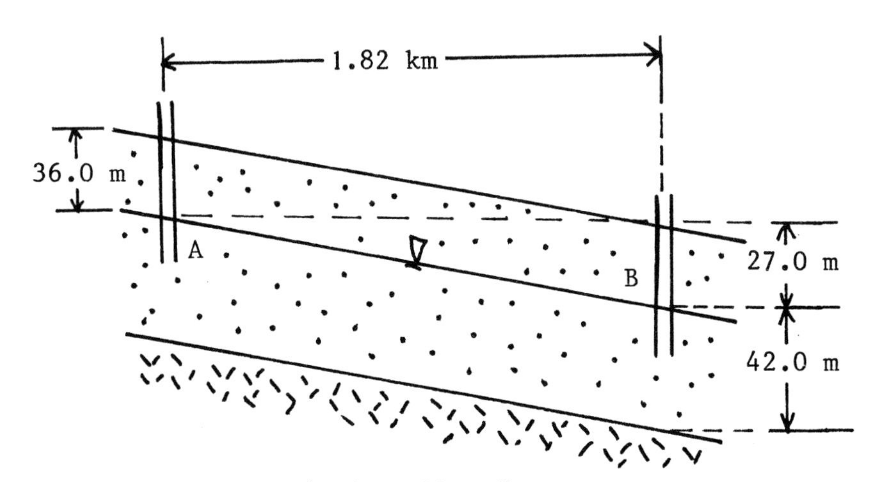

Fig.(8.7) Problem 8

9. The aquifer of Fig.(8.8) has a yield porosity of 34.0%, a retention porosity of 19.2% and a hydraulic conductivity of 160 m/d. (*a*) How long does it take for water to move from well A to well B? (*b*) What volume of water passes any cross section in 10.0 d? (*c*) Assuming that all pores below the water table are filled, find the total volume of water trapped in the rock between the wells.

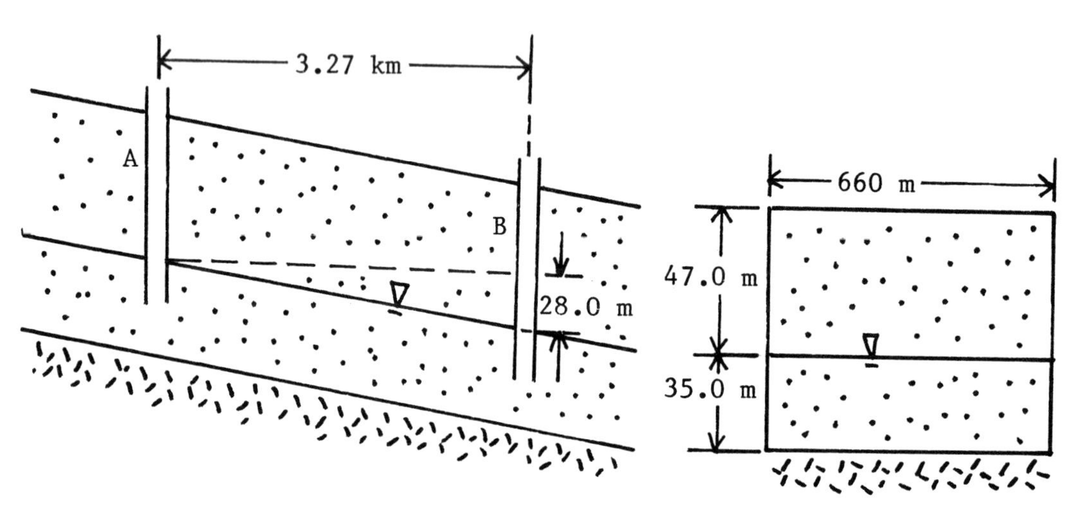

Fig.(8.8) Problem 9

10. The quantity of a chemical dissolved in groundwater is usually expressed either in units of mg/L (milligrams of chemical per liter of water), or as ppm, i.e., mass of chemical per million parts of mass of water. Show that the numerical values of these two quantities for a particular chemical are virtually the same.

11. Show that the time t required for groundwater to travel a horizontal distance L can be expressed as

$$t = \frac{n_y L^2}{KH},$$

where n_y is the yield porosity and K the hydraulic conductivity of the aquifer rocks, and H is the vertical drop of the water table over the horizontal distance L.

12. An unconfined aquifer extends to a depth of 73.0 m beneath a drainage basin of area 6.82 ha. The effective porosity of the aquifer rocks is 47.2%. Due to pumping of water, the water table falls from 12.6 m to 15.7 m beneath the ground surface. Find the volume of water pumped out of the ground.

13. An unconfined aquifer with width 420 m extends down to an impermeable rock layer at a depth of 85.0 m. The effective porosity of the aquifer rocks is 53.6%. When water moves

through the aquifer at a speed of 1.27 m/d, the aquifer discharge is found to be 2.00 X 10^7 L/d. Find the depth to the water table.

14. Calculate the Darcy speed through an unconfined aquifer 870 m wide and 106 m thick when the discharge equals 2.19 m^3/s and the water table is 18.0 m beneath the ground.

15. Over a horizontal distance of 2.87 km, the water table falls by 41.5 m. Groundwater takes 3.18 y to move the 2.87 km. The effective porosity of the aquifer is 0.640, and the aquifer's cross-sectional area is 1.80 ha. Find the hydraulic conductivity of the aquifer rocks.

16. The aquifer shown in Fig.(8.9) is 1.30 km wide. The effective porosity is 0.290 and the hydraulic conductivity equals 83.0 m/d. Water containing a dangerous radioactive isotope of plutonium is dumped (accidentally, of course) into well A. How long does it take for the plutonium to show up in well B? Assume that the plutonium is carried along at the same speed as the groundwater.

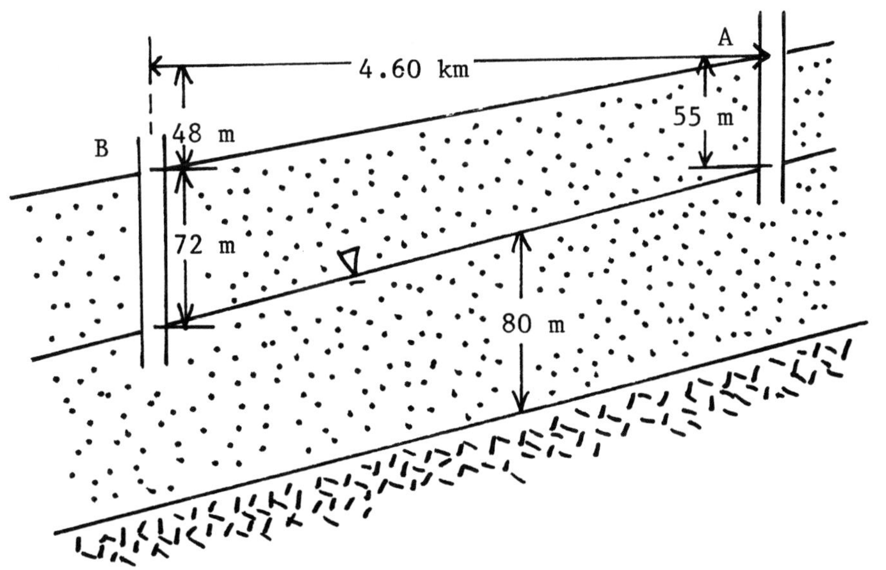

Fig.(8.9) Problem 16

17. Industrial waste containing the deadly contaminant XYX is dropped by terrorists into well A. Well B is 2.63 km from well A and at 55.0 m lower elevation. Between the wells, the depth to impermeable rock is 80.0 m, and the depth to the water table 16.7 m. The aquifer width is 1650 m. The aquifer rocks have a yield porosity of 0.364 and a hydraulic conductivity equal to 24.0 m/d. Assuming that XYX moves along with the groundwater, how long will it take for the contaminant to reach well B?

Chapter 9

Seismic Surveying

9.1 Seismic Waves

When in conversation, a person hears another by receiving and processing the sound waves that travel through the air. Sound waves can also travel through solids (such as rocks) and liquids, as well as through a gas such as air.

Sound waves traveling through a liquid or gas are longitudinal waves; i.e., as the wave passes, atoms (and/or molecules) of the gas or liquid oscillate along a line parallel to the direction of travel of the wave. Longitudinal waves can travel through a solid as well. But a solid also supports the propagation of transverse sound waves; i.e., of waves that cause the atoms (and/or molecules) to oscillate along a line perpendicular to the direction of travel of the wave.

In geology, sound waves traveling through rock often are referred to as *seismic waves.* In addition, longitudinal seismic waves are called "P waves" and transverse waves "S waves". (S waves exist in two "polarizations", corresponding to the two dimensions of the plane of oscillation perpendicular to the direction of travel of the wave; this feature of S waves can be overlooked in the present discussion.)

Seismic waves obey the same relation among frequency f, wavelength λ, and wave speed v as do light waves and sound waves in air: that is,

$$f\lambda = v. \tag{9.1}$$

In SI base units, frequency is expressed in hertz (Hz), wavelength in m, and speed in m/s. Of the three parameters appearing in Eq.(9.1), it is the wave speed that will be of primary concern in this chapter.

Seismic waves travel through homogenous and isotropic rock along straight line paths and at constant speed. (A homogenous and isotropic rock has the same properties at every location, and the properties do not vary with direction.) The shape of the wavefronts, perpendicular to the direction of travel, depends on the geometry of the source of the waves. In the laboratory, seismic waves can be produced in a block of rock by striking it with a hammer on one face; a detector attached to the opposite face can record the times required for the waves to travel through the block. By measuring the distance L between the rock faces, the velocities v_{P} and v_{S} of the P and S waves can be calculated by

$$v_{\mathrm{P}} = \frac{L}{t_{\mathrm{P}}},$$

$$v_{\mathrm{S}} = \frac{L}{t_{\mathrm{S}}}, \tag{9.2}$$

where t_{P} is the time required for the P wave to reach the opposite face, and t_{S} is the corresponding quantity for the S wave. As Eqs.(9.2) imply, the travel times, and therefore the wave speeds, are not equal. In fact, for any rock so tested, it is found that $t_{\mathrm{P}} < t_{\mathrm{S}}$, which means that $v_{\mathrm{P}} > v_{\mathrm{S}}$. P waves travel faster than S waves.

In addition to the kinematic equation for the wave speed, Eq.(9.1), formulas for the P and S wave speeds can be derived which express the wave speeds in terms of the moduli that describe the elastic properties of the rock that transmits the waves. The derivations of these formulas are somewhat tedious, and require a greater knowledge of elasticity theory and its mathematical formulation than is demanded in the present work. Consequently, the derivations will be omitted here; rather, the resulting equations will be given and their practical use described.

The formulas alluded to are:

$$v_{\mathrm{P}} = \sqrt{\frac{k + \frac{4}{3}G}{\rho}}, \tag{9.3}$$

$$v_{\mathrm{S}} = \sqrt{\frac{G}{\rho}}. \tag{9.4}$$

As introduced in Chapters 1 and 3, k is the bulk modulus, G the shear modulus, and ρ the density of the rock. Although the equations above have not been derived, the consistency of the units should be checked; that is, by substituting the units of k, G, and ρ, it can be verified that the units reduce to m/s, the SI base units for speed.

Equations (9.3) and (9.4) apply to intact rock, meaning to rock that is free of pores (porosity $n = 0$), and free of cracks (joints) or fissures. The wave speeds for porous rock are less than those calculated from Eqs.(9.3) and (9.4), but there is no simple formula for the correction due to porosity. Therefore, it will be assumed that, when Eqs.(9.3) and (9.4) are used, the porosity effects are negligible.

From Eqs.(9.3) and (9.4), two facts emerge immediately. First, since $G = 0$ for a liquid or gas, $v_{\mathrm{S}} = 0$ for these phases. The conclusion is that S waves do not travel through a liquid or a gas. Second, Eqs.(9.3) and (9.4) only differ in the numerator under the square root. Now both k and G are positive. Hence, the formulas show that $v_{\mathrm{P}} > v_{\mathrm{S}}$ for any specific rock. P waves travel faster than S waves, as noted in describing laboratory measurements of the wave speeds.

Equations (9.3) and (9.4) give the speeds of the seismic waves in terms of the moduli k and G. In Section (3.5), Eqs.(3.26) and (3.27) express these moduli in terms of the other pair of elastic moduli E and ν. If those expressions are substituted into Eqs.(9.3) and (9.4), formulas for the wave speeds in terms of E and ν (and, of course, the density ρ) will be obtained. It can be shown easily that the resulting formulas are

$$v_{\mathrm{P}} = \sqrt{\frac{E}{\rho}\frac{1-\nu}{(1+\nu)(1-2\nu)}}, \tag{9.5}$$

$$v_{\mathrm{S}} = \sqrt{\frac{E}{\rho}\frac{1}{2(1+\nu)}}. \tag{9.6}$$

Equations (9.3) and (9.4) give the wave speeds in terms of k, G, ρ, and Eqs.(9.5) and (9.6) give the speeds in terms of E, ν, and ρ.

****

EXAMPLE 1

Calculate the speeds of the P and S waves in a rock that has a bulk modulus of 31.5 GPa, a shear modulus of 37.6 GPa, and a density equal to 2.41 g/cm^3.

In SI base units the data are: $k = 31.5 \text{ X } 10^9$ Pa, $G = 37.6 \text{ X } 10^9$ Pa and $\rho = 2410 \text{ kg/m}^3$. Substituting into Eqs.(9.3) and (9.4) gives

$$v_{\mathrm{P}} = \sqrt{\frac{k + \frac{4}{3}G}{\rho}},$$

$$v_{\mathrm{P}} = \sqrt{\frac{31.5 \text{ X } 10^9 \text{ Pa} + \frac{4}{3}(37.6 \text{ X } 10^9 \text{ Pa})}{2410 \text{ kg/m}^3}},$$

$$v_{\mathrm{P}} = 5820 \text{ m/s},$$

for the P wave speed, and

$$v_{\mathrm{S}} = \sqrt{\frac{G}{\rho}},$$

$$v_{\mathrm{S}} = \sqrt{\frac{37.6 \text{ X } 10^9 \text{ Pa}}{2410 \text{ kg/m}^3}},$$

$$v_{\mathrm{S}} = 3950 \text{ m/s},$$

for the S wave speed. Values of several km/s are typical for these wave speeds in near surface rocks. By way of comparison, the speed of sound in air is about 340 m/s, and the speed of sound in water is about 1440 m/s; of course, both of these are P wave speeds.

****

9.2 Moduli from Wave Speeds

Equations (9.3) and (9.4) allow calculation of the P and S wave speeds if the values of the rock density, and bulk and shear moduli are available. Similarly, Eqs.(9.5) and (9.6) allow calculation of the wave speeds if the values of the rock density, Young's modulus and Poisson's ratio are known. But suppose instead that the values of the wave speeds are known; then, together with the density, Eqs.(9.3) and (9.4) allow for calculation of the bulk and shear moduli, and Eqs.(9.5) and (9.6) permit calculation of Young's modulus and Poisson's ratio. In fact, it is these "inverse relations" that are of great engineering significance: knowledge of the wave speeds allows calculation of the elastic moduli without subjecting the rock to actual loading and deformation in the laboratory, as required for evaluating the moduli from their defining equations as given in Chapter 3. (The values of the moduli obtained by these two methods do not always agree within the precision of the procedures used, but this difference will be overlooked in this work.)

These inverse relations, then, are significant enough to justify doing the algebra explicitly here. Start with Eq.(9.4); square both sides and multiply by the density ρ to get

$$G = \rho v_{\mathrm{S}}^2. \tag{9.7}$$

Similarly, square both sides of Eq.(9.3), multiply by ρ, and solve for k to find that

$$k = \rho v_{\mathrm{P}}^2 - \frac{4}{3}G.$$

Substitute Eq.(9.7) for G and factor out the density ρ to obtain

$$k = \rho(v_{\mathrm{P}}^2 - \frac{4}{3}v_{\mathrm{S}}^2). \tag{9.8}$$

Equations (9.7) and (9.8) are the relations sought that allow calculation of G and k from ρ, v_{P} and v_{S}.

Something can be learned about the relative values of v_{P} and v_{S} from Eq.(9.8). Since $k > 0$ and $\rho > 0$, that equation requires that

$$v_{\mathrm{P}}^2 - \frac{4}{3}v_{\mathrm{S}}^2 > 0,$$

which leads to

$$v_{\mathrm{P}} > \sqrt{\frac{4}{3}}v_{\mathrm{S}}. \tag{9.9}$$

For example, if $v_{\mathrm{S}} = 2000$ m/s, then $v_{\mathrm{P}} > 2309$ m/s.

A "sharper" lower bound on v_{P} can be found by invoking "Eq.(3.31)": $k > \frac{2}{3}G$. Substituting this into Eq.(9.3) gives

$$v_{\mathrm{P}} > \sqrt{\frac{2G}{\rho}}.$$

Comparing this result with Eq.(9.4) yields

$$v_P > \sqrt{2} v_S. \tag{9.10}$$

Since $\sqrt{2} > \sqrt{\frac{4}{3}}$, it can be said that "Eq.(9.10)" is a stronger, or sharper, lower bound to v_P than "Eq.(9.9)"; i.e., "Eq.(9.10)" excludes more values of v_P than does "Eq.(9.9)". Considering the previous example, i.e., $v_S = 2000$ m/s, "Eq.(9.10)" requires that $v_P > 2828$ m/s, considerably greater than the lower limit of 2309 m/s allowed by "Eq.(9.9)".

Turn now to Eqs.(9.5) and (9.6) to find the inverse relations for E and ν. Square both sides of both equations and then divide Eq.(9.5) by Eq.(9.6) to find that

$$(v_P/v_S)^2 = \frac{2(1-\nu)}{1-2\nu}.$$

This can be solved for Poisson's ratio ν, resulting in

$$\nu = \frac{\frac{1}{2}(v_P/v_S)^2 - 1}{(v_P/v_S)^2 - 1}. \tag{9.11}$$

Evidently, ν can be determined from the wave speeds alone, without knowledge of the rock density ρ. Note that, since $\nu > 0$, Eq.(9.11) requires that $\frac{1}{2}(v_P/v_S)^2 > 1$ or $v_P > \sqrt{2} v_S$, the same lower limit as expressed in "Eq.(9.10)".

Only an equation for Young's modulus E remains to be found. To accomplish this, substitute Eq.(9.11) into Eq.(9.5) and solve for E to find

$$E = 2\rho v_S^2(1+\nu),$$

$$E = (\rho v_S^2)\frac{3(v_P/v_S)^2 - 4}{(v_P/v_S)^2 - 1}. \tag{9.12}$$

Equations (9.11) and (9.12) are the inverse relations allowing for calculation of E and ν once the wave speeds v_P and v_S and the density ρ are known. The density is not needed to calculate Poisson's ratio.

****

EXAMPLE 2

Find the density of a rock that has wave speeds of 1.74 km/s and 2.93 km/s, and a Young's modulus of 19.5 GPa.

Since $v_P > v_S$, the wave speeds are $v_S = 1.74$ km/s and $v_P = 2.93$ km/s. This means that $(v_P/v_S)^2 = 2.836$. Equation (9.12) becomes

$$19.5 \text{ X } 10^9 \text{ Pa} = \rho(1740 \text{ m/s})^2[\frac{3(2.836) - 4}{2.836 - 1}],$$

$$\rho = 2620 \text{ kg/m}^3.$$

****

9.3 Reflection and Refraction

Seismic waves travel on straight line paths as long as the elastic moduli and/or density of the rock through which they travel do not change; that is, as long as the speed of the wave, as calculated from the appropriate formula given in Section (9.2) remains constant, the wave travels in a straight line at constant speed. If the wave encounters rock in which the wave speed differs, then the direction of travel as well as the speed of the wave will change. If the wave speed changes slowly along the wave path, then the direction of travel will change slowly and the wave follows a curved path. But, if the wave speed changes abruptly, then the direction of travel changes abruptly too. It is this latter situation that is examined here.

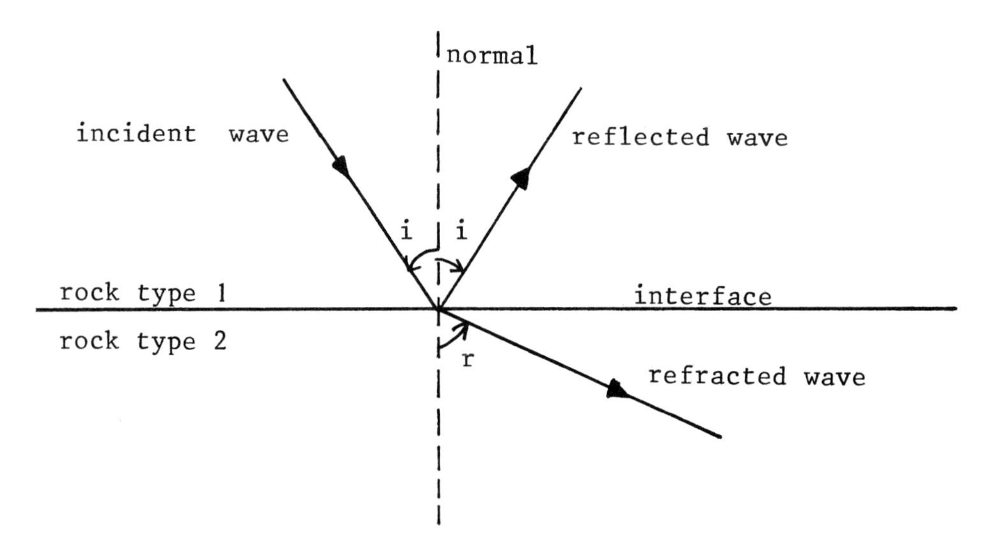

Fig.(9.1) Reflected and Refracted Waves

In Fig.(9.1), a layer of rock sits on top of a layer of a different kind of rock. These different rocks are called simply rock type 1 (upper layer) and rock type 2 (lower layer). The line (really a plane) drawn where the different rocks touch is given various names: interface, boundary, discontinuity, for example. A seismic wave moving through rock type 1 is shown approaching and striking this interface. At the interface, the wave "splits" into two waves (not usually of the same intensity): one wave appears to "bounce off" the interface, the other passes through the interface into rock type 2.

The wave striking the interface is called the *incident wave*; the *reflected wave* is the wave that appears to bounce off the interface; the wave that passes through the interface to rock type 2 is called the *refracted wave.* The incident and reflected waves travel in rock type 1; call their speeds v_1. The refracted wave travels in rock type 2; call its speed v_2. The three waves travel in the same plane; in Fig.(9.1), this is the plane of the page.

In Fig.(9.1), all of the waves shown are presumed to be of the same type. That is, if the incident wave is a P wave, then the reflected and refracted waves also are P waves. In this case, v_1 is the speed of P waves in rock type 1 and v_2 is the speed of P waves in rock type 2. On the other hand, if the incident wave is an S wave, then the reflected and refracted waves both are S waves. In this event, v_1 is the speed of S waves in rock type 1, and v_2 the speed of S waves in rock type 2. (The possibility of wave conversions, P to S, and vice versa, upon reflection or refraction, is ignored.)

The directions of propagation of the waves are important. In a plane, directions can be specified by giving an angle from a reference direction. The interface itself forms a natural reference line, and this sometimes is used, but it is more common to use the normal (perpendicular) to the interface as the reference line from which to express the angles. This line is shown dashed in Fig.(9.1), passing through the intersection of the incident wave with the interface, and continuing into rock type 2.

The direction of the incident wave is specified by giving the angle i (angle of incidence) that it makes with that part of the normal that is in rock type 1. It is found that the reflected wave makes the same angle i with this normal, but on the other side, as shown in Fig.(9.1).

The direction of travel of the refracted wave is specified by giving the angle r (angle of refraction) that it makes with that part of the normal that is in rock type 2. It also is on that side of the normal opposite to the incident wave: the refracted wave crosses the normal.

The angles of incidence and of refraction are related to the wave speeds v_1 and v_2 by the same equation, known as *Snell's law*, or the *law of refraction*, that governs the refraction of light; to wit,

$$\frac{\sin i}{v_1} = \frac{\sin r}{v_2}. \tag{9.13}$$

In Eq.(9.13), the speeds v_1 and v_2 must be expressed in the same units; the angles usually are expressed in degrees.

****

EXAMPLE 3

A P wave strikes an interface between two rock types at an angle of incidence of 63.0°. The speed of the incident wave is 4.73 km/s. The P-wave speed in the rocks beyond the interface is 2.94 km/s. Find the angle of refraction.

In this situation, $v_1 = 4.73$ km/s, $i = 63.0°$, $v_2 = 2.94$ km/s, and the angle of refraction r is sought. By Eq.(9.13),

$$\frac{\sin i}{v_1} = \frac{\sin r}{v_2},$$

$$\frac{\sin 63.0^\circ}{4.73\ \text{km/s}} = \frac{\sin r}{2.94\ \text{km/s}},$$

$$\sin r = 0.5538,$$

$$r = 33.6^\circ.$$

****

In this example, the angle of refraction r is less than the angle of incidence. This is unlike the situation displayed in Fig.(9.1), where $r > i$. Each case occurs in nature.

Now suppose that the wave speed in the upper layer is $v_1 = 2.30$ km/s and the wave speed in the lower layer is $v_2 = 5.42$ km/s. Find the angle of refraction for a wave incident at 30.0° on the interface from the upper layer. Upon substituting the data into Eq.(9.13) and solving for r, a calculator gives

$$\frac{\sin 30.0^\circ}{2.30 \text{ km/s}} = \frac{\sin r}{5.42 \text{ km/s}},$$

$$\sin r = 1.178,$$

$$r = -\text{E}-,$$

or some other error message. (No error message will appear if the calculator displays complex numbers; examination of the display in this case will show that no real angle is indicated.)

The source of the error message can be identified if Eq.(9.13) is solved algebraically for $\sin r$:

$$\sin r = \left(\frac{v_2}{v_1}\right) \sin i.$$

The greatest possible value of $\sin r$ is one (1), so that

$$\left(\frac{v_2}{v_1}\right) \sin i \leq 1.$$

Suppose that $v_2 > v_1$; then $v_2/v_1 > 1$. For angles of incidence from 0° to 90°, $\sin i$ ranges from 0 to 1. It follows that there is a particular angle of incidence, which will be called I, for which

$$\left(\frac{v_2}{v_1}\right) \sin I = 1,$$

$$\sin I = \frac{v_1}{v_2}. \tag{9.14}$$

This special angle of incidence I is called the *critical angle.* Waves striking the interface at angles of incidence i greater than I will generate an error message when an attempt is made to find the angle of refraction since, for these angles, $\sin i > \sin I$, making $(v_2/v_1) \sin i > 1$, $\sin r > 1$, $r =$ error.

What is the meaning of this result that displays "error"? It can only signify that, if no angle of refraction can be calculated, then there is no refracted wave. (There still is a reflected wave.) For the situation that led to the "error" above,

$$\sin I = \frac{2.30 \text{ km/s}}{5.42 \text{ km/s}},$$

$$I = 25.1^\circ.$$

No critical angle is found if $v_1 > v_2$; regardless of the value of the angle of incidence, all incident waves produce a refracted wave along with a reflected wave.

Figure (9.2) summarizes the situation when $v_2 > v_1$. Waves incident with $i < I$ produce refracted waves. The wave incident with $i = I$ produces the *critical wave*: the wave incident at the critical angle together with its refracted wave at $r = 90°$. Waves incident with $i > I$ produce no refracted wave. A reflected wave appears in all cases.

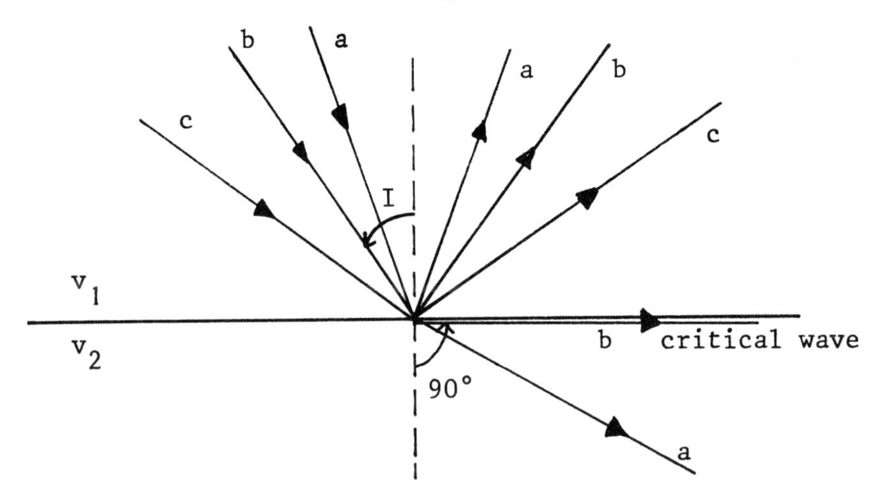

Fig.(9.2) Refraction with $v_2 > v_1$.

****

EXAMPLE 4

In Fig.(9.3), find the wave speed in the upper layer.

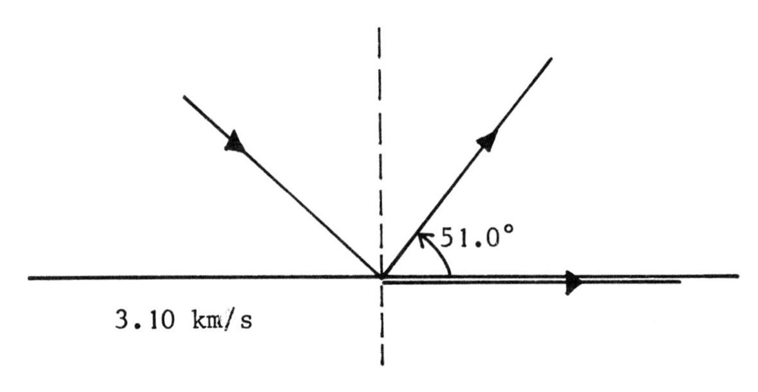

Fig.(9.3) Example 4

Since the reflected wave makes an angle of 51.0° with the interface, it makes an angle of $90.0° - 51.0° = 39.0°$ with the normal. As the angle of reflection equals the angle of incidence, $i = 39.0°$. The angle of refraction, from the diagram, evidently is 90.0°, so the wave is the critical wave. Therefore $I = 39.0°$. By Eq.(9.14),

$$v_1 = v_2 \sin I,$$

$$v_1 = (3.10\ \text{km/s}) \sin 39.0°,$$

$$v_1 = 1.95\ \text{km/s}.$$

****

9.4 Seismic Surveying Principles

The term "surveying" usually conjures up images of surveyors sighting through instruments to measure angles and/or distances, or dragging tape measures to construct or verify a plot plan of a certain parcel of land. This procedure is of no value, however, if a survey of the region beneath the ground is needed, perhaps to locate the onset of bedrock beneath a layer of soil, to find the depth of the water table, or to locate a layer of mineral bearing rock. Earth, unlike air, is not transparent: underground features cannot be sighted visually. So totally different techniques must be devised for a subsurface survey.

Digging a hole into the ground to a sufficient depth (borehole) to obtain actual samples of the subsurface rocks seems an obvious technique. But the very act of drilling, with the associated high pressure and contamination by the fluid needed for lubrication and removal of cuttings, can distort the removed rock samples, in that they may no longer be identical with the undisturbed "in situ" rock. Boreholes are not "nondestructive". Also, unless exploratory instruments are positioned in the hole, the borehole only provides data on the rocks at the site of the borehole. If information over a wide area is needed, many boreholes must be drilled.

A method similar to the use of ultrasound in medicine can quickly yield a picture of the arrangement of rocks beneath the ground over a wide area. The detonation of a small amount of explosive, such as TNT, near the surface produces seismic waves which travel into the Earth. The interfaces between layers, or formations, of different rocks reflect and refract the waves. Some are thereby redirected back to the surface, where they are detected by small, portable instruments called *geophones* (an odd name, perhaps: no one can place a call, or speak, on a geophone!). The geophones record the arrival of waves, and the time of their arrival (geophones contain, or are connected to, a real time clock). Since the time of detonation of the explosive is known, the travel times of the waves can be obtained. From the travel times and the distances of the geophones from the detonation point, the arrangement of underground rocks can be inferred.

The procedure is illustrated by considering a very simple, but not uncommon, arrangement of subsurface rocks, shown on Fig.(9.4). Immediately beneath the ground are rocks of type 1, in which the wave speed is v_1. A horizontal interface is at depth h. Below the interface are rocks of rock type 2, in which the wave speed is v_2.

Geophones can detect both P and S waves. The analysis given below applies to both types of waves (except that possible wave conversions are ignored). If P waves are being considered, then v_1 and v_2 are P wave speeds; similarly if S waves are being considered. (Generally, P waves are easier to excite than S waves.) Interfaces deeper than the one shown on Fig.(9.4) are ignored, or it can be assumed that the source of the seismic waves is not strong enough to produce waves that can penetrate to a second interface. Seismic waves "die out" by giving up their energy as thermal energy to the rocks through which they pass.

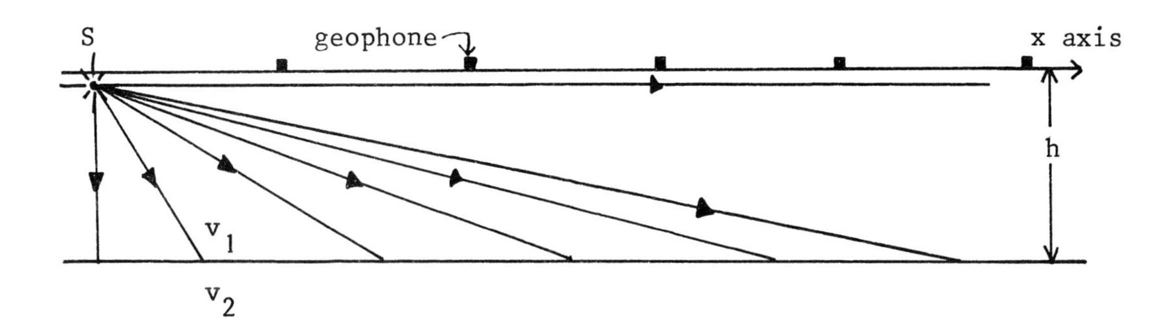

Fig.(9.4) Two Layers of Rock

The term "rock type" is generic, and may not represent actual rock. For example, rock type 1 could actually be soil and rock type 2 bedrock. Or, both rock type 1 and rock type 2 could be soil, with the interface being the water table separating dry and saturated soil.

The source of the seismic waves is located at point S, called the *shot point.* This is where the small amount of explosive is placed. The geophones, represented in Fig.(9.4) by the small, dark, squares, usually are laid out equally spaced on a straight line from S. If, in fact, the seismic wave production is by explosive (other methods, such as mechanical hammering, or acoustic pulsation, are available), then there is not likely to be a geophone at the shot point itself. If seismic wave production is by one of the other methods just mentioned, then there could be a geophone virtually at the shot point. In any event, call the line through S containing the geophones the x-axis, with the origin at S, so that the distance of any geophone from S is the value of x for that geophone.

The measured data consists of the travel times for all the waves recorded by each geophone, together with the value of x for each geophone. The quantities sought are the wave speeds v_1 and v_2, and the depth h to the interface.

It is convenient to separate the analysis of the reflected waves received by the geophone from the analysis of the refracted wave data. This dichotomy of consideration arises not only from the fact that these waves follow different paths, but also because the chosen geophones may not permit reception of all the refracted waves.

9.5 Seismic Reflection

First, then, consider only the reflection of waves from the interface. Seismic waves travel outward from the shot point S in all downward directions. On Fig.(9.5) are shown only waves that, on reflection, travel to the geophones shown. Also on Fig.(9.5) are two additional waves: First, the reflected wave that travels vertically downward and back up to the shot point; Second, the wave that moves parallel to, and just beneath, the ground surface, never striking the interface, but whose passage still is recorded by the geophones. This latter wave is called the *direct wave.*

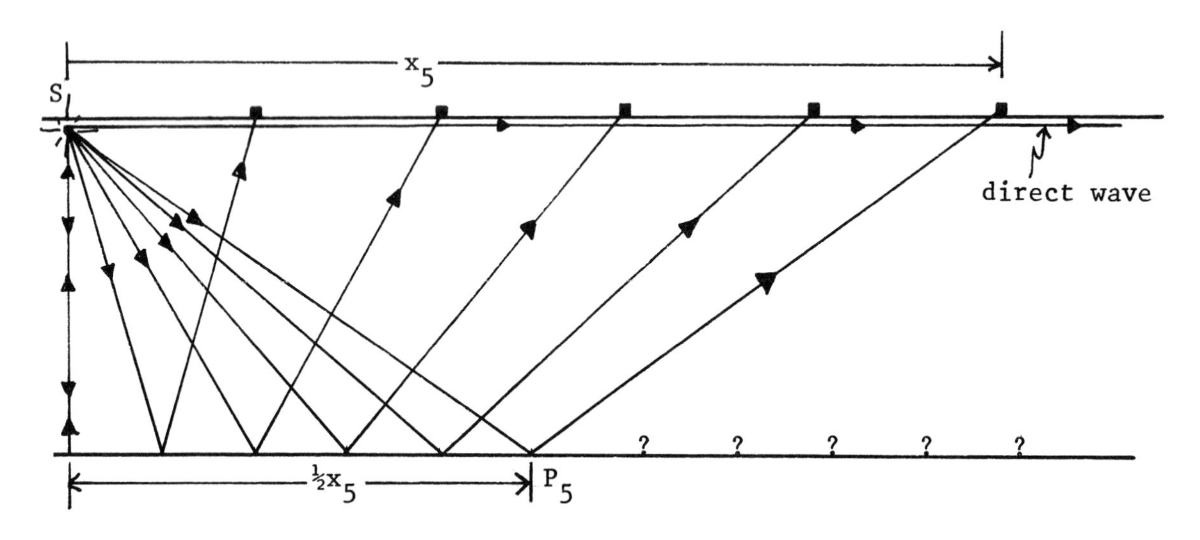

Fig.(9.5) Reflected and Direct Waves

Several conclusions can be drawn from Fig.(9.5). First, any geophone, regardless of its distance from shot point S, can, as far as the geometry is concerned, receive a reflected wave. Even a geophone at S itself can receive a reflected wave. (Of course, if the geophone is very far from S, the wave may die out before reaching it.) Second, the geophones, by virtue of their necessarily finite number, only "sense" the interface at certain regions. For example, geophone #5 at distance x_5 from S senses the interface in a region about point P_5, at distance $\frac{1}{2}x_5$ from the point on the interface directly below S. The factor $\frac{1}{2}$ occurs because of the symmetry imposed by the angle of incidence equalling the angle of reflection. If geophone #5 is the last geophone of the array, then no conclusions can be drawn about the nature of the interface beyond point P_5; hence, the ? marks attached to the presumed horizontal extension of the interface beyond P_5 in Fig.(9.5). (Possible diffraction effects are ignored.)

Suppose, then, that each geophone records the passage of the direct and reflected waves; the travel times of these waves are thereby recorded; the geophone-shot point distances are recorded as the geophones are laid out. To see if the data are consistent with the presence of a horizontal interface, it is necessary to obtain equations for the travel times based on that assumption, and see if the field data fit the equations.

Refer to Fig.(9.6). The shot point is at S. A single geophone is shown. Since this geophone represents any individual geophone, its distance from S is called simply x (rather than, say,

x_5, for the 5th geophone).

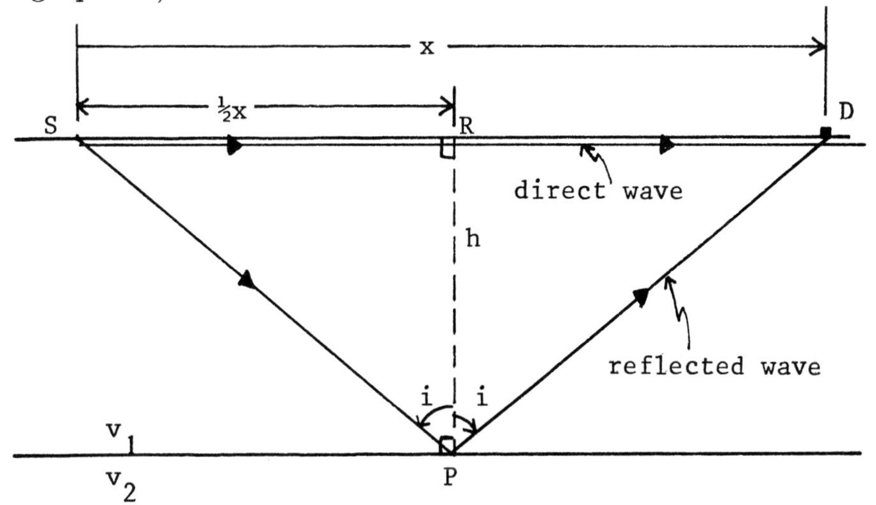

Fig.(9.6) To Find Travel Times

An equation for the travel time t_d of the direct wave is very easy to come by. This wave travels the distance x in reaching the geophone and the speed of the wave is v_1, since its path is entirely in the upper layer. Since this speed is constant, it follows that

$$t_d = \frac{x}{v_1}. \tag{9.15}$$

For the reflected wave the calculation is not much more difficult. Since the angle of incidence equals the angle of reflection, the distances SP = PD, where P is the interface reflection point and D is the location of the detector (geophone). Since the reflected wave remains in the upper layer, its speed also is v_1 throughout its trip. Hence, the travel time t_r of the reflected wave is given from

$$t_r = \frac{2(SP)}{v_1}.$$

Projecting the normal to the interface at P up to the surface at R, it is seen that SR = $\frac{1}{2}x$, again by virtue of the equality of the angles of incidence and of reflection. Using the Pythagorean theorem,

$$\mathrm{SP} = \sqrt{(\frac{1}{2}x)^2 + h^2},$$

so that

$$t_r = \frac{\sqrt{x^2 + 4h^2}}{v_1}. \tag{9.16}$$

Equation (9.16) is the equation for the time required for a reflected wave to travel from the shot point S down to the interface and be reflected back up to a point on the surface a distance x from the shot point. Notice that the equation has been algebraically simplified, with two 2's replaced, in effect, with one 4.

****

EXAMPLE 5

In a reflection survey, the geophones are 135 m apart, with no geophone at the shot point. The interface depth is 360 m. A reflected wave takes 287 ms to reach the first geophone, 135 m from S. How much later does the second geophone receive a reflected wave?

Use Eq.(9.16) to solve for the speed of the waves, employing the travel time to the first geophone:

$$v_1 = \frac{\sqrt{x^2 + 4h^2}}{t_{\mathrm{r}}},$$

$$v_1 = \frac{\sqrt{(135 \text{ m})^2 + 4(360 \text{ m})^2}}{0.287 \text{ s}},$$

$$v_1 = 2552 \text{ m/s}.$$

For the second geophone, $x = 270$ m. Again, apply Eq.(9.16), but this time solve for the travel time to this geophone using the speed of the waves just found:

$$t_{\mathrm{r}} = \frac{\sqrt{x^2 + 4h^2}}{v_1},$$

$$t_{\mathrm{r}} = \frac{\sqrt{(270 \text{ m})^2 + 4(360 \text{ m})^2}}{2552 \text{ m/s}},$$

$$t_{\mathrm{r}} = 0.301 \text{ s}.$$

Therefore, the second geophone receives a reflected wave at a time 301 ms $-$ 287 ms $=$ 14 ms after the first geophone records a reflected wave.

****

Data from a seismic survey often is presented graphically. To see if survey results so presented match Eqs.(9.15) and (9.16), it is necessary to present these equations in a graphical format. Both equations are plotted on the same graph, with travel time t on the "y" axis (ordinate) and horizontal distance from shot point x on the x axis (abscissa).

It is fairly easy to see how Eq.(9.15) should appear on such a graph. If that equation is rewritten as

$$t_{\mathrm{d}} = (\frac{1}{v_1})x,$$

then it can be seen to parallel the equation of a straight line, usually written in algebra books as

$$y = mx + b,$$

where m is the slope and b the y-intercept. For Eq.(9.15), in which t_d replaces y, the slope is $1/v_1$, and the t_d-intercept is zero. The direct wave takes zero time to travel to a geophone placed at the shot point.

The graph of Eq.(9.16) is not so obvious. First, the equation does not match that of a straight line. This means that the equation plots as a curved line. Second, the curved line does not go through the origin, since the $x = 0$ travel time for a reflected wave, $t_{r,0}$ is, from Eq.(9.16) with $x = 0$ substituted,

$$t_{r,0} = \frac{2h}{v_1}. \tag{9.17}$$

This makes sense, since the $x = 0$ reflected wave must arrive back at the surface right at the shot point S. The wave travels a distance h vertically down to the interface, and the same distance h back up to the surface, for a total distance $2h$, the wave moving at speed v_1 throughout. See the wave on the extreme left of Fig.(9.5).

Fig.(9.5) shows the reflected waves propagating to the right from the shot point. But waves also propagate to the left of S in exactly the same way. This means that the graph of t_r vs. x must be symmetrical about $x = 0$, which in turn implies that the curve of t_r vs. x must have a "horizontal" tangent at $x = 0$ (i.e., a tangent parallel to the x axis). This can also be proven by taking the derivative.

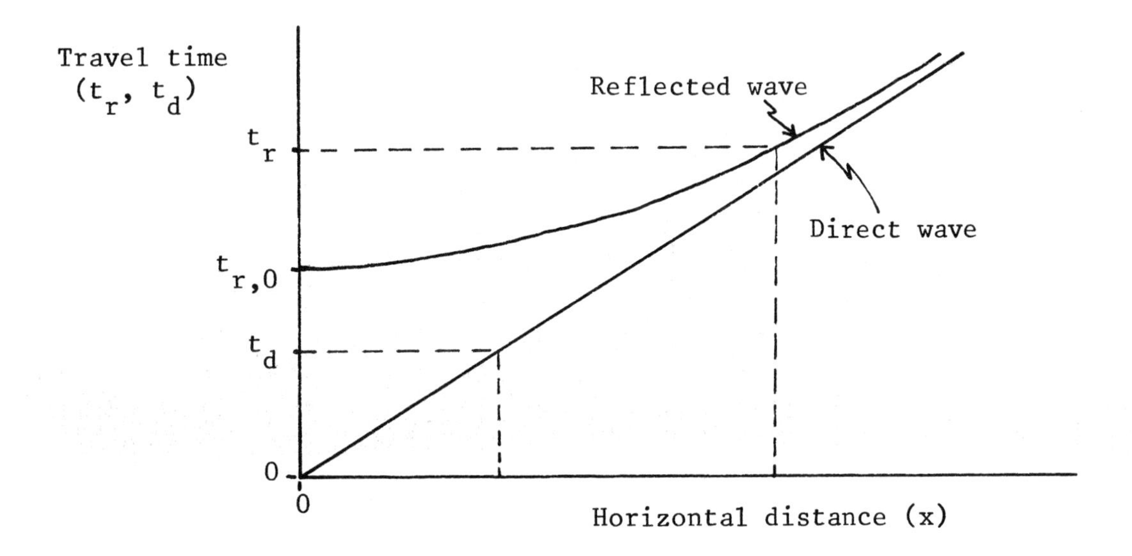

Fig.(9.7) Graphical Direct and Reflected Times

With this background, examine Fig.(9.7). Equation (9.15), the travel time for the direct wave, is a straight line through the origin, with slope $1/v_1$. Equation (9.16), the travel time for the reflected wave, is the curve that intercepts the time axis at $2h/v_1$. This curve turns upward at increasing values of x, since the travel time is greater for the greater distances.

Two additional features of Fig.(9.7) are noteworthy. At any x, the travel time for the reflected wave is greater than the travel time for the direct wave (the reflected wave travels a greater distance than the direct wave to reach the same geophone). The two travel lines do not intersect, for this would mean that the reflected and direct waves take the same time to travel from the shot point to the geophone at the intersection location x. But, the direct wave always travels a shorter distance to every geophone, no matter where it is placed. Since the two waves travel at the same speed, they cannot arrive at the same time.

For geophones very far from the shot point ($x >> h$), the difference in travel times between the reflected and direct waves becomes very small, since the difference in distances traveled becomes very small. Mathematically speaking, the direct wave line in Fig.(9.7) is an asymptote to the reflected wave curve, which itself is a hyperbola.

Each geophone will yield two data points, one for the direct wave and one for the reflected wave. (There may be other waves recorded, but for now these are ignored.) If the data points appear to fall along two lines looking like, and positioned as, those in Fig.(9.7), then it is fair to conclude that a horizontal interface exists at some depth. If the data fall along lines other than those described above, then a different arrangement of subsurface rocks is indicated.

Assume, then, that the lines drawn averaging the data points do form a picture like Fig.(9.7). The graph must be analyzed to extract, if possible, the desired quantities which, be it reminded, are the depth h to the interface and the wave speeds in the upper and lower levels v_1 and v_2. Unlike the generic diagram of Fig.(9.7), a graph drawn from real data will have numbers along the axes, so the values of the travel times t_d and t_r can be read for any chosen value of x.

First, from the direct wave line, pick a value of x and its corresponding time t_d. Then, calculate the wave speed in the upper layer v_1 from Eq.(9.15). Next, read the value of the $x = 0$ reflected wave travel time from the reflected wave curve. There may not have been a geophone at the shot point, represented by $x = 0$. In this case, the $x = 0$ value of t_r is obtained by an extrapolation, or extension, of the reflected wave curve from the closest geophone to $x = 0$, forming a horizontal tangent there. This travel time is $t_{\mathrm{r},0}$. Since v_1 has been found from Eq.(9.15), the depth h to the interface can now be found from Eq.(9.17). If the travel time for $x = 0$ is not so used, but rather the reflected wave travel time from another value of x, then Eq.(9.16) must be used to find the interface depth.

The wave speed v_2 in the lower layer cannot be determined from analysis of these reflected and/or direct waves, because these waves never penetrate into the lower layer.

****

EXAMPLE 6

P wave reflection data taken in alluvium over chalk is shown on Fig.(9.8) as travel time in ms vs. distance in m. Find (*a*) the wave speed in the upper layer (alluvium), and (*b*) the depth to the alluvium-chalk interface.

(*a*) To find v_1, use, for example, the point $x = 10$ m, $t_\mathrm{d} = 35$ ms on the direct wave (straight) line. By Eq.(9.15),

$$v_1 = \frac{x}{t_\mathrm{d}},$$

$$v_1 = \frac{10 \text{ m}}{0.035 \text{ s}},$$

$$v_1 = 290 \text{ m/s}.$$

Such a low seismic wave speed, less even than the speed of sound in air, indicates very loosely-consolidated material.

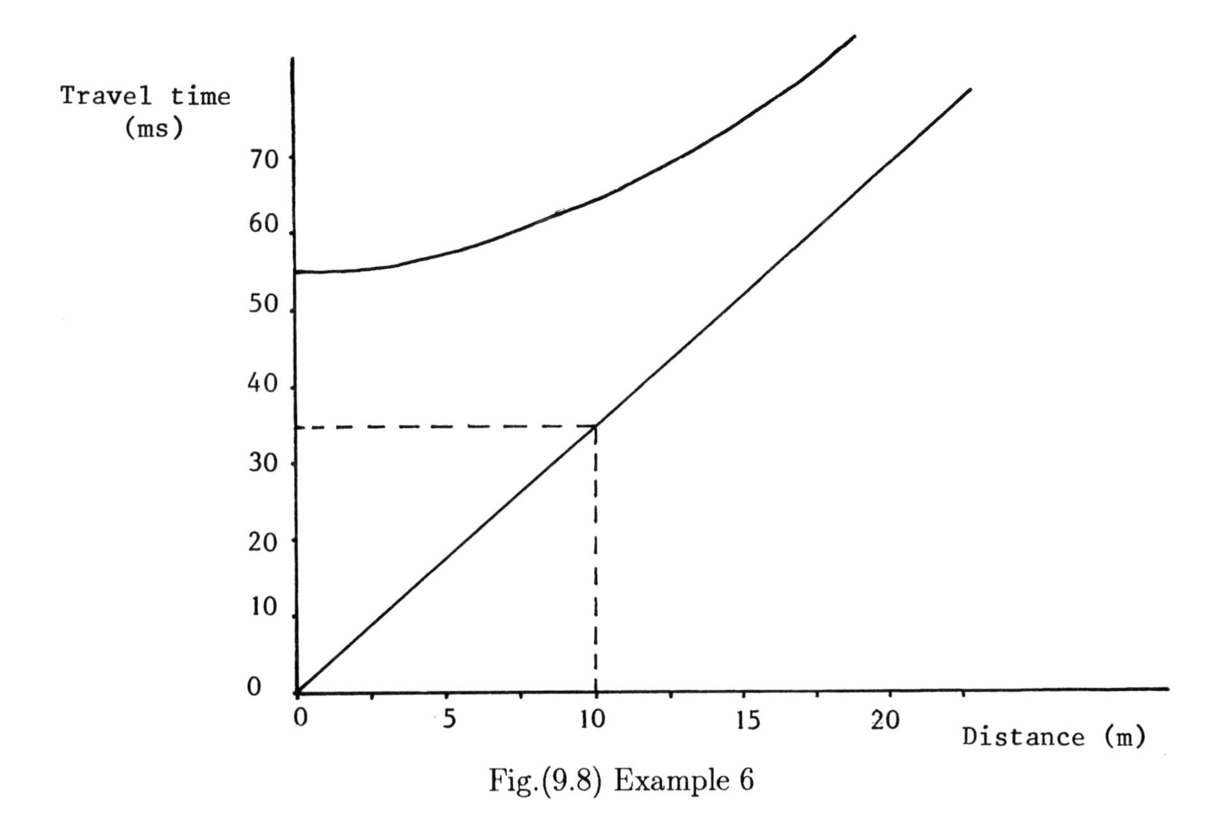

Fig.(9.8) Example 6

(*b*) The reflected wave curve intersects the time axis at $t_{r,0} = 55$ ms. Using the value of v_1 found in (*a*), Eq.(9.17) yields

$$h = \frac{1}{2} v_1 t_{r,0},$$

$$h = \frac{1}{2}(290 \text{ m/s})(0.055 \text{ s}),$$

$$h = 8.0 \text{ m}.$$

At such a shallow depth, it should be easy to verify this result with a bore hole.

****

9.6 Seismic Refraction

In Example 6 above, the only wave speed that could be found is v_1, the speed in the upper layer. The reflection method does not identify the value of v_2, the wave speed in the lower layer. The reason is that the reflected waves never penetrate into the lower layer, so the speed in that layer does not enter into the calculation of the travel time. Of course, the same applies to the direct wave.

The situation is different when refracted waves are considered, for these waves cross the interface and penetrate into the lower layer. In the lower layer, they travel at the speed of waves in that layer, which is v_2. Hence, their travel times depend on v_2 as well as v_1, so that measurements of the travel times may allow discovery of the value of v_2.

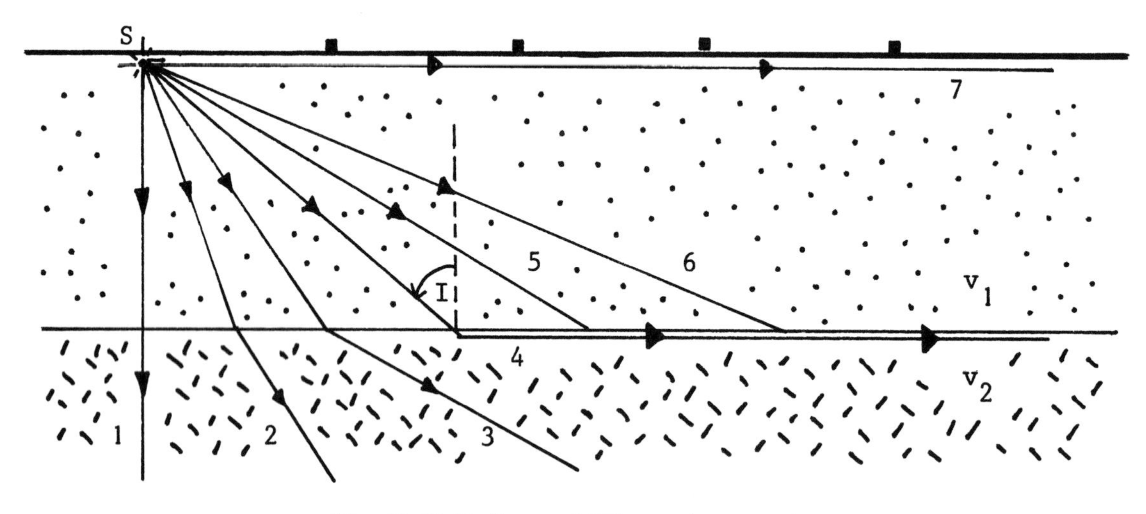

Fig.(9.9) Refracted Waves ($v_2 > v_1$)

Figure (9.9) shows the refracted waves produced by waves reaching the interface from the shot point. The refracted waves are present along with the reflected waves discussed above, but the reflected waves are not shown on Fig.(9.9). It is assumed in Fig.(9.9) that the wave speed in the lower layer is greater than the wave speed in the upper layer; i.e., $v_2 > v_1$. This is evident in Fig.(9.9) in that, for all the waves except the one at normal incidence (wave 1), the angle of refraction is greater than the angle of incidence. The principles of seismic refraction described below depend on the wave speeds being ordered in this way.

In Fig.(9.9), wave 1 strikes the interface at an angle of 0°; by Snell's law, the angle of refraction is 0° also, so the refracted wave continues along the same straight line as the incident wave but into the lower layer. Waves 2 and 3 each impinge on the interface at an angle of incidence sufficiently small to generate a refracted wave. Although these refracted waves enter rock type 2, they are useless for surveying purposes since they propagate deeper into the Earth, whereas all the geophones are on the surface. (These waves could be returned to the surface by other interfaces at depths greater than h, but this possibility is not considered here.)

Wave 4 strikes the interface at the critical angle I; the refracted wave it produces travels parallel to and just below the interface at the wave speed v_2 in the lower layer.

Neither of waves 5 or 6 generates a refracted wave, since these waves have angles of incidence greater than the critical angle. Wave 7 is the direct wave.

Focus attention on wave 4. At first sight, it would appear that this wave also is useless for surveying purposes, for the same reason that waves 1, 2, and 3 are useless: the refracted waves do not return to the surface. But wave 4 has a special property that does make it useful: this critically refracted wave, as it travels just below the interface, continually generates new waves, called *secondary waves*, that travel upward, through rock type 1, to the surface. These secondary waves leave the interface at the critical angle I with the normal, on the side of the normal opposite to that of the incident wave.

Figure (9.10) shows the critical wave and some of the secondary waves the critical wave produces as it propagates "along" the interface, ultimately to "die out" as its energy is dissipated into the secondary waves and as thermal energy to the rocks through which it passes. (Since the "useless" waves 1, 2, 3, 5, 6 are not shown, there being no need to show them, it is no longer necessary to refer to the critical wave as wave 4, nor the direct wave as wave 7.) A reminder: in Fig.(9.10), as in the other figures, the waves shown are either all P waves, or all S wave.

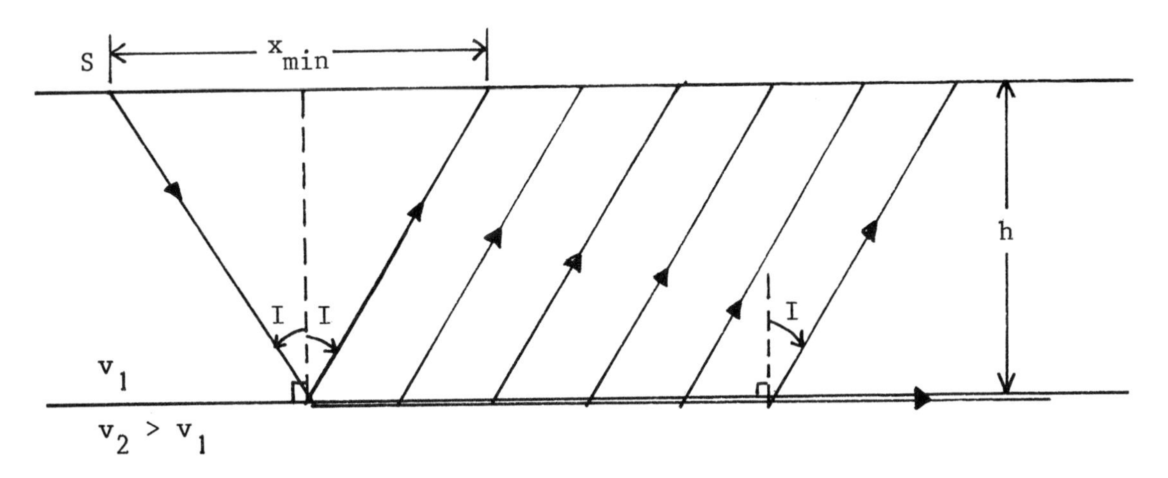

Fig.(9.10) Critical and Secondary Waves

No secondary waves are received at points on the surface at distances from the shot point less than $x_{\min}$, as shown on Fig.(9.10). The secondary wave that does reach the surface at this distance looks like a reflected wave for which the angles of incidence and reflection equal the critical angle. From Fig.(9.10), it can be seen that

$$\tan I = \frac{\frac{1}{2}x_{\min}}{h},$$

$$x_{\min} = 2h \tan I. \tag{9.18}$$

As in the seismic reflection surveying procedure, it is necessary to derive a formula for the time required for a wave to travel from the shot point to a geophone placed on the surface. By "a wave" is now meant a disturbance that propagates from the shot point to the interface, "along" the interface a certain distance, and then proceeding as a secondary wave back to the surface. A name must be given to a disturbance that makes this trip. There is no general agreement on a name: the names refracted wave, head wave, critical wave, are used in engineering literature. Since the disturbance strikes and leaves the interface at the critical angle, the name "critical wave" will be used in this work.

Figure (9.11) shows the path of a typical critical wave. The wave travels at speed v_1 from the shot point S to the interface, arriving at the point labelled A at an angle of incidence I equal to the critical angle. The wave then travels in the lower layer at speed v_2 to a point labelled B, and finally travels upward to geophone G at speed v_1 in the upper layer, leaving the interface at angle I from the normal.

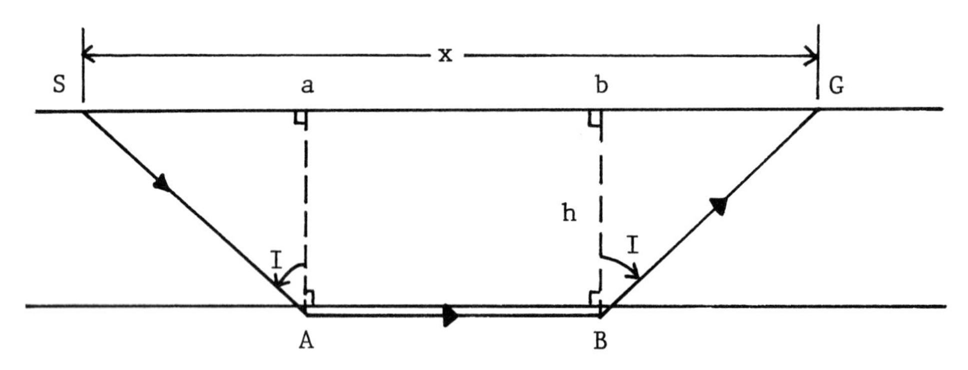

Fig.(9.11) Representative Critical Wave

The formula for the travel time of this wave must relate the quantities that can be measured, the travel time t_c itself to a geophone at distance x from the shot point, to the quantities it is desired to find, the depth h to the interface and the wave speeds v_1 and v_2 in the two rock layers. The derivation of the travel time formula is a little more involved than the derivation of the reflected wave travel time formula, Eq.(9.16), because, unlike that wave, the critical wave does not travel at the same speed over its entire journey.

Because of the symmetry of the situation, it is evident that the distances SA and BG are equal. Hence, the travel time can be expressed as

$$t_c = \frac{2(\mathrm{SA})}{v_1} + \frac{\mathrm{AB}}{v_2}. \tag{9.19}$$

From the right triangle SaA,

$$\mathrm{SA} = \frac{h}{\cos I}. \tag{9.20}$$

Also, from Fig.(9.11)

$$\mathrm{AB} = x - \mathrm{Sa} - \mathrm{bG}.$$

But note from Fig.(9.11) that Sa = bG. Using the same right triangle as before.

$$\mathrm{Sa} = h \tan I. \tag{9.21}$$

Combining Eqs.(9.19), (9.20), (9.21) yields

$$t_\mathrm{c} = \frac{2h}{v_1 \cos I} + \frac{x - 2h \tan I}{v_2}.$$

It is convenient to split-off the term involving x; i.e., rewrite the equation above as

$$t_\mathrm{c} = \frac{x}{v_2} + 2h \left[\frac{1}{v_1 \cos I} - \frac{\tan I}{v_2} \right]. \tag{9.22}$$

The last part of the derivation is to replace $\cos I$ and $\tan I$ with equivalent expressions involving the wave speeds (the angle I cannot be directly measured). To do this, recall from Eq.(9.14) that $\sin I = v_1/v_2$. Since $\sin^2 I + \cos^2 I = 1$, it follows that

$$\cos I = \sqrt{1 - (v_1/v_2)^2}. \tag{9.23}$$

Also, $\tan I = \sin I/\cos I$, which indicates that

$$\tan I = \frac{v_1/v_2}{\sqrt{1 - (v_1/v_2)^2}}. \tag{9.24}$$

Substitute Eqs.(9.23) and (9.24) into Eq.(9.22) and rearrange to show that

$$t_\mathrm{c} = \frac{x}{v_2} + \frac{2h\sqrt{v_2^2 - v_1^2}}{v_1 v_2}. \tag{9.25}$$

As with the reflection method, it is helpful to draw a graph of travel time t_c versus the distance x. At first sight, Eq.(9.25) looks very formidable as far as graphing is concerned. However, it should be noted that the last term contains neither of the variables being plotted. Hence, Eq.(9.25) does fit the generic straight line equation

$$t_\mathrm{c} = mx + b.$$

The slope of the line is $m = 1/v_2$, and

$$b = \frac{2h\sqrt{v_2^2 - v_1^2}}{v_1 v_2}, \tag{9.26}$$

is the t_c-intercept.

Unlike the time intercepts for the reflected and direct waves, the time intercept of the critical wave line has no direct physical significance. It is not possible for a critical wave to return to the surface at the shot point. See Eq.(9.18) for the smallest distance between the shot point and a returning critical wave. The existence of the time intercept is a mathematical manifestation of the fact that a straight line can be extended indefinitely, regardless of whether the extension corresponds to anything that is real in a particular situation.

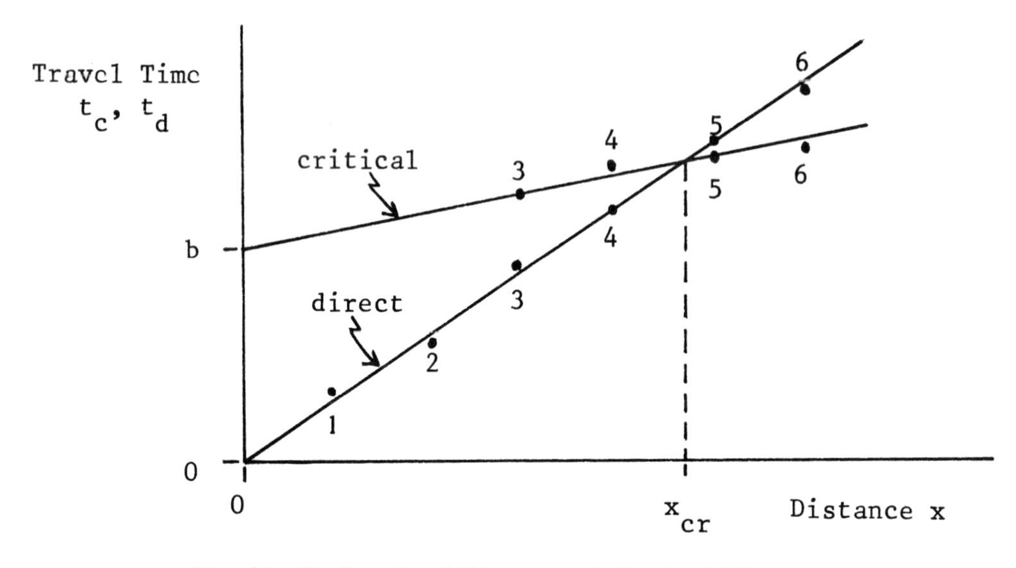

Fig.(9.12) Graph of Direct and Critical Waves

Passage of the direct wave and the recording of its travel time also is needed. This travel time is given by Eq.(9.15), which plots as a straight line through the origin with slope $1/v_1$. Since $v_2 > v_1$, it follows that $1/v_1 > 1/v_2$; that is, the slope of the direct wave line is greater than the slope of the critical wave line. This guarantees that, on Fig.(9.12), the two straight lines representing these waves will intersect; the point at which this occurs is called the *cross-over point*, and its distance from the shot point is written x_{cr}. A geophone fortunately located at the cross-over point records the direct and critical waves as arriving at the same time.

Also shown on Fig.(9.12) are the data points from geophones 1-6 (#1 closest to the shot point) from which the lines are drawn. Geophones 1 and 2 do not record the critical wave; evidently, they are positioned too close to the shot point. Hence, the portion of the critical wave line to the left of the geophone #3 data point is simply an extension of the straight line averaging the data points of geophones 3-6. The direct wave line must pass through the origin.

From the graph, the speed v_1 in the upper layer can be found from Eq.(9.15), using values of x and t_{d} from a point on the direct wave line, as is done in the reflection surveying procedure. To find the speed v_2 in the lower layer, select two points on the critical wave line (not necessarily data points). Since the slope of this line equals $1/v_2$, and the slope, by definition, is $\Delta t_{\mathrm{c}}/\Delta x$, it follows that

$$v_2 = \frac{x_2 - x_1}{t_{\mathrm{c}2} - t_{\mathrm{c}1}}. \tag{9.27}$$

In Eq.(9.27), the subscripts 1 and 2 refer to the two selected points on the critical wave line, and not to particular geophones.

The fact that v_2, the wave speed in the lower rock layer, can be found is the vital feature of the refraction method. (If $v_1 > v_2$, the refraction method does not work since no critical wave is generated.)

The last quantity to be calculated, after the wave speeds have been found, is the depth h to the horizontal interface. There are several ways to do this.
(i) Read the value b of the time intercept of the critical wave line from the graph; then substitute the values of b, v_1, v_2 into Eq.(9.26) and solve for h.
(ii) From the critical wave line, read off the values of t_c and x at one point on the line; substitute these values, along with those of v_1 and v_2 into Eq.(9.25) and solve for h.
(iii) From the graph, read off the value of x_{cr} at the cross-over point, the point where the direct and critical lines intersect. At $x = x_{cr}$, $t_d = t_c$. By Eqs.(9.15) and (9.25), this means that at $x = x_{cr}$,

$$t_d = t_c,$$

$$\frac{x_{cr}}{v_1} = \frac{x_{cr}}{v_2} + \frac{2h\sqrt{v_2^2 - v_1^2}}{v_1 v_2}.$$

This equation can be rearranged to yield

$$h = \frac{1}{2} x_{cr} \sqrt{\frac{v_2 - v_1}{v_2 + v_1}}, \quad (9.28)$$

and therefore h can be found from the values of x_{cr}, v_2, and v_1.

In displaying data on the graph of Fig.(9.12), parts of the lines are sometimes omitted. Often, only the "first arrivals" are plotted. This means that for $0 < x < x_{cr}$, the critical wave line segment is not shown, since the direct wave arrived first. For $x > x_{cr}$, the critical wave is the first arrival so it is shown, but the direct wave line is not drawn.

****

EXAMPLE 7

A seismic surveying team takes the first arrival refraction data shown in Fig.(9.13). Only P wave data is shown. (*a*) Find the speed of the P waves in the upper layer of rock. (*b*) Find the speed of the P waves in the lower layer. (*c*) Find the depth to the horizontal interface.

(*a*) The cross-over point is on both line segments and therefore on the direct wave segment. Using its coordinates in Eq.(9.15) gives

$$v_1 = x/t_d,$$

$$v_1 = (1200 \text{ m})/(0.4 \text{ s}),$$

$$v_1 = 3000 \text{ m/s}.$$

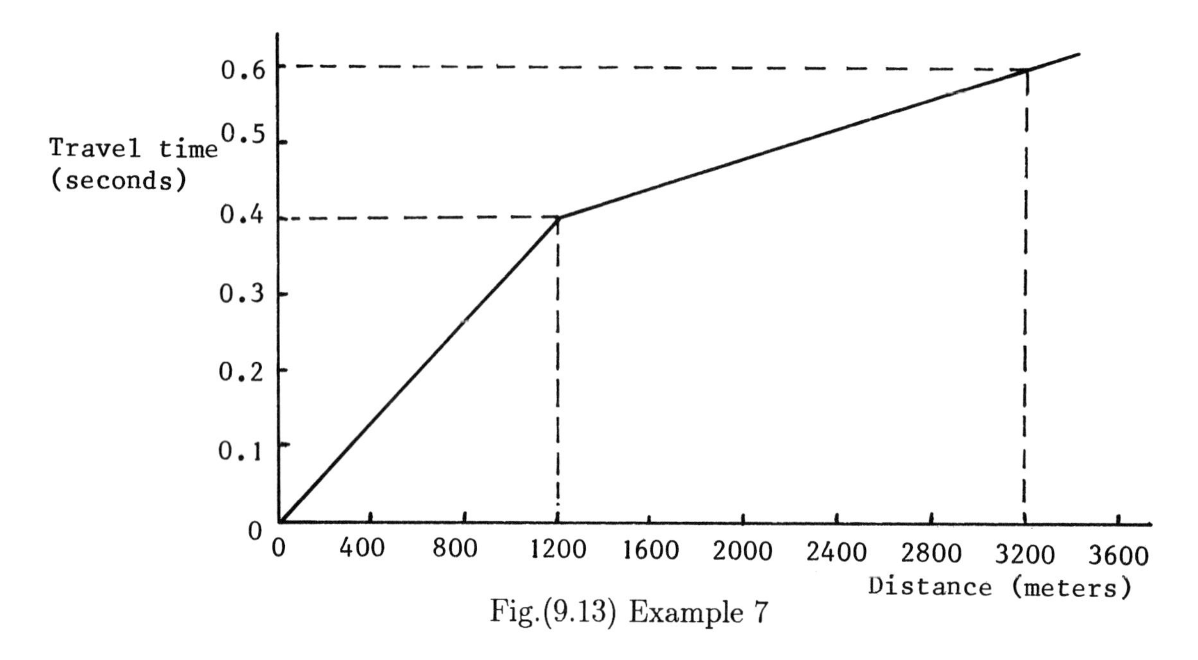

Fig.(9.13) Example 7

(*b*) The speed in the lower layer is found from Eq.(9.27). The cross-over point is on the critical wave line segment as well as on the direct wave segment, so it can be used as one point. The other point can be the upper point marked by dashed lines. Hence,

$$v_2 = \frac{x_2 - x_1}{t_{c2} - t_{c1}},$$

$$v_2 = \frac{3200 \text{ m} - 1200 \text{ m}}{0.6 \text{ s} - 0.4 \text{ s}},$$

$$v_2 = 10,000 \text{ m/s}.$$

(*c*) Suppose that method (iii) is used to find h. The value of x at the cross-over point is, as seen above, $x_{\text{cr}} = 1200$ m. When the values of the wave speeds are substituted into Eq.(9.28), their units cancel and need not be written out; the depth is found to be

$$h = \frac{1}{2} x_{\text{cr}} \sqrt{\frac{v_2 - v_1}{v_2 + v_1}},$$

$$h = \frac{1}{2}(1200 \text{ m}) \sqrt{\frac{10,000 - 3000}{10,000 + 3000}},$$

$$h = 440 \text{ m}.$$

This result can be checked by using methods (i) and (ii). For method (i), the critical wave line segment must be extended to the time axis; the intercept is $b \approx 0.28$ s. In method (ii), the coordinates of any point on the critical wave line can be used in Eq.(9.25).

****

9.7 Combined Methods

In the two previous sections, the reflection method and the refraction method of seismic surveying have been analyzed separately, and applied to the simple geometry of a single, horizontal, interface as an illustration of their use. Only the properties of the direct wave are common to both procedures. This dichotomy of presentation aids in clarifying the features of each method. In the field, however, the geophones may record direct, critical, and reflected waves on the same output, so it is appropriate to understand how the reflection and refraction analyses fit together, if only so that the signals can be sorted-out.

This synthesis can be carried out by drawing one time-distance graph that shows the direct, refracted, and reflected waves together; that is, to merge, as it were, Fig.(9.7) and Fig.(9.12). Since both of these figures show the direct wave line, all that has to be decided is where the reflected wave curve fits relative to the critical wave line.

To do this, examine first the time intercepts of these two lines. The time intercept of the reflected wave curve, $t_{\mathrm{r},0}$, is given by Eq.(9.17); the critical wave time intercept b by Eq.(9.26). But, comparing these two equations, it can be seen that

$$\frac{2h}{v_1} > \frac{2h}{v_1}\left(\frac{\sqrt{v_2^2 - v_1^2}}{v_2}\right),$$

since the term in parenthesis is always less than 1. Hence $t_{\mathrm{r},0} > b$; the reflected wave intercept is "higher" on the graph than the critical wave intercept.

The next matter to be decided is whether or not the critical line and the reflected line intersect. If they do, their travel times at the intersection point(s) must be equal. To find the value or values of x at which this happens, if any, set $t_{\mathrm{r}} = t_{\mathrm{c}}$ and solve for x. By Eqs.(9.16) and (9.25), this means finding the values of x that solves

$$\frac{\sqrt{x^2 + 4h^2}}{v_1} = \frac{x}{v_2} + \frac{2h\sqrt{v_2^2 - v_1^2}}{v_1 v_2}.$$

Square both sides of the equation; it can be confirmed that the resulting equation can be written in the form

$$\left[x - \frac{2hv_1}{\sqrt{v_2^2 - v_1^2}}\right]^2 = 0.$$

The solution (really, and this is important, two equal solutions) of the quadratic is

$$x = \frac{2hv_1}{\sqrt{v_2^2 - v_1^2}},$$

$$x = 2h \tan I.$$

This result is anticipated, since $2h \tan I = x_{\mathrm{min}}$, the horizontal distance traveled by the very first critical wave; see Fig.(9.10). This one critical wave looks like a reflected wave; hence, its travel time should plot on the reflected wave line on the graph. The calculation above shows that this is the only critical wave that has a travel time equal to that of a reflected wave traveling the same horizontal distance x. Since the slope of the critical wave line is less than the slope of the direct wave line, and the direct wave line is asymptotic to the reflected wave curve, it follows that the critical wave line is tangent to the reflected wave curve at x_{min}, for otherwise there would be two separate solutions for their intersection.

Putting all the conclusions reached together allows for drawing the complete time-distance graph, showing all three types of waves. In Fig.(9.14), the direct wave line goes through the origin, the reflected wave line is a curve asymptotic to the direct wave line. The reflected wave line is tangent to the critical wave line at x_{min}. If the critical wave line is extended from x_{min} to the time axis, its intercept b is found to be less than the time intercept of the reflected wave curve; this also follows from their tangency. The direct and critical wave lines intersect at the cross-over distance x_{cr}.

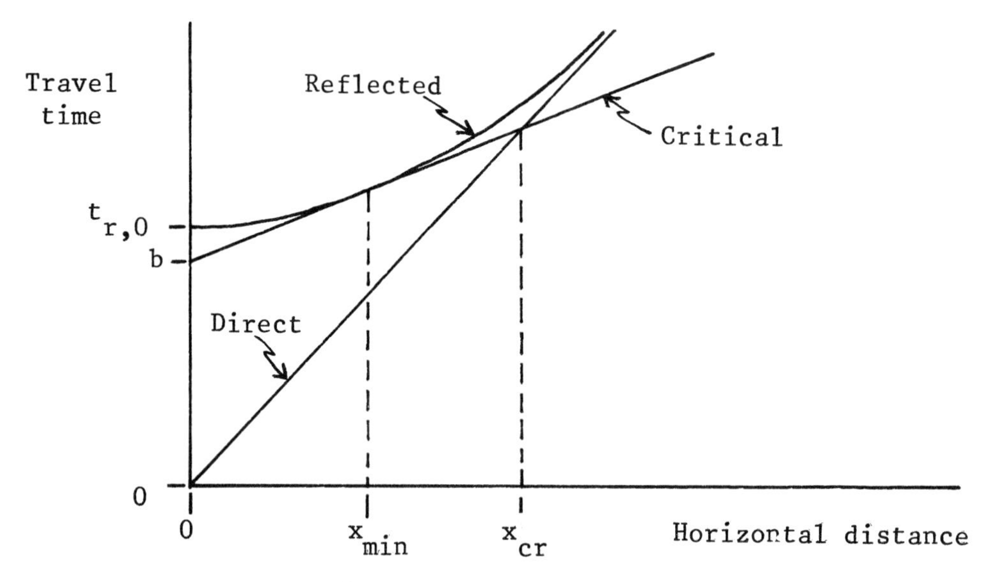

Fig.(9.14) Combined Time-Distance Graph

9.8 Problems

1. When one face of a slab of rock is struck with a hammer, a detector at the opposite face 85.0 cm away receives waves 230 μs and 425 μs later. Find the speeds of (*a*) the S waves, and (*b*) the P waves in this rock.

2. A particular rock of density 3.42 g/cm^3 has the following elastic moduli: Young's modulus = 58.2 GPa, shear modulus = 20.3 GPa, bulk modulus = 138 GPa. Calculate the speed of P waves in the rock.

3. Certain rocks have density 2.72 g/cm^3, Young's modulus = 47.4 GPa, and Poisson's ratio = 0.413. Calculate the speed of S waves in the rock.

4. A certain rock has an S wave speed of 2.11 km/s and a Poisson's ratio of 0.322. Find the speed of P waves in this rock.

5. A particular rock has a shear modulus of 12.6 GPa and seismic wave speeds of 2.25 km/s and 3.87 km/s. Find the bulk modulus of the rock.

6. Calculate the shear modulus of a rock with density 2.70 g/cm^3 and wave speeds of 2.90 km/s and 4.60 km/s.

7. Calculate Poisson's ratio for a rock with density 3.12 g/cm^3 and wave speeds of 3.40 km/s and 5.90 km/s.

8. Find Poisson's ratio for a rock in which the P waves travel twice as fast as the S waves.

9. A rock has an S wave speed of 3.40 km/s, a density equal to 2.70 g/cm^3, and a Poisson's ratio of 0.360. (*a*) Find the speed of the P waves in the rock. (*b*) Calculate the values of all the elastic moduli that can be found from the given data.

10. A particular rock has seismic wave speeds of 6.73 km/s and 3.87 km/s. The rock has a compressibility of 0.0130 (GPa)$^{-1}$. Find (*a*) the unit weight of the rock, and (*b*) its shear modulus.

11. Rewrite Snell's law in a version appropriate when the angles are measured from the interface, rather than from the normal.

12. Show that the largest possible angle of refraction r_{max}, for a wave incident from rock

type 1 and $v_1 > v_2$, is given by

$$\sin r_{\text{max}} = v_2/v_1.$$

13. Find the angle of incidence of a wave from the upper layer if the wave speed in the lower layer is twice the wave speed in the upper layer and the angle between the reflected and refracted waves is 90.0°.

14. A wave, originating in the lower layer, is moving upward and strikes the interface between the lower and an upper layer of rock. The wave speed in the lower layer is 4760 m/s and in the upper layer is 2230 m/s. (*a*) Find the angle of refraction if the angle of incidence is 59.0°. (*b*) Find the critical angle, if one exists.

15. Where possible, supply the missing quantity in each case in the following table, where i = angle of incidence, r = angle of refraction, v_1 = wave speed in the upper layer, and v_2 = wave speed in the lower layer. The wave originates in the upper layer.

Case	i (°)	v_1 (km/s)	r (°)	v_2 (km/s)
(*a*)		4.36	72.5	3.20
(*b*)	33.3	2.55		3.82
(*c*)	58.4	1.82	77.2	
(*d*)	26.0		61.2	4.48
(*e*)		3.26	90.0	5.10

16. Calculate the difference in travel times between the reflected and direct waves for horizontal distances of $x = 0$, h, $2h$, $5h$, and $10h$, where h is the depth of the interface. Express the answers in terms of the $x = 0$ reflected wave travel time.

17. Explorers are conducting a seismic reflection survey in a region in which a horizontal interface between two rock types is known to exist at some depth. A geophone 186 m from the shot point receives P waves 62.0 ms and 87.7 ms after detonation. Find the depth to the interface. (Any critical waves received have been filtered out.)

18. A seismic reflection survey is being conducted in a region with a horizontal interface between two kinds of rock. The interface is at a depth of 340 m. A seismic wave detector is placed very close to the shot point. This detector receives a P wave echo 184 ms after detonation and an S wave echo 309 ms after detonation. The upper layer rocks have density 2.80 g/cm^3. Find the shear modulus of these rocks.

19. In Fig.(9.15), wave (a) takes 259 ms to reach detector D and wave (b) takes 325 ms to

reach D. Calculate the travel time for wave (c).

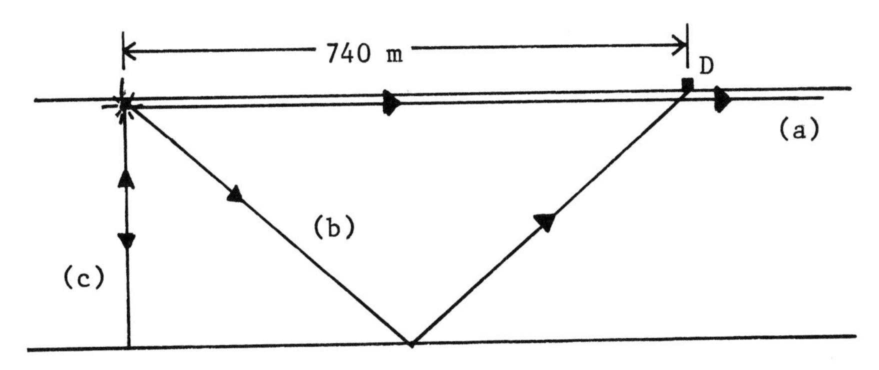

Fig.(9.15) Problem 19

20. Beneath the ground in a particular region, two horizontal interfaces at different depths meet at a vertical fault; see Fig.(9.16). Find the distance Δx within which no reflected waves will be recorded in the direction normal to the fault. (Geophones in this region will detect waves generated by diffraction processes not discussed in this book.)

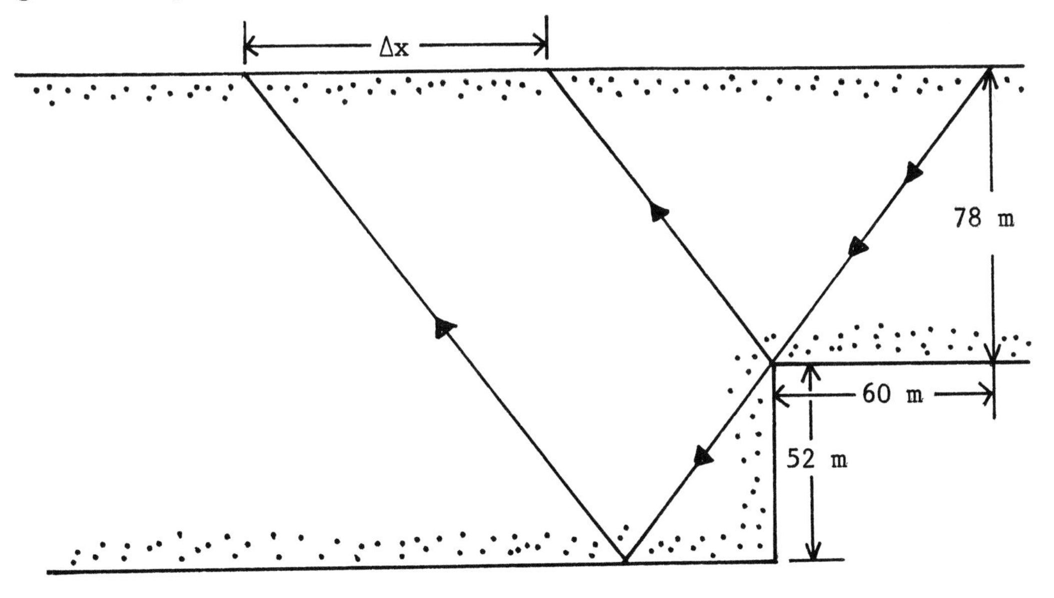

Fig.(9.16) Problem 20

21. A reflection survey is being conducted with a mechanical hammer, rather than an explosive, at the shot point to produce the seismic waves. Hence, a geophone can be placed at the shot point. Both P and S direct and reflected waves are recorded. The geophone at the shot point receives signals 56.0 ms and 100 ms after a hammer blow. A geophone

placed 350 m from the shot point receives signals 140 ms, 151 ms, 250 ms, and 269 ms after a hammer blow. (*a*) Find the speeds of the P and S waves. (*b*) Find the depth to the horizontal interface. (*c*) At what distance from the shot point do the direct S and reflected P waves arrive at the same time?

22. The reflected wave takes 117 ms to reach geophone G_1, which is 120 m from the shot point, as shown on Fig.(9.17). How long does it takes for the reflected wave to reach geophone G_2, which is 340 m from the shot point in the other direction?

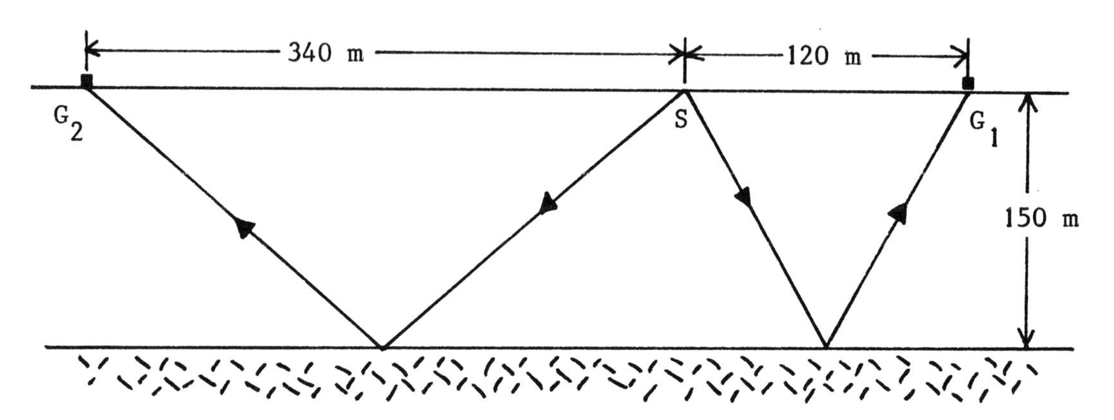

Fig.(9.17) Problem 22

23. In a reflection survey, the explosive is placed at a depth d in the ground. A horizontal interface exists at depth h, where $h > d$. Show that the travel time t_r for a reflected wave to reach a geophone on the surface at a distance x from the vertical hole into which the explosive is placed, is given by

$$t_r = \frac{\sqrt{x^2 + (2h - d)^2}}{v_1},$$

where v_1 is the wave speed in the upper layer of rock.

24. In parts of Antarctica, the ice floats on a layer of liquid water. Evidently, heat escaping upward through the bedrock on which the ice rests has, in these regions, melted ice at the base of the ice sheet. To determine the thicknesses of the ice layer and the water layer, explorers conduct a seismic reflection survey. The detector is very close to the shot point, so that the reflections at the two interfaces are essentially at normal incidence. Echoes are received 3.36 s and 3.53 s after detonation of 5 kg of TNT. The P wave speed in ice is 2200 m/s and in water is 1440 m/s. Find the thickness of (*a*) the ice sheet, and (*b*) of the water layer.

25. A seismic survey yields the results shown on Fig.(9.18). Find the depth to the horizontal interface.

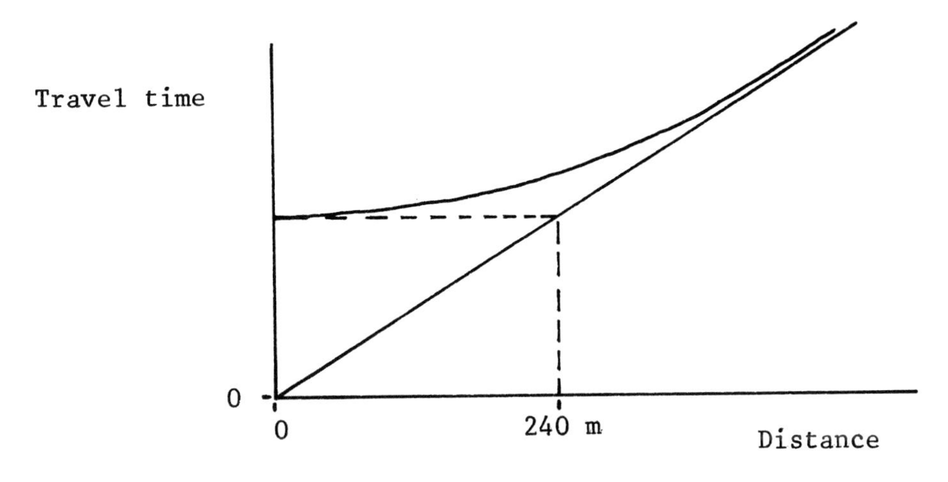

Fig.(9.18) Problem 25

26. A seismic survey yields the results shown on Fig.(9.19). Travel times are in ms and distances in m. Find the depth to the horizontal interface.

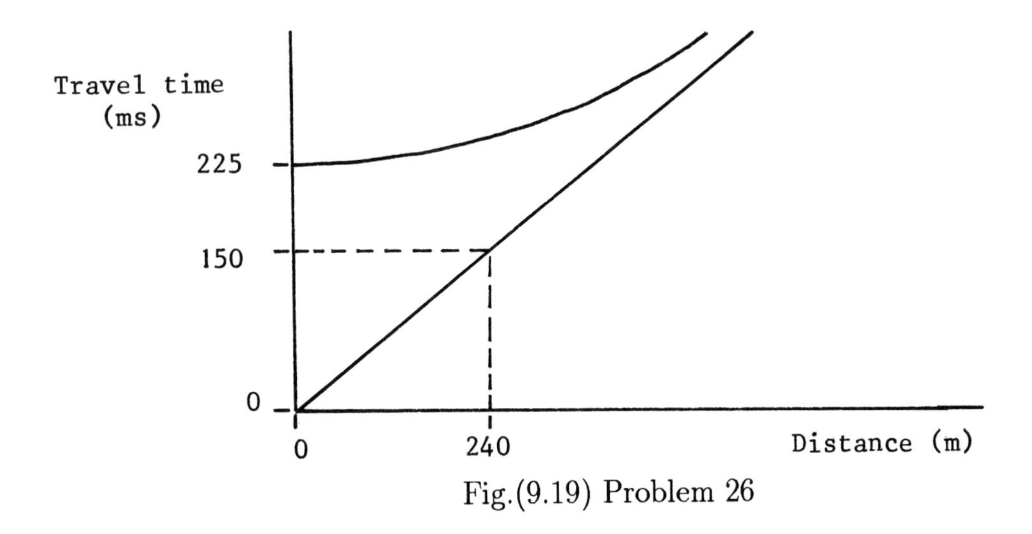

Fig.(9.19) Problem 26

27. In a seismic refraction survey, no detector within a distance of 288 m from the shot point receives a critical wave. A detector at a distance of 420 m from the shot point receives the direct wave 130 ms after detonation, and receives the critical wave 97.0 ms after the direct wave. Find (a) the depth to the horizontal interface, and (b) the wave speed in the lower layer.

28. A seismic refraction survey results in data fitting the lines shown on the graph of Fig.(9.20). Find the depth to the horizontal interface.

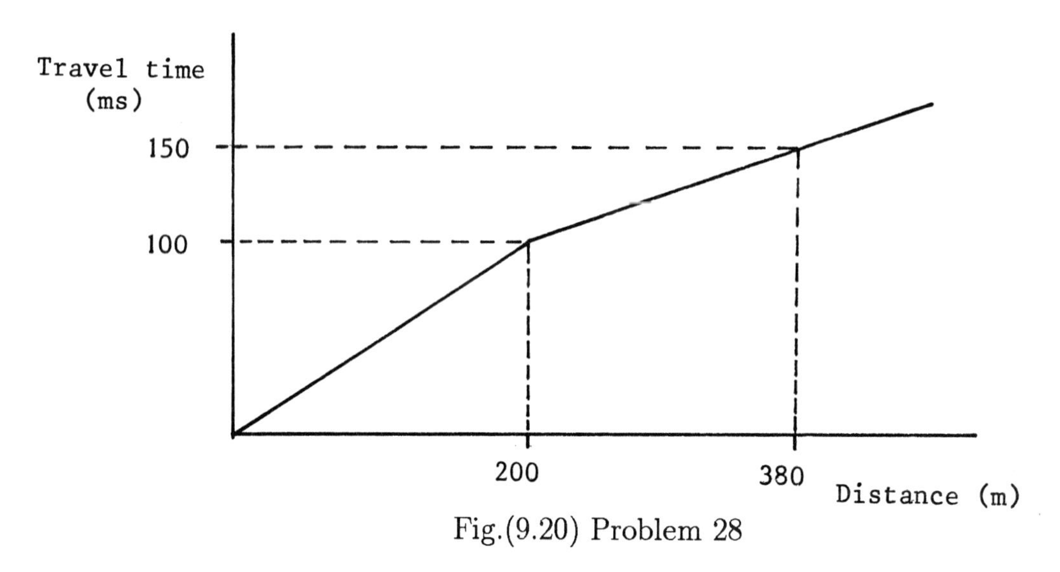

Fig.(9.20) Problem 28

29. Seismic survey P and S wave data are shown on Fig.(9.21). (*a*) Find the depth to the horizontal interface. (*b*) Find the greatest distance inside which no critical waves of either kind are detected.

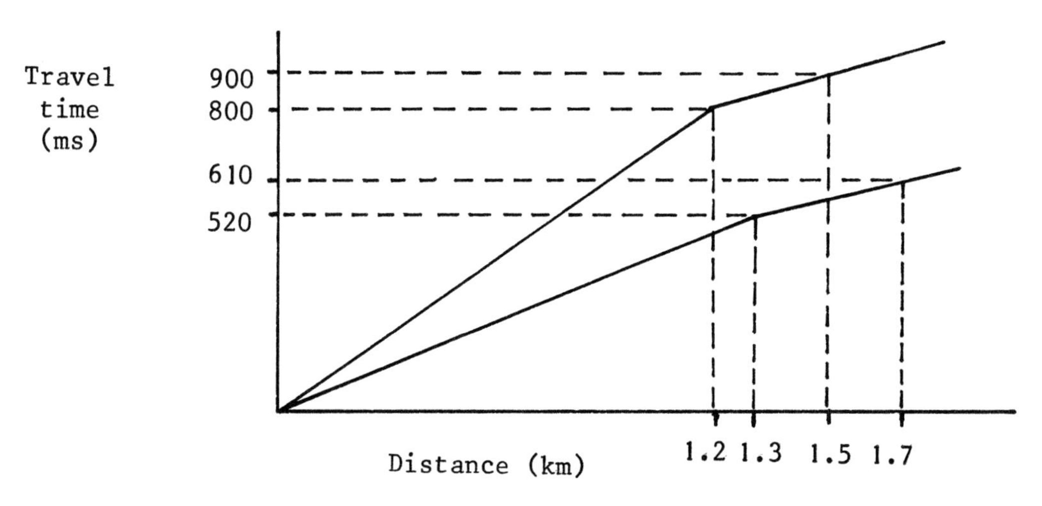

Fig.(9.21) Problem 29

30. A horizontal interface exists at a depth of 78.5 m. In the upper layer S waves travel at 1.22 km/s and P waves at 2.16 km/s. In the lower layer, S waves travel at 1.95 km/s and P waves at 4.07 km/s. Find the area of the region within which (*a*) no critical S waves, and (*b*) no critical P waves, are detected.

31. Can the cross-over point be "inside" the minimum distance for detection of a critical wave; i.e., is $x_{\text{cr}} < x_{\text{min}}$ possible? Answer this query by showing that

$$x_{\text{cr}} - x_{\text{min}} = 2h \sec I.$$

32. The table shows recorded travel times for the direct, reflected, and critical P waves versus distance from shot point for a certain seismic survey. Graph the data and find(*a*) the wave speeds in the two layers of rock, and (*b*) the depth to the horizontal interface.

x (m)	t_{d} (ms)	t_{r} (ms)	t_{c} (ms)
0			
50	29	115	
100	56	124	
150	83	139	139
200	111	157	155
250	139	178	171
300	167	200	187
350	195	224	203
400	222	248	219
450	250	274	236
500	278	299	252
550	306	325	268
600	333	351	284
650	361	378	300
700	389	404	316

33. Show that the derivative dt_{r}/dx of the reflected wave travel time curve (*a*) at $x = 0$ equals zero, and (*b*) at $x = x_{\text{min}}$ equals $1/v_2$. The second result shows that the reflected and critical wave lines are tangent at x_{min}, since they have the same slope there.

34. P waves travel from a shot point to a geophone that is 1500 m away. The speed of the P waves in the upper layer of rock is 4.70 km/s and in the lower layer is 5.80 km/s. The horizontal interface is at a depth of 330 m. (*a*) Find the angle made by the reflected wave with the interface. (*b*) Find the travel times of the direct wave, the reflected wave, and the critical wave. (*c*) How much time does the critical wave spend in traveling along the interface before emitting the secondary wave that reaches the geophone?

35. Seismic P wave refraction and reflection data over the same region are shown together in Fig.(9.22). Assume a horizontal interface between two rock types. (*a*) Find the speed of the P waves in the upper rock layer. (*b*) Find the speed of the P waves in the lower rock

layer. (*c*) Find the depth to the interface. (*d*) Calculate the numerical value of the time T shown on the travel time axis. (*e*) Find the distance marked X on the distance axis.

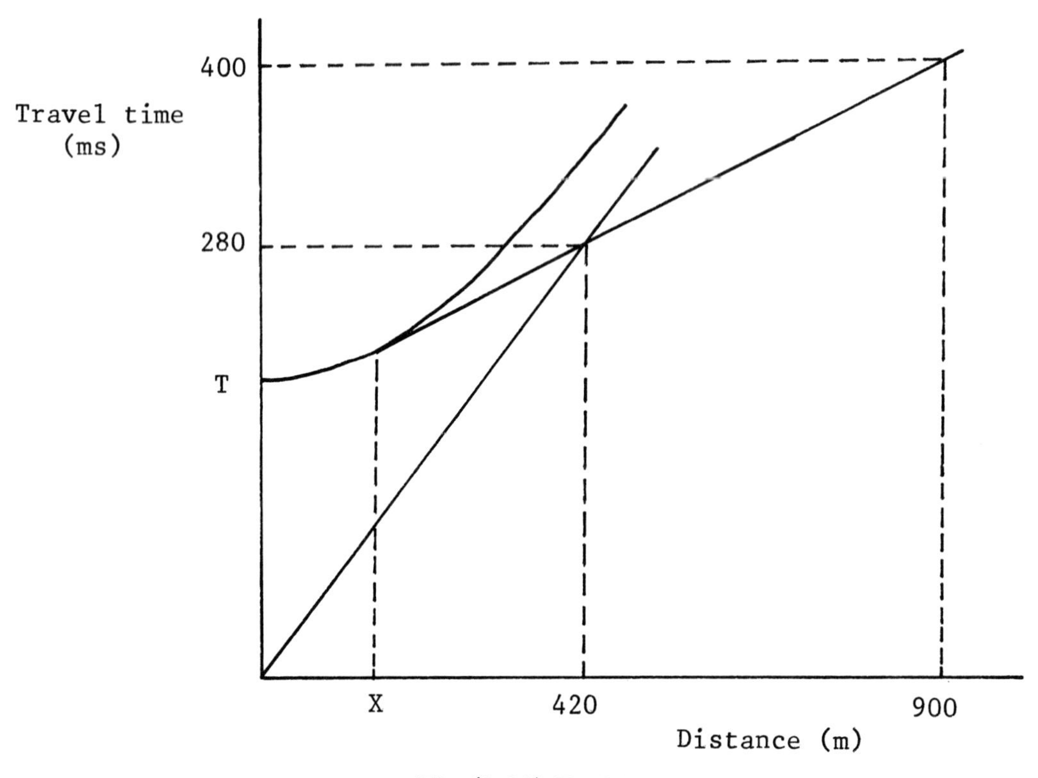

Fig.(9.22) Problem 35

Chapter 10

Seismology

10.1 Earthquake Waves

Usually, the ground, the land surface of the Earth, is a fixed, stable and secure surface. Occasionally, however, the ground shakes for several seconds, or even longer, for no apparent reason. Sometimes, the shaking is violent enough to cause extensive damage to buildings, freeways, and other large structures. Loss of life is not unusual. Often, over the next few days, weeks, or even months, the shaking recurs many times, although with less intensity. The size of the region affected can be hundreds of square kilometers, including an entire city.

It is reported that an *earthquake*, followed by several *aftershocks*, has taken place.

The shaking is only the surface manifestation of the earthquake. This shaking is the result of the passage of seismic waves. If the seismic waves in a particular earthquake are traced backwards, they are found to originate at a point in the Earth; this point is called the *focus* of the earthquake. Most earthquakes have a focus at a depth of less than 100 km; none are at depths greater than about 700 km. The point on the Earth's surface directly above the focus is called the *epicenter*.

The focus cannot really be literally a point; however, the region in the Earth within which the earthquake originates, although perhaps several tens of kilometers in size, is very small compared to the radius of the Earth (6370 km), so that for many purposes the focus can be considered to be a point.

Waves carry energy, the amount being proportional to the square of the amplitude of the wave. Some earthquakes can cause severe damage to large parts of a city, even after the seismic waves have traveled through tens of kilometers of rock, and thereby suffered some attentuation of energy. It can be concluded that an earthquake is associated with the sudden release of large amounts of energy at the focus. At present, there is no complete, consistent, theory for the mechanism by which this energy release takes place.

Figure (10.1) shows a part of the Earth and some seismic waves associated with an earthquake. The focus of the earthquake is F; the center of the Earth is at C, so the epicenter of the earthquake is at E, where the line CF intersects the surface. From the focus,

P and S seismic waves travel away in all directions; only four are shown on Fig.(10.1). These waves are also referred to as *body waves*, since they travel through the body of the Earth. When a body wave encounters the Earth's surface, seismic waves are generated that travel along the surface of the Earth; these are known, naturally enough, as *surface waves*, also as L waves. In Fig.(10.1), only the surface waves asociated with the body waves that travel the shortest path from the focus to the Earth's surface are shown.

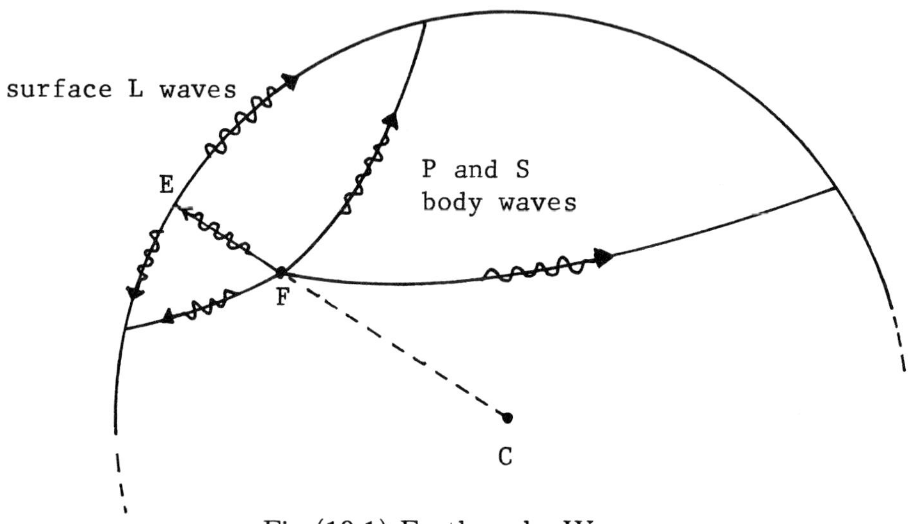

Fig.(10.1) Earthquake Waves

On Fig.(10.1), the paths of the seismic waves are shown as curved, not straight, lines. In describing the principles of seismic surveying in Chapter 9, the paths of "artifically produced" seismic waves traveling relatively short distances (compared to the size of the Earth) are examined. In such local situations, the waves travel though uniform homogenous rock until they encounter a different rock type at an interface or discontinuity. The waves travel on straight line paths in each rock type, but their direction of travel changes abruptly at the discontinuity.

But Fig.(10.1) displays a global picture: the waves travel distances comparable to the size of the Earth. Local variations are too small to show on this scale. As the body waves travel through the Earth, they travel, at first, through material whose properties change gradually and continually, not abruptly, along the wave path. As a result, the wave direction changes gradually and continually, yielding a curved path that is convex toward the center of the Earth. However, if the wave travels deep enough into the Earth, then real discontinuities are encountered, and at these discontinuities, reflected and refracted waves are produced, these waves moving off in directions demanded by Snell's law.

Instruments called *seismographs* record the passage of seismic waves. A seismograph records in detail the displacement of the near-surface rocks at the location of the seismograph as the waves pass. (Seismographs are usually constructed at a fixed location, and are much more accurate than the small portable geophones used in seismic surveying.) Since the occurrence of earthquakes cannot be predicted (at present, and probably not for a long time to come), these instruments are left running continuously. The Earth is never completely calm or static, so a background noise of small, random, vibrations is always present on the instrument's graphical record (*seismogram*). As waves from an earthquake pass, then deflections are recorded. The amplitude (size) of these deflections depends on the strength of the earthquake and the distance of the epicenter from the seismograph (which may be many thousands of kilometers). If the epicenter is close to the instrument (within several hundred kilometers), then the depth to the focus is also important in affecting the amplitude, unless the focus is very close to the surface. In what follows it is assumed that the focus is at a shallow depth, so that the seismograph-focus distance is virtually the same as the seismograph-epicenter distance.

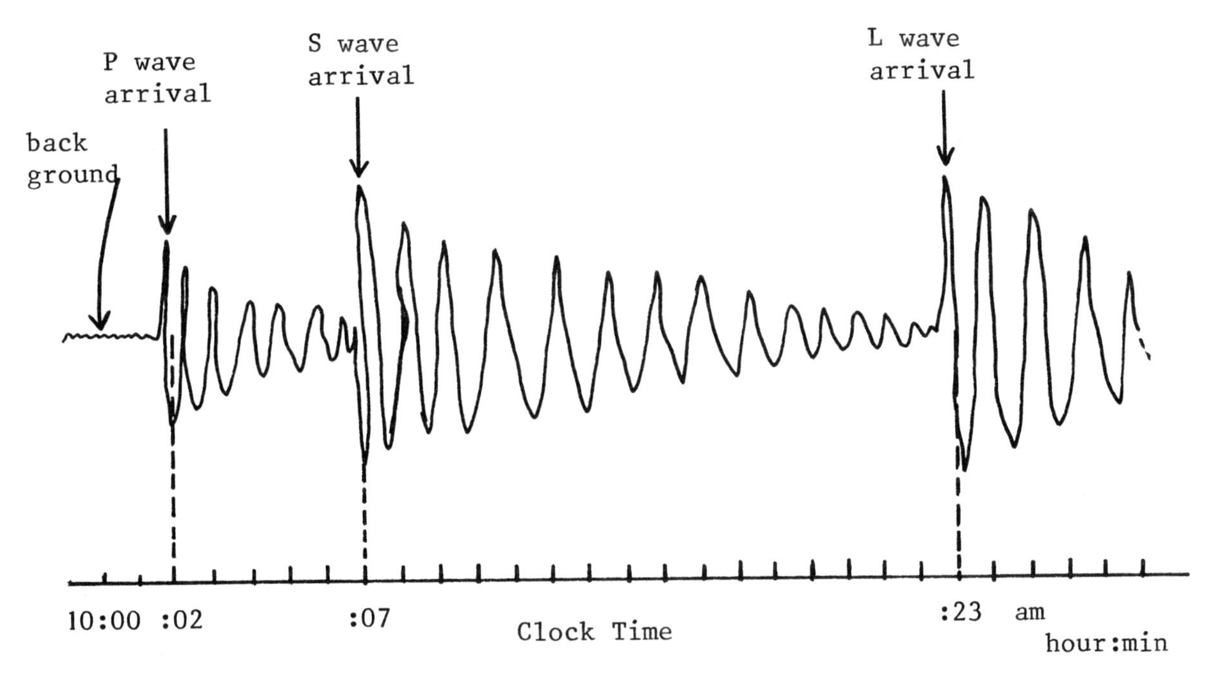

Fig.(10.2) Schematic Seismogram

Figure (10.2) shows a schematic seismogram of earthquake waves. The seismograph is connected to a clock, so the local (or the Greenwich) times of all events are recorded. On Fig.(10.2), before 10:02 am all is quiet; only background deflections from many microscopic rock movements are arriving. But then at 10:02 am, P waves from an earthquake begin arriving. At 10:07 am the first S waves arrive, and finally at 10:23 am the first L waves. At any time, the deflection shown is the net displacement due to all the waves that are passing at that time. For example, between 10:07 am and 10:23 am, the deflections are the sum of the P and S wave deflections; since the waves "die out" gradually after first arrival, most of these deflections are probably due to the S wave. L waves have a more complicated amplitude-time dependence.

The order of arrival shown (P, then S, then L) is always observed. In Section (9.1) it is revealed that, through any kind of rock, P waves travel faster than S waves. The first P and S waves are generated simultaneously, at the instant the earthquake begins. Hence, regardless of the different rocks encountered on their travel from focus to seismograph, the first P wave, moving faster, always arrives before the first S wave. The surface waves move slower than S waves, and cannot be generated until a body wave reaches the surface of the Earth. They arrive last.

10.2 Global Travel Times

Concentrate now on the body waves, i.e., the P and S waves. Sensitive seismographs started to become available early in the twentieth century, and "seismic observatories" housing these instruments eventually were established at many locations around the world (the seismic network). Over the succeeding decades, each observatory recorded the passage of seismic waves from hundreds of earthquakes. For many of these earthquakes (not all), the location of the epicenter and the time of occurrence soon became known. From the time of occurrence and the recorded times of arrival of the P and S waves at the observatory, the travel times of the P and S waves can be calculated by subtraction. From the latitude and longitude of the epicenter and of the observatory, the distance between the epicenter and observatory, measured on the surface of the Earth along the great circle arc connecting them, can be calculated. Over the years, then, each observatory accumulated values of the travel times for the P and S waves as a function of the epicenter-observatory distance. Two striking features emerged from this data.

First, at any given observatory, the travel times for the P waves, say, from all epicenters at the same distance are virtually equal. The same applies to the S waves. This may not seem extraordinary until it is realized that specifying just a distance only places the epicenter somewhere on the circumference of a circle with that distance as radius. Regions of very different local geology are likely to be on that circumference.

Second, all seismic observatories report the same travel times for the same epicenter-observatory distance. Again, this applies both to the P waves and to the S waves. This holds in spite of the fact that the different observatories themselves are in regions of different local geology.

Taken together, these two features mean that the travel time for P waves, and the travel time for S waves, depend only on the distance between the epicenter and the seismic observatory, not on their specific locations. Evidently, local geological conditions at the epicenter and the seismic observatory are not important in determining the travel times of the body waves over long distances. This characteristic of earthquake body wave propagation can be understood if it is supposed that, except near the surface of the Earth, the nature of the interior (composition, pressure, temperature, etc.) possesses "spherical symmetry" to a high degree; that is, although the properties vary with depth, they are virtually invariant for all points at the same depth below the surface of the Earth. Differences in near-surface features are unimportant in affecting body wave travel times over large distances, because these waves spend most of their time in traveling at great depth.

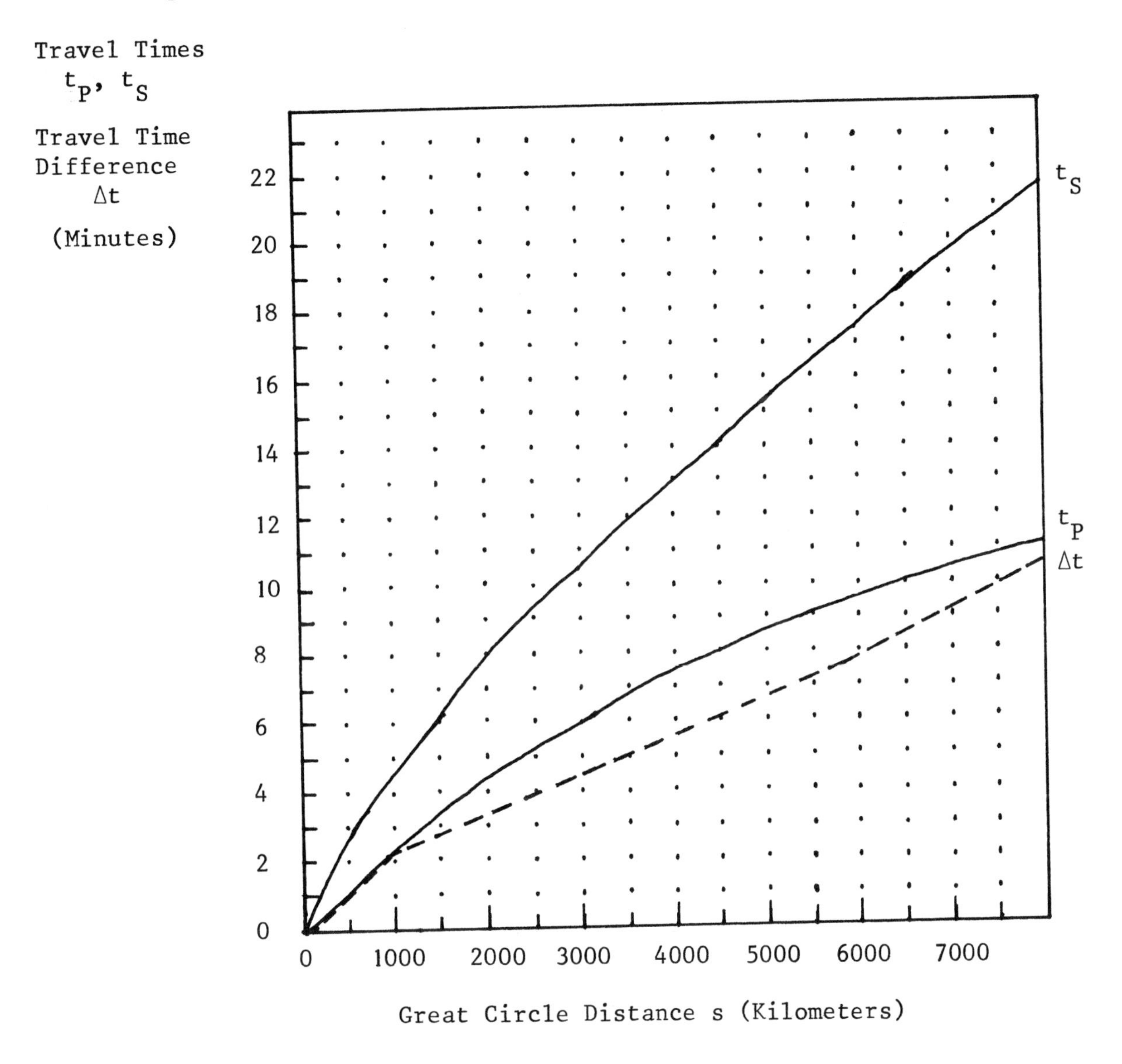

(after Rahn, P.H., 1996, p.439; Δt curve added.)

Fig.(10.3) Global Travel Times

Figure (10.3) shows the global P and S wave travel times graphically. The travel times, in minutes, are plotted on the ordinate, and the great circle distance in kilometers between epicenter and observatory is plotted on the abscissa. These curves apply globally, i.e., to all observing stations and epicenters. (The only restriction is that the observatory not be very close to the epicenter. Just what "very close" means depends on the depth of the focus;

most earthquakes occur at focal depths of less than 100 km.) The dashed line on Fig.(10.3) is discussed below.

In the formalism of one-dimensional kinematics, in which the motion of an object moving along a straight line (usually called the x axis) is analyzed, the instantaneous velocity v of the object is given by $v = dx/dt$. However, for seismic waves in which s is the great circle distance measured on the surface of the Earth and t the corresponding travel time, then it is not true that the speed of the wave is given by $v = ds/dt$. One reason for this is that waves that travel different great circle distances s from the same epicenter travel along geometrically and physically different paths; see Fig.(10.4). Also, even if the paths were straight lines and the Earth's interior strictly homogenous in its properties, the fact that s is measured on the Earth's surface, and not along the path of the waves, would invalidate the $v = ds/dt$ relation. All this means that the slope at any point on a travel time curve in Fig.(10.3) cannot be directly related to an instantaneous wave speed.

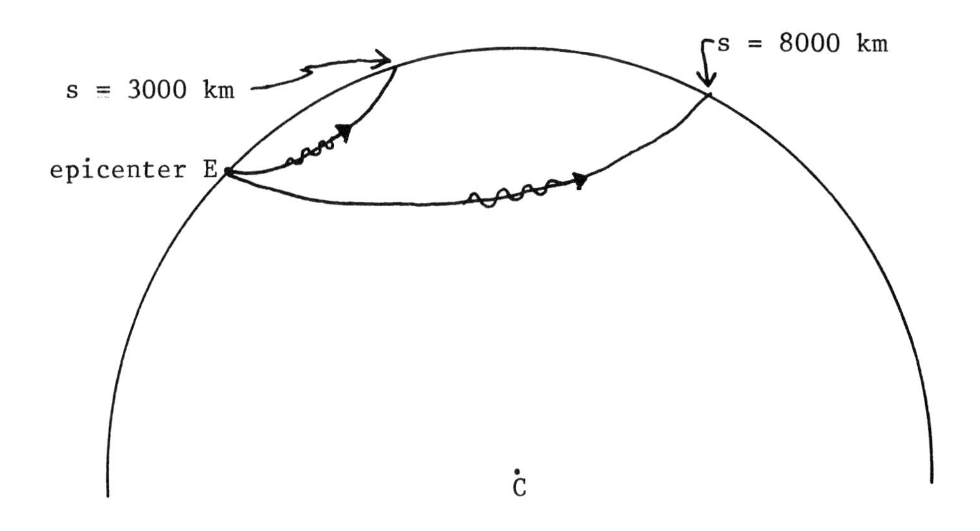

Fig.(10.4) Two Body Waves

The travel times in Fig.(10.3) are for waves, like those shown on Fig.(10.4), that travel from the epicenter to the seismograph without encountering any discontinuities that would generate reflections and refractions. [Of course, the waves really originate at the focus, but this is usually so close to the surface that the distinction between epicenter and focus can be ignored within the accuracy of the travel times shown in Fig.(10.3).] For P waves this kind of path is usually called the "P phase"; for S waves, the path is the "S phase".

Figure (10.5) shows the "PP phase", in which the P wave from epicenter E reaches the seismic observatory at O after a reflection at the Earth's surface. The surface of the Earth is a real discontinuity. The arrival of these waves can be seen on the seismogram recorded at O, and must be properly identified to avoid errors in reading the seismogram record. For example, the arriving PP phase must not be mistaken for the arriving S wave.

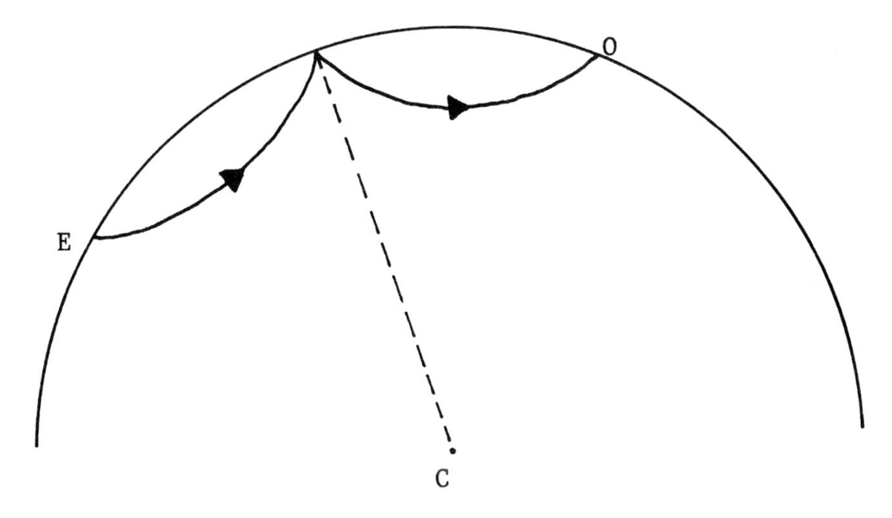

Fig.(10.5) The PP (or SS) Phase

****

EXAMPLE 1
Use the global travel time curve of Fig.(10.3) to calculate the travel time of the PP phase wave shown in Fig.(10.5). The epicenter to observatory great circle distance is 6000 km. What is the order in which the P, S, PP waves are received at O?

The Earth's surface is a rock discontinuity, so the reflection therefrom must obey Snell's law: the angle of incidence equals the angle of reflection. This means that the point at which the reflection takes place is midway between E and O. Therefore, the travel time t_{PP} of the PP phase wave for the distance s = 6000 km is given by

$$t_{PP}(6000 \text{ km}) = 2t_P(3000 \text{ km}),$$

$$t_{PP}(6000 \text{ km}) = 2(6 \text{ min}),$$

$$t_{PP}(6000 \text{ km}) = 12 \text{ min},$$

where the travel time t_P of the P wave for s = 3000 km is read from Fig.(10.3).
At s = 6000 km (the location of O), Fig.(10.3) indicates that $t_P \approx 9.5$ min and $t_S \approx 17.5$ min. Therefore, the order of arrival is P, PP, S.

****

10.3 Locating the Epicenter

Suppose that an earthquake (or a banned nuclear test, for that matter) takes place with the epicenter in a remote, little populated area, or under the ocean. How can the location of the epicenter, or test site, be identified from reading seismograms recording the passage of the waves generated by either of these events?

P and S waves are produced at the very instant that the earthquake begins. These are the first to arrive at a particular seismograph. If t_{CP} is the time shown on the seismograph clock at the instant the first P wave arrives, and t_{CS} the time on the seismograph clock when the first S wave arrives, then the travel time t_{P} of the first P wave and the travel time t_{S} of the first S wave, are given by

$$t_{\mathrm{P}} = t_{\mathrm{CP}} - t_{\mathrm{CE}}, \tag{10.1}$$

$$t_{\mathrm{S}} = t_{\mathrm{CS}} - t_{\mathrm{CE}}, \tag{10.2}$$

where t_{CE} is the time on the seismograph clock when the earthquake takes place. This time is, so far, unknown. Subtracting Eq.(10.1) from (10.2) gives

$$t_{\mathrm{S}} - t_{\mathrm{P}} = t_{\mathrm{CS}} - t_{\mathrm{CP}},$$

$$\Delta t = \Delta t_{\mathrm{C}}. \tag{10.3}$$

That is, the difference in arrival times of the P and S waves shown on the seismograph clock equals the difference in travel times. But the travel times t_{P} and t_{S} are shown on the global travel time curves of Fig.(10.3) as a function of epicenter to observatory distance. By subtracting the travel times shown by the solid lines the dashed line is obtained; this dashed line represents $\Delta t = t_{\mathrm{S}} - t_{\mathrm{P}}$. This difference in travel times dashed line is not a smooth curve, but closely forms a sequence of three straight line segments.

To find the distance s to the epicenter from a seismograph station, then, read t_{CS} and t_{CP} from the seismogram. By subtraction, obtain $\Delta t_{\mathrm{C}} = t_{\mathrm{CS}} - t_{\mathrm{CP}}$. This equals Δt, where $\Delta t = t_{\mathrm{S}} - t_{\mathrm{P}}$. Find the value of Δt on the time axis of Fig.(10.3) and read the corresponding distance s from the dashed line.

This puts the epicenter somewhere on the circumference of a circle with the distance s as the radius and the observatory at the center. The intersection of this circle with similar circles obtained from two other seismographic observatories locates the epicenter.

****

EXAMPLE 2

A seismic observatory receives P and S body waves from an earthquake as indicated on the seismogram of Fig.(10.6). (*a*) How far is the observatory from the epicenter? (*b*) When did the earthquake occur? (*c*) At what time did the first P wave reach a city 3700 km from the epicenter? (*d*) At what time did the first S wave reach a resort 6200 km from the epicenter? (*e*) The first S wave reached your home at 3:06 am. How far from the epicenter do you live? Assume that all times given, or asked for, are in Greenwich Mean Time (GMT), so that no time zone corrections need be made. Also, assume that only direct P and S wave phases are displayed on the seismogram. (That is, any PP phase, for example, has been filtered out.)

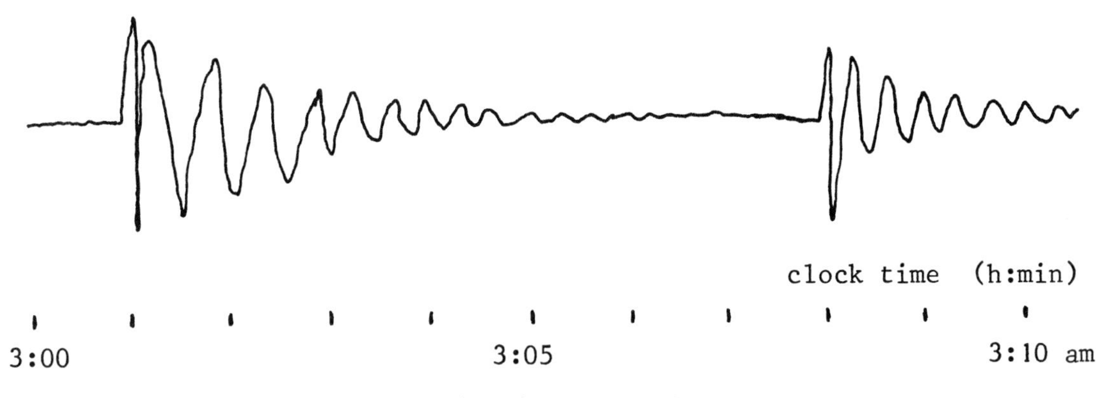

Fig.(10.6) Example 2

(*a*) From the seismogram, t_{CP} = 3:01 am and t_{CS} = 3:08 am, so that $\Delta t_C = 7$ min. By Eq.(10.3), $\Delta t = 7$ min also. On Fig.(10.3), 7 min on the time axis indicates a distance from the dashed line of $s = 5400$ km.
(*b*) Either the P or S wave data can be used to find when the earthquake took place. Using the P wave data, the seismogram gives t_{CP} = 3:01 am, and Fig.(3.10) for $s = 5400$ km yields a travel time of $t_P = 9$ min. By Eq.(10.1), the earthquake occurred at the time t_{CE} given by

$$t_{CE} = t_{CP} - t_P,$$

$$t_{CE} = 3:01 - 0:09,$$

$$t_{CE} = 2:52 \text{ am}.$$

The same result will be obtained using the S wave data.
(*c*) From Fig.(10.3), the travel time for P waves to a point $s = 3700$ km from the epicenter is $t_P = 7$ min. Since the earthquake occurred at t_{CE} = 2:52 am, the P waves will arrive at the city at time t_{CP} given by

$$t_{CP} = t_P + t_{CE},$$

$$t_{CP} = 0:07 + 2:52,$$

$$t_{CP} = 2:59 \text{ am}.$$

(*d*) By Fig.(10.3), S waves require 18 min to reach the resort. The earthquake took place at 2:52 am, so the first S wave reached the resort at $t_{CS} = t_S + t_{CE}$, t_{CS} = 0:18 + 2:52, t_{CS} = 3:10 am.
(*e*) When the first S wave reaches your home, it has been traveling for $t_S = t_{CS} - t_{CE}$, t_S = 3:06 − 2:52, t_S = 14 min. On Fig.(10.3), 14 min on the t_S curve indicates a distance of s = 4500 km from the epicenter.

****

10.4 Internal Structure of the Earth

Body waves generated from even only moderate earthquakes can travel through the Earth several times before dying out by dissipation of their energy as thermal energy imparted to the media through which they pass. This means that they can be used to map the interior of the Earth, somewhat like the use of X rays to image the internal structure of the human body.

For example, consider one observational feature of S wave propagation in the Earth: beyond an angular distance of 105° from an epicenter, the angle measured at the center of the Earth, no direct S waves are seen. The region in which no S waves are seen is called the S wave *shadow zone.*

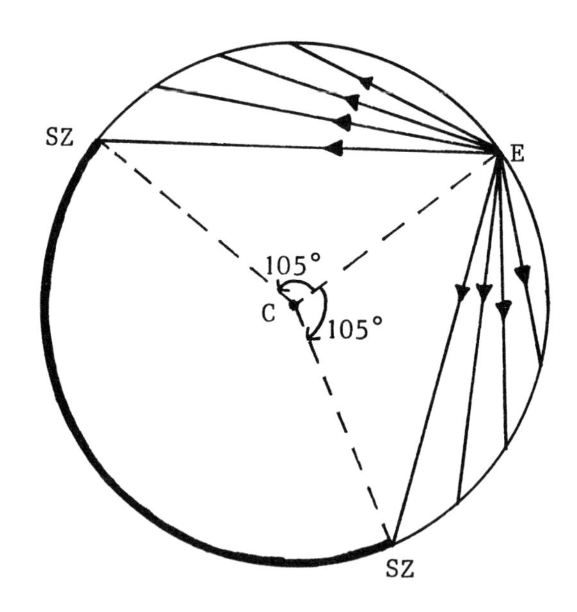

Fig.(10.7) Direct S Wave Shadow Zone

Figure (10.7) shows the shadow zone for an epicenter located at E. In drawing Fig.(10.7), the geometry has been simplified in that it is pretended that the S waves travel in straight lines, avoiding the complexity of dealing with the actual curved paths. This simplification does not materially affect the conclusions drawn about the internal structure of the Earth implied by the existence of the shadow zone. As suggested in Fig.(10.7), no S waves from epicenter E reach the indicated region between the points marked SZ. (S waves can reach this region from earthquakes with epicenters in different locations.) It should be remembered that Fig.(10.7), and similar diagrams, are two-dimensional representations of the three-dimensional Earth; the actual shadow zone is an area, not a length. Diffraction effects are ignored.

For a direct S wave from E to reach a point on the Earth's surface within the shadow zone, it would have to pass closer to the center of the Earth C than does the wave that travels from E to one of the points labelled SZ that is at the edge of the shadow zone. But suppose that below this greatest depth reached by a wave from E to SZ the Earth is liquid, not solid. Recall from Chapter 9 that S waves cannot pass through a liquid, and the existence of the shadow zone is thereby explained.

An S wave shadow zone is observed, starting at 105° from E, regardless of the location of the epicenter E around the Earth. This implies that the outer boundary of the liquid region is spherical and centered on the center of the Earth. From the presence of the shadow zone, nothing can be said about the inner boundary (if any) of the liquid region; the liquid region need not extend to the center of the Earth to explain the suppression of the S waves. In fact, apparently it does not extend to the center. There is a solid, spherical, region centered at the center of the Earth, within the liquid. By virtue of this geometry, the liquid region is called the *outer core*, and the solid region the *inner core* of the Earth, the two components making up the *core* of the Earth.

The radius of the core can be calculated, since it is also the radius of the outer liquid core. On Fig.(10.8) the radius of the core is R_{C} and the radius of the Earth is R. The inner core is not shown. The angle θ is the 105° shown on Fig.(10.7). The line marked R_{C} is perpendicular to the wave from E to SZ, since this wave is tangent to the spherical outer boundary of the core. Therefore

$$\cos\frac{\theta}{2} = \frac{R_{\mathrm{C}}}{R},$$

$$R_{\mathrm{C}} = R\cos\frac{\theta}{2}, \tag{10.4}$$

$$R_{\mathrm{C}} = (6370\ \mathrm{km})\cos 52.5^\circ,$$

$$R_{\mathrm{C}} = 3880\ \mathrm{km}.$$

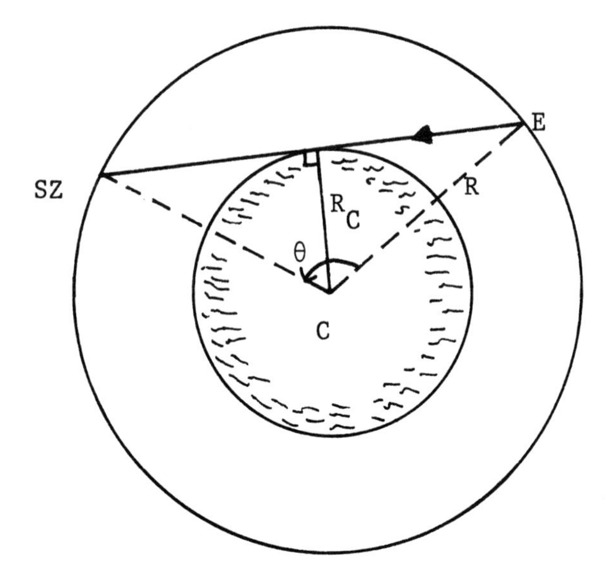

Fig.(10.8) The Outer Core

The actual value of the core radius, found by taking account of the real curvature of the S wave paths, is $R_C = 3480$ km. The discontinuity, at the outer boundary of the liquid outer core, is called the *Gutenberg discontinuity.*

P waves can pass through a liquid. Nevertheless, P waves feel the presence of the liquid core since, at the Gutenberg discontinuity, they must undergo reflection and refraction. Curiously, the result is the creation of a "shadow band" for P waves: P waves are not recorded at angles between 105° and 140° from the epicenter, but are recorded at all other locations. For example, P waves are recorded at 180°, the point on the Earth's surface directly opposite the epicenter. This 180° wave travels in a straight line along the diameter of the Earth that passes through the epicenter. (No S waves are seen at 180°.) See Fig.(10.9). Diffraction effects are ignored.

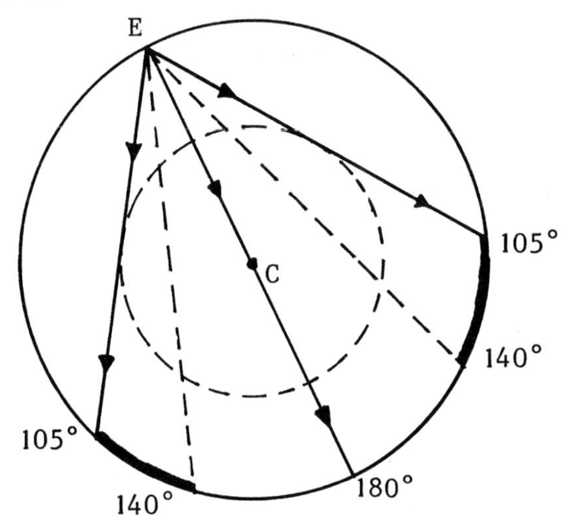

Fig.(10.9) The P Wave Shadow Band

The 140° line from E to the inner edge of the P wave shadow band is shown as a dashed line in Fig.(10.9) because that is not the path of the P wave. This wave undergoes refraction as it enters and leaves the core so that, even ignoring the curvature of the wave paths, its path is a sequence of three non-colinear straight lines. The 105° wave path is tangent to the outer core.

The methods of surveying described in Chapter 9 (Seismic Surveying) have been used to detect and locate other discontinuities in the Earth. The reflection method, using waves produced by nuclear weapons undergoing testing as the explosive, confirmed the presence of a solid core centered at the center of the Earth and inside the liquid outer core. Timing the P wave echoes reflected from the solid inner core yielded its radius as 1220 km. (The existence of the solid inner core had been suggested earlier by Lehmann to explain the presence of very weak P waves in the P wave shadow band.) The discontinuity at the outer surface of the solid inner core is named the *Lehmann discontinuity.*

The refraction method showed the existence of a very shallow discontinuity. Unlike the two discontinuities discussed above in association with the core region, this shallow discontinuity is between two solid rock layers. This *Mohorovičić discontinuity* is found at a depth that varies between about 5 km below the ocean floor and 35 km beneath the surface in some continental areas; the average depth is about 20 km.

Figure (10.10) summarizes pictorially the description of the Earth's interior outlined above. This picture is simplified in that not all of the discovered discontinuities are shown. Also, the actual small departures from sphericity are ignored: for example, the Mohorovičić discontinuity is shown as having a uniform depth, equal to its average depth, rather than with the actual distribution of depths. Nontheless, Fig.(10.10) adequately portrays the main features of the Earth's interior.

The adoption of a spherically symmetric picture as being sufficiently accurate (circularly symmetric on a two-dimensional cross section) allows for the drawing of a wedge-shaped section as being representative of the entire interior. The diagram is not entirely to scale, especially in that the Mohorovičić discontinuity appears to be deeper than it actually is. The center of the Earth is at the apex of the wedge. Distances from the center of the Earth are on the left of the wedge and depths below the surface on the right.

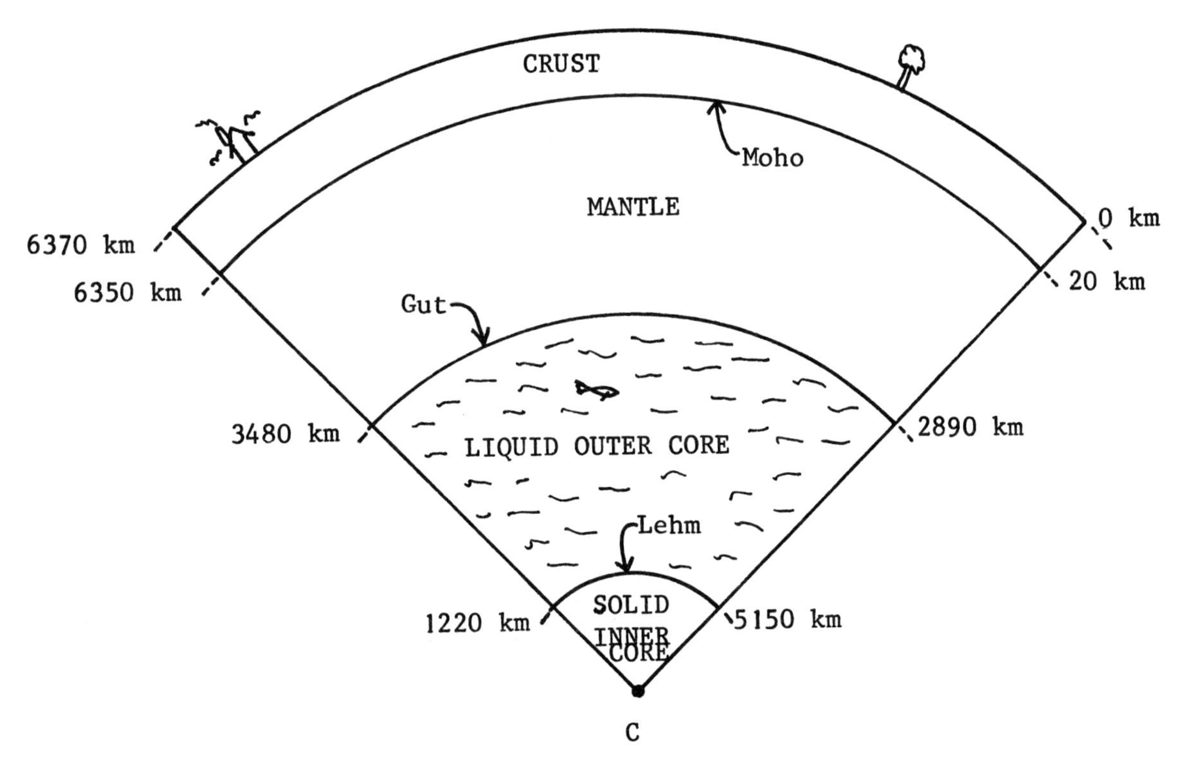

Fig.(10.10) Interior of the Earth

The Earth's solid surface forms the outer surface of the *crust* of the Earth; the lower surface of the crust is the Mohorovičić discontinuity. The average thickness of the crust, then, is about 20 km. Many of the rocks making up the crust are directly accessible and can be studied; their average density is about 2.80 g/cm^3.

The region between the Mohorovičić and Gutenberg discontinuities is called the *mantle.* The vents of some volcanoes reach down into the upper mantle; the density of the rocks that such volcanoes eject during eruptions is about 4.64 g/cm^3. Although the density of the upper mantle is probably less than the density of the lower mantle, the value of 4.64 g/cm^3 is probably fairly close to the average density of the mantle. The greater density of the mantle compared to the crust is due, in part, to the presence of proportionally more of the heavier elements in the mantle rocks compared with the crustal rocks.

Beneath the mantle is the core which, as discussed above, has a liquid outer region and a solid inner part.

****

EXAMPLE 3

Using the dimensions shown on Fig.(10.10), and the values of 2.80 g/cm^3 and 4.64 g/cm^3 for the average densities of the crust and mantle, respectively, find (*a*) the average density of the entire core (i.e., the region within the Gutenberg discontinuity), and (*b*) the average density of the Earth. The mass of the Earth is M = 5.98 X 10^{24} kg and its radius is R = 6370 km.

(*a*) The average density ρ_{core} of the core is given by $\rho_{\text{core}} = M_{\text{core}}/V_{\text{core}}$. The volume V_{core} of the core can be found from the dimensions shown on Fig.(10.10). The masses of the core, mantle, and crust must add to the mass of the Earth, so the mass of the core M_{core} can be obtained by subtraction. The crust occupies a region between two spheres, the outer sphere being the surface of the Earth and the inner sphere being the Mohorovičić discontinuity. Therefore

$$M_{\text{crust}} = \rho_{\text{crust}} V_{\text{crust}},$$

$$M_{\text{crust}} = \rho_{\text{crust}} (\frac{4\pi}{3}) [R^3 - R^3_{\text{Moho}}],$$

$$M_{\text{crust}} = (2800 \text{ kg/m}^3)(\frac{4\pi}{3})[(6370 \text{ X } 10^3 \text{ m})^3 - (6350 \text{ X } 10^3 \text{ m})^3],$$

$$M_{\text{crust}} = 2.847 \text{ X } 10^{22} \text{ kg}.$$

Similarly, the mantle is the region between two spheres, the outer sphere being the Mohorovičić discontinuity and the inner sphere being the Gutenberg discontinuity. Hence,

$$M_{\text{mantle}} = \rho_{\text{mantle}} V_{\text{mantle}},$$

$$M_{\text{mantle}} = \rho_{\text{mantle}} (\frac{4\pi}{3}) [R^3_{\text{Moho}} - R^3_{\text{Gut}}],$$

$$M_{\text{mantle}} = (4640 \text{ kg/m}^3)(\frac{4\pi}{3})[(6350 \text{ X } 10^3 \text{ m})^3 - (3480 \text{ X } 10^3 \text{ m})^3],$$

$$M_{\text{mantle}} = 4.157 \text{ X } 10^{24} \text{ kg}.$$

Subtracting the masses of the mantle and crust from the mass of the Earth gives for the mass of the core $M_{\text{core}} = 1.795 \text{ X } 10^{24}$ kg. The complete core is a sphere bounded by the Gutenberg discontinuity, so that

$$\rho_{\text{core}} = \frac{M_{\text{core}}}{V_{\text{core}}},$$

$$\rho_{\text{core}} = \frac{M_{\text{core}}}{\frac{4\pi}{3} R_{\text{Gut}}^3},$$

$$\rho_{\text{core}} = \frac{1.795 \text{ X } 10^{24} \text{ kg}}{\frac{4\pi}{3}(3480 \text{ X } 10^3 \text{ m})^3},$$

$$\rho_{\text{core}} = 1.017 \text{ X } 10^4 \text{ kg/m}^3,$$

$$\rho_{\text{core}} = 10.2 \text{ g/cm}^3.$$

It is likely that the density of the outer liquid core is a little less than this, and the density of the solid inner core a little greater.

(*b*) The average density ρ of the Earth as a whole (its bulk density) is

$$\rho = \frac{M}{V},$$

$$\rho = \frac{M}{\frac{4\pi}{3} R^3},$$

$$\rho = \frac{5.98 \text{ X } 10^{24} \text{ kg}}{\frac{4\pi}{3}(6370 \text{ X } 10^3 \text{ m})^3},$$

$$\rho = 5.52 \text{ g/cm}^3.$$

Note that

$$\rho \neq \frac{1}{3}[\rho_{\text{crust}} + \rho_{\text{mantle}} + \rho_{\text{core}}],$$

$$\rho \neq \frac{1}{3}[2.80 \text{ g/cm}^3 + 4.64 \text{ g/cm}^3 + 10.2 \text{ g/cm}^3],$$

$$\rho \neq 5.88 \text{ g/cm}^3,$$

because the crust, mantle, and core do not occupy equal volumes.

****

10.5 Earthquake Energy and Magnitude

The seismic P and S waves radiate from the focus of the earthquake. It is not possible at present to describe just what happens at the focus to produce these waves; i.e., the precise mechanism that is an earthquake is not yet understood. It seems clear that an earthquake involves the release, over a very short period of time (seconds, perhaps a few minutes) of potential energy stored in the rocks at the focus. This energy accumulated over long periods of time as the rocks deformed under stress. An applied stress means that forces are exerted on the rock; the rock deforms, and this deformation involves displacement (strain). The energy stored is the work (force times displacement) done in deforming the rock. The ultimate origin of the stress is probably global ("tectonic") in character.

Whatever the details of its origin, the total amount of energy released in an earthquake (its *strength*) is one of the important parameters that determines whether or not a specific site on the Earth's surface will see significant damage to the structures built thereon. Other parameters affecting this issue are the depth of the focus, the distance to the epicenter, and the local geological conditions at the site.

It has been shown how a seismogram can be used to find the distance to the epicenter by noting the times of arrival of the P and S waves. But another important feature of a seismogram consists in the values of the largest recorded deflections of the P, S, and surface waves: the *amplitudes* of the waves.

The power (rate of energy transport) carried by a mechanical wave is proportional to the square of the amplitude. By measuring the amplitude, the rate at which energy is delivered by the wave to the site of the seismograph can be calculated (other properties of the wave have to be measured). To find the energy released by the earthquake, two additional factors must be taken into account: dilution of the waves through geometrical spreading and the absorption of energy by the rocks through which the waves pass on their journey from the focus. The distance and the shape of the wavefront (only approximately spherical) are needed to calculate the dilution; the absorption of energy can only be estimated, for the detailed properties of deep rocks are not known to great precision.

There are other complications in obtaining a value of the earthquake energy release. For example, it is not clear which wave amplitude best represents the energy produced. It is probable that different wave amplitudes best represent earthquakes of different strengths. In fact, it appears that amplitudes alone are not sufficient: detailed knowledge of the earthquake mechanism may be necessary.

All this means is that, at present, only an approximate value of the energy liberated in a particular earthquake can be found. However, it is clear from the analysis of many earthquakes that the range of energy released is enormous, so much so that a logarithmic scale is usually used to express this energy (much like the use of the logarithmic decibel scale to represent acoustic wave intensities). Several different scales are in use, depending on the particular wave amplitude employed. Also, as energy estimates improve over time, the various energy scale equations are appropriately modified.

One very commonly reported scale is the *Richter scale.* The strength of the earthquake is represented by a dimensionless number M, called the *magnitude*; the energy E in joules released by the earthquake is given from the magnitude by the equation

$$E = (25.1 \text{ kJ})(31.6)^M. \tag{10.5}$$

For positive integral values of M up to $M = 5$, the energies calculated from Eq.(10.5) are listed in Table(10.1).

M	E
0	25.1 kJ
1	793 kJ
2	25.1 MJ
3	792 MJ
4	25.0 GJ
5	791 GJ

Table (10.1) Energy Released in Small Earthquakes

Notice that the energy increases very rapidly with magnitude. Between $M = 0$ and $M = 4$ the energy increases by a factor of $1 \text{ X } 10^6$.

The joule is a very small unit of energy as far as earthquakes are concerned. Sometimes the energy is expressed instead in terms of the number of megatons (1 Mton = $1 \text{ X } 10^6$ tons) of TNT that give the same energy release. The equivalence is

$$1 \text{ Mton of TNT} = 4.18 \text{ X } 10^6 \text{ GJ}.$$

This is the same energy unit used to express the yield of nuclear weapons, but an earthquake should not be considered as strictly analogous to the detonation of a nuclear bomb in the Earth. Some differences are: in earthquakes, temperatures do not reach millions of degrees and no fissionable material is involved; earthquakes take several seconds, or even minutes, to transpire, while a nuclear device requires only a few microseconds; a nuclear explosion is primarily "radial" in nature, resulting in the production mainly of P waves, whereas an earthquake excites both P and S waves.

The energy that is released in an earthquake is the potential energy of deformation accumulated by, and stored in, the involved rocks. There is an upper limit to the amount of energy that can be so stored, since rock will rupture if the applied stress is too large. This means that there is an upper limit to the Richter magnitude scale; this maximum possible magnitude seems to be about $M = 9$.

Equation (10.5) can be inverted to yield the magnitude corresponding to a given energy. Taking the base-10 logarithm of both sides of Eq.(10.5) yields

$$\log E = \log(25.1) + M \log(31.6),$$

$$\log E = 1.40 + 1.50M. \tag{10.6}$$

In Eq.(10.6), the energy E must be expressed in kJ.

****

EXAMPLE 4
Find the magnitude of an earthquake that releases an amount of energy equivalent to 1.5 megatons of TNT.

The energy released in kJ is

$$E = (1.5 \text{ Mtons})(4.18 \text{ X } 10^6 \text{ GJ/Mton}),$$

$$E = 6.27 \text{ X } 10^6 \text{ GJ},$$

$$E = 6.27 \text{ X } 10^{12} \text{ kJ}.$$

By Eq.(10.6),

$$\log E = 1.40 + 1.50M,$$

$$\log(6.27 \text{ X } 10^{12}) = 1.40 + 1.50M,$$

$$12.8 = 1.40 + 1.50M,$$

$$M = 7.6.$$

No attempt is made to trace through the units, since they are suppressed in Eq.(10.6). This is often the case with logarithmic (exponential) equations.

****

10.6 Structural Damage

Nearly all of the structural damage done by an earthquake is a result of the ground shaking associated with the passage of the seismic waves generated by the earthquake. The body waves and the surface waves are implicated in this shaking. Although assigning the damage to structures, buildings and freeways for example, to ground shaking seems obvious, it is important to discover the specific features of the shaking that are most involved; armed with such knowledge, construction techniques can be devised to render a building, say, safe against collapse, safeguarding its occupants, even if the building suffers damage severe enough to require eventual reconstruction.

Now a seismogram is a graph of ground displacement versus time at the site of the seismograph. If x is the displacement, then the second derivative of x with respect to time t yields the ground acceleration a:

$$a = \frac{d^2x}{dt^2}.$$

The ground acceleration is important because the force F that will act on the structure if it follows the ground is given by $F = Ma$, where M is the structure's mass. P waves, being longitudinal, cause ground acceleration along the path of travel of the wave, but the body S waves and the surface waves provide transverse, horizontal and vertical, displacements. Hence, the actual ground displacement is three-dimensional in nature, with a vertical and two horizontal components.

Extensive damage is not normally caused by the vertical component of the ground acceleration. The reason is that buildings must be self-supporting against gravity. The effect of a vertical ground acceleration is to increase (acceleration up) or decrease (acceleration down) the effective weight of the material located above a supporting member. For example, consider a single pillar supporting a structural component (block) of weight W. If the ground suffers an upward vertical acceleration a, the pillar must exert a normal force R on the block, given from Newton's second law:

$$F = Ma,$$

$$R - W = (\frac{W}{g})a,$$

$$R = W(1 + \frac{a}{g}). \tag{10.7}$$

By Newton's third law, the block exerts an equal and opposite force R on the pillar. Hence, R is the effective weight of the block on the pillar.

An earthquake-induced vertical ground acceleration of $0.20g$ would be quite large. In this case, $R = 1.20W$; i.e., the effective weight to be supported is 20% greater than the normal load. The compressive strength of building materials, and the tolerances normally allowed for in building design, usually permits for this extra load to be safely supported.

In contrast with the vertical situation, the usual external horizontal load on a building is near zero. A horizontal ground acceleration of $0.20g$ can be very destructive since buildings not seismically-engineered are not specifically designed to support such horizontal loads. (Gravity acts vertically.) Even if the structure "holds together", it could slide or tip.

Consider a building that simply sits on the ground (a house with no foundation, or a gravity dam, for example). If the ground undergoes a horizontal acceleration a, then in order for the building to remain at rest relative to the ground, a static friction force f_s must act, so that $f_s = Ma$, M the mass of the building. But, the force of static friction has as its maximum possible value $f_{s,max} = \mu_s R$. On level ground $R = Mg$. Therefore, with a coefficient of static friction μ_s between ground and building, the maximum horizontal ground acceleration that can be supported is

$$f_{s,max} = Ma,$$

$$\mu_s Mg = Ma,$$

$$a = \mu_s g. \tag{10.8}$$

Hence, for a horizontal ground acceleration of $0.2g$, a coefficient of friction $\mu_s > 0.2$ will prevent sliding, with its attendant problems of broken gas and water pipes, etc. Ordinarily, dry ground will provide this required coefficient of static friction. But the passage of seismic waves can drastically alter the character of certain ground materials resulting in, among other effects, a substantial reduction in the value of the coefficient of friction. Also, horizontal ground accelerations of about $1g$ have been reported in some earthquakes, putting very severe requirements on ground friction to prevent sliding.

The nature of the ground material on which the structures rest can substantially affect the amplitude of the resulting ground vibrations. Loose, unconsolidated, material tends to amplify the wave displacements. Also, the formation of standing waves by reflection also increases the vibration amplitudes in regions near the antinodes, just as for standing waves on a string, or sound waves in an air column. If the frequency of the seismic waves is close to the natural oscillation frequency of a building, the amplitude of the building's shaking can be increased by resonance. Finally, it is clear that another important parameter in determining the amount of structural damage is simply the time-duration of the shaking due to the passing waves; this time can vary from only a few seconds to three or four minutes in severe earthquakes.

10.7 Problems

1. An earthquake takes place at 1:16 pm. The first S wave reaches your home at 1:32 pm. When does the first P wave reach your home?

2. From the seismogram record of an earthquake, it is concluded that the distance to the epicenter is 7500 km. Find the length of the circumference of the circle, centered on the seismograph, that passes through the epicenter.

3. A seismic observatory records P and S waves from an earthquake as shown on Fig.(10.11). (*a*) Find the distance between the seismic observatory and the epicenter. (*b*) When did the earthquake occur? (*c*) At what time did the first S wave reach a city 7700 km from the epicenter?

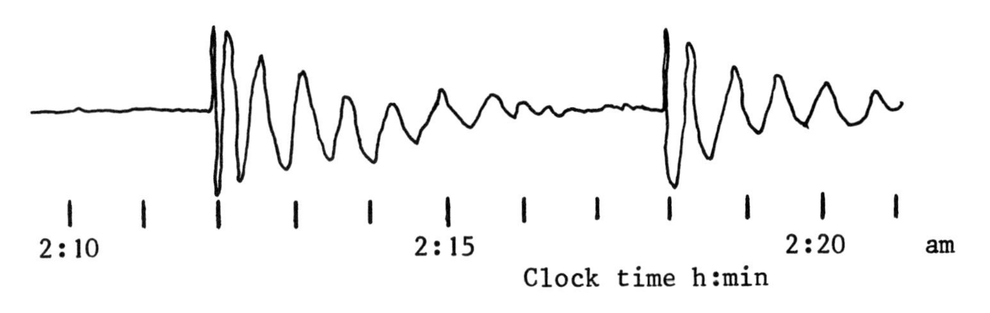

Fig.(10.11) Problem 3

4. A seismic station received the first P wave from an earthquake at 4:02 pm; the first S wave arrived 10 minutes later. At what time did the first P wave reach a resort 3600 km from the epicenter?

5. On a recently discovered satellite, astronauts note that no S waves are received beyond an angle of 120° from the shot point in a seismic exploration survey. Show that the volume of the presumably wholly liquid core is one-eighth the total volume of the satellite.

6. (*a*) Show that the average speed $\overline{v}$ of the seismic body wave that travels a great circle distance s in time t is given by

$$\overline{v} = (\frac{2R}{t})\sin(\frac{s}{2R}),$$

where R is the radius of the Earth. Assume that the focus of the earthquake is very close to the surface of the Earth. (*b*) Suppose that the waves travel along straight line paths; show that the greatest depth h reached by such a wave is

$$h = R[1 - \cos(\frac{s}{2R})].$$

(c) For P waves, use values of s and t from Fig.(10.3) to calculate the average speed and greatest depth. Graph the results to obtain an estimate of the variation of wave speed with depth in the mantle. Repeat for S waves.

7. Find the area of the S wave shadow zone, in terms of the area A of the Earth.

8. Calculate the travel times for the PP phase for enough distances to draw a PP travel time curve on Fig.(10.3).

9. Find the travel time for the SSS phase wave (two reflections from the Earth's surface) that reaches a seismic observatory directly opposite the epicenter on the other side of the Earth. See Fig.(10.12).

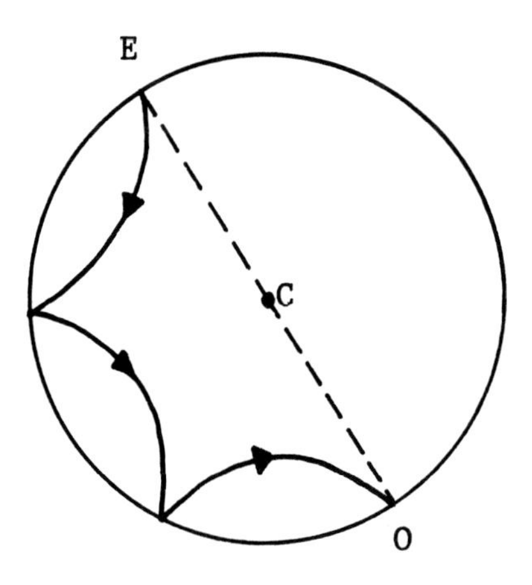

Fig.(10.12) Problem 9

10. Find the area of the P wave shadow band, in terms of the area A of the Earth.

11. At the bottom of the mantle, the P wave speed is about 13 km/s and the S wave speed about 7.2 km/s. Assuming that the mantle material at that location behaves like an elastic solid, find the value of Poisson's ratio implied by the wave speed data.

12. Suppose that the internal structure of the Earth can be approximated as follows: no crust; a homogenous solid mantle; a homogenous wholly-liquid core of radius R_C. Also suppose that, as in the actual Earth, the P wave speed v_1 in the mantle at the core-mantle interface is greater than the P wave speed v_2 in the core at the same interface. In the model of the Earth in this problem, these wave speeds are constant throughout their respective regions. Hence, the waves travel in straight line paths in the mantle and in the core. Show

angular distance of travel Δ of a P wave from an epicenter very near the Earth's surface that refracts through the core to reach the surface again, is given by

$$\Delta = 180° + 2\left[i - \sin^{-1}\left(\frac{R_C}{R}\sin i\right) - \sin^{-1}\left(\frac{v_2}{v_1}\sin i\right)\right],$$

where i is the angle of incidence at the core-mantle interface. Make a graph of Δ vs. i; is there a P wave shadow band? See Fig.(10.13). Use $v_1 = 13$ km/s, $v_2 = 8.0$ km/s.

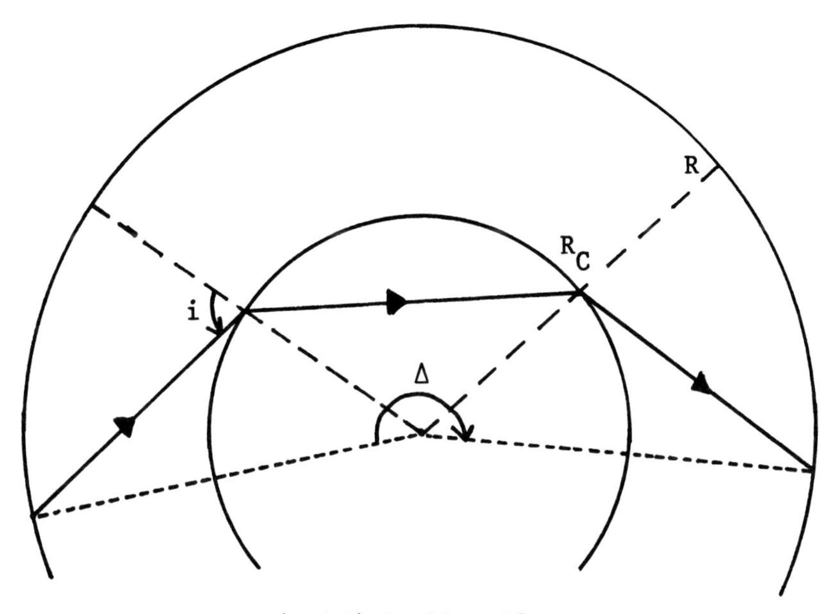

Fig.(10.13) Problem 12

13. Suppose that the internal structure of a planet is as shown on Fig.(10.14). Find the average density of the planet.

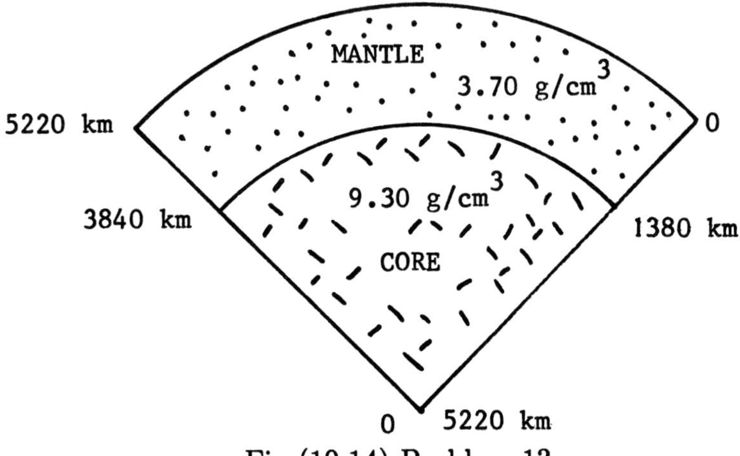

Fig.(10.14) Problem 13

14. A hypothetical planet has the internal structure shown on Fig.(10.15). The average density of the planet is 3.22 g/cm^3. Find the density of the core.

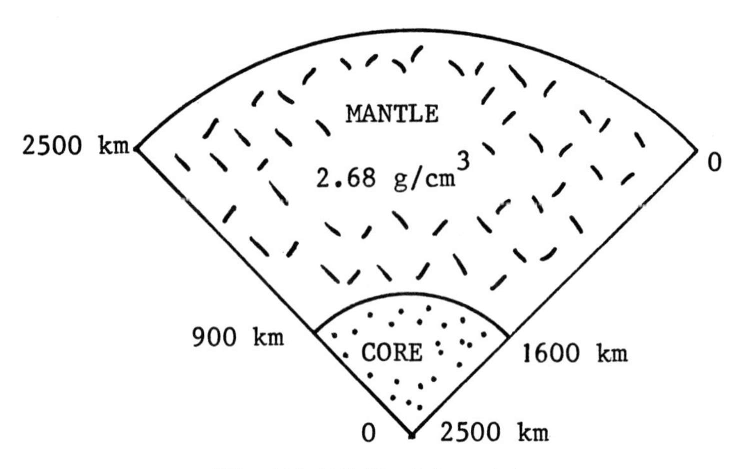

Fig.(10.15) Problem 14

15. Planet Strange has just been discovered (maybe). It has a radius of 7600 km. There is no crust. The mantle has a density of 2700 kg/m^3. The core has a radius of 4400 km and is empty! Find the mass of planet Strange.

16. The density of the solid inner core of the earth is thought to be 13.5 g/cm^3. Find the density of the liquid outer core. (Refer to Example 3.)

17. Find the magnitude of an earthquake with an energy yield of 20 kilotons of TNT (about the yield of the World War II atomic bombs).

18. Show that the energy release E in megatons of TNT is related to the magnitude M of the earthquake by

$$E = (6.00 \text{ X } 10^{-12} \text{ Mton})31.6^M.$$

19. (*a*) How much energy, in megatons of TNT, was released in the 1989 Loma Prieta earthquake, which had a magnitude of 7.1? (*b*) What is the magnitude of an earthquake that releases an energy equivalent to 2.0 megatons of TNT? (*c*) The 1985 Mexico City earthquake had a magnitude of 8.1 (about what is anticipated for the "big one" expected anytime in California). Find the energy release, in megatons of TNT.

20. Find the magnitude of an earthquake with an energy release that is 190 times that of a magnitude 6.50 earthquake.

21. Suppose that a seismic wave can be approximated as the sinusoidal wave

$$x = A\sin(\frac{2\pi t}{T}),$$

where x is the ground displacement, A the amplitude of the wave, and T its period. Find the amplitude of such a wave that will give a maximum horizontal ground acceleration of $0.75g$ if the period of the wave is 1.20 s.

22. Because of a nearby earthquake, an urban location experiences the passage of seismic waves for 1.50 min. The waves have wavelength 5.40 km and wave speed 3.72 km/s. How many time does a building in the area undergo the maximum ground acceleration due to the waves? (Ignore diminution of amplitude with time and any possible resonance effects.)

23. A freeway pillar 47.0 cm in diameter is responsible for supporting a section of freeway 5.60 m wide and 24.5 m long. The roadbed is 9.20 cm thick and is made of concrete with density 2.38 g/cm^3. Find the effective stress on the top of the pillar if, due to an earthquake, the ground undergoes an upward acceleration of $0.160g$.

Chapter 11

Gravity Surveying

11.1 Newton's Law of Gravitation

The seismic surveying techniques presented in Chapter 9 make possible the determination of the P and S wave speeds in a subsurface rock layer. As described in Section (9.2), knowledge of the values of these wave speeds, along with that of the density of the rock, allows for calculation of the elastic moduli. The elastic moduli, in turn, permit evaluation of the rock deformation to be expected under a specified applied load. It is the aim of gravity surveying to supply information on the density of rock layers and formations.

By *gravity* is meant the force of attraction that exists between any two objects. The value, or strength, of the force depends on the each object's size, shape, and internal distribution of matter; it also depends on the distance between the objects. Simple algebraic formulas for the strength of the force exist only for the cases where both (finite) objects have one of two very basic shapes. The first case, and the one for which Newton originally wrote down the law of gravitation, is that in which both objects are "points"; i.e., their dimensions are so very small compared to the distance between them that, in any diagram drawn to scale, the objects appear as geometrical points.

In this event, Newton's *Law of Universal Gravitation* ("universal" because it applies to all such pairs of objects, no matter where in the universe they be) states that the magnitude of the force is given by

$$F = \frac{GMm}{r^2}. \qquad (11.1)$$

In Eq.(11.1), F is the magnitude of the force, M and m are the masses of the two objects, and r is the distance between them. The symbol G stands for the "universal gravitational constant", the value of which depends on the units used; however, in a given set of units, G has the same value everywhere in the universe (presumably).

In the SI base units, F has units of newtons (N), M and m have units of kilograms (kg), and r must be in meters (m). The corresponding units of G can be found from Eq.(11.1). The value of G is found by experiment (quite difficult experiments), and, to three significant figures, is

$$G = 6.67 \text{ X } 10^{-11} \text{ N} \cdot \text{m}^2/\text{kg}^2.$$

The gravitational force, which is always attractive, acts with the same magnitude on each object and is directed along the line defined by their positions, the force on each object pointing toward the other object; see Fig.(11.1).

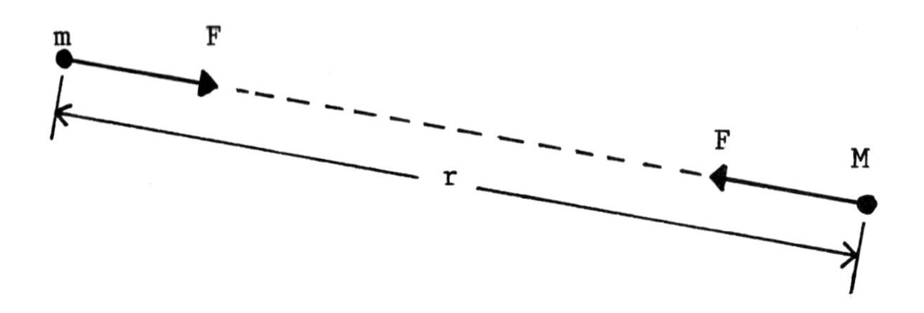

Fig.(11.1) Gravity Between Point Objects

The other object shape for which the strength of the force can be easily calculated is that of a sphere within which the density of matter is a function only of distance to the center of the sphere (a spherically symmetric distribution of matter). Specifically, there is a simple formula for the force of attraction between two such spheres, and between one such sphere and a point object. It is this latter configuration that is of immediate interest.

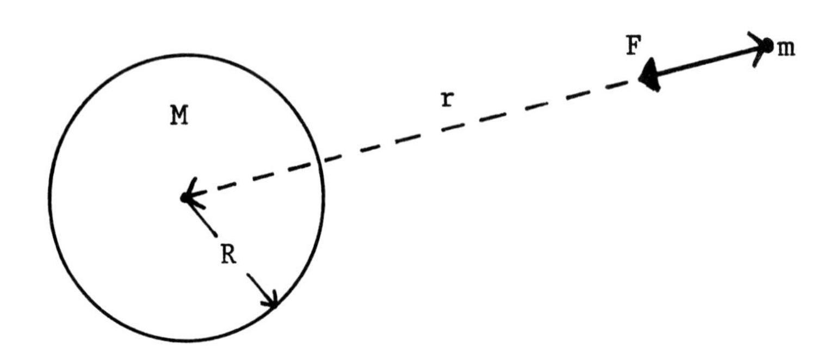

Fig.(11.2) Sphere and Point Objects

Figure (11.2) shows the two objects, a sphere of total mass M and radius R, and a point object of mass m. (The choice of symbols for the masses is not meant to convey the impression that $M > m$ necessarily.) The distance r between the objects is measured between the point object and the center of the sphere. By a remarkable fortuity, it turns out that the formula for the gravitational force F on the point object looks just like Eq.(11.1) for the force between two points; i.e.,

$$F = \frac{GMm}{r^2}, \tag{11.2}$$

except that r has the meaning given to it in Fig.(11.2) and stated above. The force is directed toward the center of the sphere. There is an equal, but oppositely-directed, force on the sphere itself (as required by Newton's third law), but this force on the sphere will not be of concern.

Equation (11.2) is valid when the point object is located "outside" the sphere, as in Fig.(11.2), or on the surface of the sphere. Equation (11.2) is not valid if the point object is inside the sphere. Put another way, Eq.(11.2) holds for values of r in the range $R \leq r \leq \infty$.

Now suppose that the sphere is fixed in position and that the gravitational force F due to the sphere is the only force acting on the point object. Then the acceleration a of the point object can be found from Newton's second law. When the only force acting on an object is gravity, the resulting acceleration is called the free-fall acceleration, or the *acceleration due to gravity*, often contracted to just *gravity*. Also, it is very common to use the symbol g, rather than a, for this special acceleration. Putting all this together with Eq.(11.2) leads to

$$\Sigma F = ma, \tag{11.3}$$

$$F = mg,$$

$$\frac{GMm}{r^2} = mg,$$

$$g = \frac{GM}{r^2}. \tag{11.4}$$

The value of g does not depend on the value of the mass m of the point object, and so has the same value for any point object placed at the same position. Hence, g is really a function of position. In geological applications of gravity, the sphere is the Earth itself. Now the Earth is not a perfect sphere, and its internal distribution of matter is not precisely spherically symmetric. However, these two conditions for the validity of Eq.(11.2), and therefore of Eq.(11.4), hold to a high-enough degree of precision for the purposes of this chapter, so that Eq.(11.4) will be considered correct as it is applied to the Earth.

One obviously important "location" is the surface of the Earth. Using Eq.(11.4), the value of the acceleration due to gravity at the surface g_S is found by putting $r = R$, where R is the radius of the Earth, so that

$$g_S = \frac{GM}{R^2}, \tag{11.5}$$

$$g_S = \frac{(6.67 \text{ X } 10^{-11} \text{ N} \cdot \text{m}^2/\text{kg}^2)(5.98 \text{ X } 10^{24} \text{ kg})}{(6.370 \text{ X } 10^6 \text{ m})^2},$$

$$g_S = 9.82987858 \text{ m/s}^2.$$

The value of g_S as written above contains (many!) more significant figures than the quoted values of G, M, and R allow. This is to provide a value of g_S that reflects the accuracy needed for gravity surveying purposes. To obtain a value of g_S to nine significant figures while observing the rules of significant figures in its calculation, would require a discussion of a great many effects that are outside the scope of the present work. [For example, Eq.(11.5) could not be used at all, since a much more accurate representation for the shape of the Earth than is provide by a sphere would have to be used.] A necessarily detailed treatment of these effects is sidestepped by suspending the rules of significant figures in the calculation of the surface value of gravity g_S.

The SI base unit of acceleration, m/s^2, is a large unit of acceleration as far as gravity surveying is concerned. Gravity measurements usually are reported in *milligalileos* (mGal), where the *galileo* (Gal) is defined by

$$1 \text{ Gal} = 1 \text{ cm/s}^2,$$

which means that

$$1 \text{ mGal} = (1 \text{ X } 10^{-3})(1 \text{ X } 10^{-2} \text{ m/s}^2),$$

$$1 \text{ mGal} = 1 \text{ X } 10^{-5} \text{ m/s}^2. \quad (11.6)$$

Therefore, the value of g_S found above is equivalent to

$$g_S = 982987.858 \text{ mGal}. \quad (11.7)$$

This value of surface gravity is based on a calculation that assumes the Earth to be a perfect sphere with a spherically symmetric distribution of matter. Since the Earth does not meet these conditions precisely, actual measured values of gravity at the surface will be different at different locations. That is, the measured value of g_S varies with location on the surface of the Earth. The value given in Eq.(11.8) can be considered as a global average of actual measured values.

There are many types of instruments, gravity meters, that measure the acceleration due to gravity. Most actually measure the vertical component of this acceleration. They can be based on: measuring the time of fall of an object through a known vertical distance; measuring the period of a pendulum; measuring the extension of a spring carrying a known load. Although simple in principle, it is not easy to obtain extremely accurate values of the acceleration due to gravity. Most physics textbooks describe each of these systems and derive and display the relevant formulas, so the details of gravity meter construction will not be treated here. However, see the problems for a few examples.

****

EXAMPLE 1

"Eight Mountain" is 320 m high. As shown on Fig.(11.3), the mountain consists of two spheres of rock, arranged with their centers on a vertical line, so that one is balanced on top of the other. (Evidently, there have been no earthquakes in the vicinity of Eight Mountain for a long time.) The rock of which the mountain is composed has density 2.94 g/cm^3.

Find the value of that part of the acceleration due to gravity at point P at the peak of the mountain that can be attributed to the mountain itself (i.e., with the contribution from the rest of the Earth ignored).

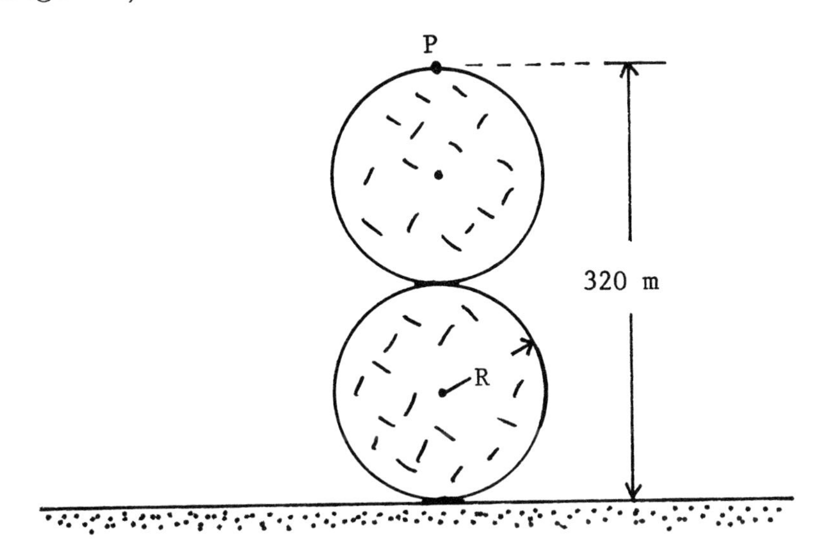

Fig.(11.3) Example 1

Since the spheres are identical, the radius of each is $\frac{1}{4}$(320 m), or $R = 80$ m. The acceleration due to the gravity of each sphere is directed to the center of that sphere. Because of the collinear arrangement of P and the centers of the spheres, the two contributions to g at P both are directed vertically down, and therefore their magnitudes add. Note that P is at a distance $r = R$ from the center of the top sphere and $r = 3R$ from the center of the bottom sphere. Applying Eq.(11.4) to each sphere as seen from P, the resultant acceleration due to gravity from the two spheres at P is, with M the mass of each sphere,

$$g = \frac{GM}{R^2} + \frac{GM}{(3R)^2},$$

$$g = (\frac{10}{9})\frac{GM}{R^2}.$$

But $M = \frac{4}{3}\pi R^3 \rho$, where ρ is the rock density; hence,

$$g = (\frac{40}{27})\pi G R \rho.$$

Substitution of the numerical values $G = 6.67 \text{ X } 10^{-11}$ N·m^2/kg^2, $\rho = 2940$ kg/m^3 and, as found above, $R = 80$ m, gives $g = 7.30 \text{ X } 10^{-5}$ m/s^2. Then converting the result from m/s^2 to mGal by Eq.(11.6) leads to $g = 7.30$ mGal.

****

11.2 Elevation Correction

To obtain information on the density distribution of the subsurface rocks in a region where the terrain is not flat, gravity measurements must be made at several locations over the region that are not at the same elevation. For example, in Fig.(11.4), readings may be made at a point A on the surface and at point b at the top of a hill. These gravity readings cannot immediately be compared to yield data on the subsurface rocks until the reading at b has been "reduced" or "corrected" to indicate the value that would have been recorded at the surface at B, vertically below b, if the hill was not present. The reading at A does not have to be corrected. Then, any variation between the reading at A and the corrected reading at B would be attributed to variations in the density of the rocks beneath the surface.

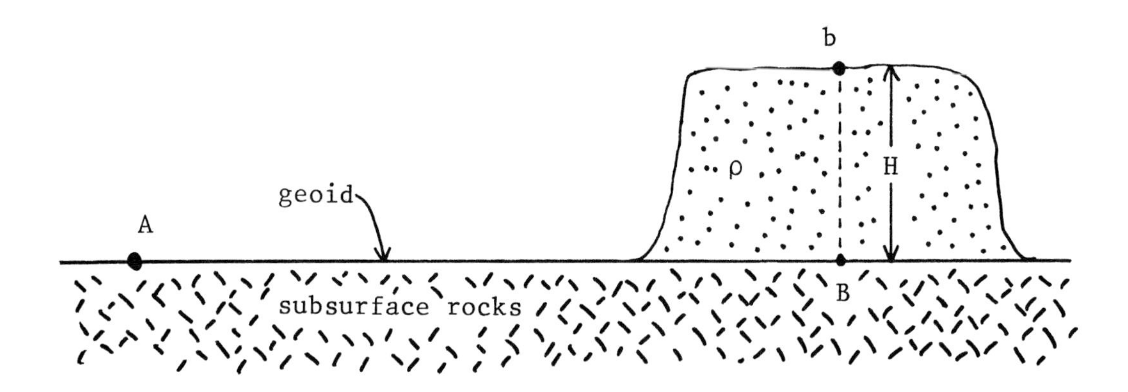

Fig.(11.4) Gravity and Elevated Terrain

By the "surface" of the Earth is meant the surface of a sphere, centered on the center of the Earth, that has a radius of 6370 km. This sphere represents the average radius of the Earth, as indicated by the mean global sea level, suitably extrapolated through land areas. This average sea-level surface is called the *geoid*. The surface in Fig.(11.4) is part of the geoid. (When effects due to the rotation of the Earth are taken into account, an ellipsoid becomes a better representation of the geoid, but this complication, and others, will not be considered here.)

To reduce the gravity reading at b to the reading that would be obtained at B if the hill was not there, two effects must be accounted for: first, point b is at an elevation H above the surface of the Earth and, second, between b and the surface is hill rock that contributes to the reading at b, and this contribution must be subtracted away.

Consider the elevation, or altitude effect, alone; i.e., ignore the presence of the hill but keep point b at altitude H, as though the gravity reading is taken from a balloon; see Fig.(11.5). By Eq.(11.4), the gravity reading g at an altitude H above the geoid is given by

$$g = \frac{GM}{(R+H)^2}, \tag{11.8}$$

since a point at an altitude H is at a distance $r = R + H$ from the center of the Earth.

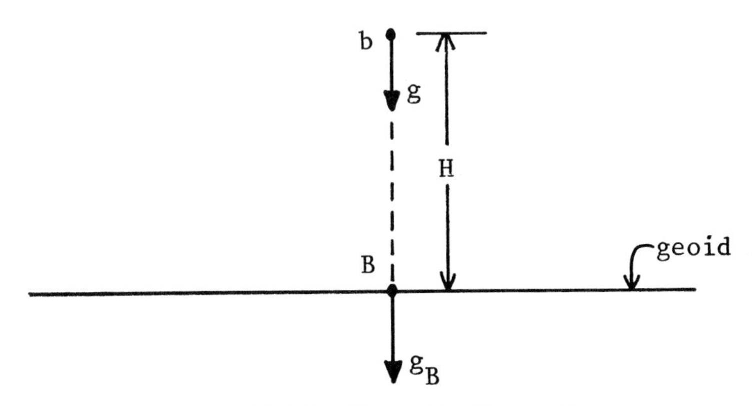

Fig.(11.5) The Free-Air Correction

Now, in any practical engineering situation, $H \ll R$, so that use of the approximation

$$\frac{1}{(1+x)^2} \approx 1 - 2x, \tag{11.9}$$

valid for $x \ll 1$ is appropriate. To apply "Eq."(11.9), factor out R from the parenthesis in Eq.(11.8) for g to obtain

$$g = \frac{GM}{R^2} \frac{1}{(1+H/R)^2},$$

$$g \approx \frac{GM}{R^2} \left[1 - \frac{2H}{R}\right],$$

$$g \approx g_{\rm B} \left[1 - \frac{2H}{R}\right],$$

$$g \approx g_{\rm B} - g_{\rm B} \left(\frac{2H}{R}\right),$$

$$g_{\rm B} \approx g + g_{\rm B} \left(\frac{2H}{R}\right). \tag{11.10}$$

Since point B is on the geoid, little error is committed if, for $g_{\rm B}$ on the right-hand side of Eq.(11.10) the average surface value $g_{\rm S}$ is substituted. Doing this, and replacing the inequality with an equality valid for practical purposes, gives

$$g_{\rm B} = g + \left(\frac{2g_{\rm S}}{R}\right) H. \tag{11.11}$$

The correction that must be applied to g due to the altitude of the observation point is the last term in Eq.(11.11); this term is called the *free-air correction* $\Delta g_{\rm fa}$:

$$\Delta g_{\rm fa} = \left(\frac{2g_{\rm S}}{R}\right) H. \tag{11.12}$$

It is usual to evaluate numerically the quantity in parenthesis, since the value found holds everywhere. This is

$$\frac{2g_S}{R} = \frac{2(982987.858 \text{ mGal})}{6370 \text{ X } 10^3 \text{ m}},$$

$$\frac{2g_S}{R} = 0.309 \text{ mGal/m}. \tag{11.13}$$

When putting this into Eq.(11.12), it is conventional to not write out the units in the final formula, but with the understanding that in writing the free-air correction as

$$\Delta g_{\text{fa}} = 0.309H, \tag{11.14}$$

H is expressed in meters and Δg_{fa} in milliGal. This free-air correction must be added to a reading taken at altitude H above the geoid to compensate for the elevation itself.

The second correction that must be applied to the reading taken at point b on the hill is that arising from the gravitational effects of the earth materials between point b and the geoid. It is not possible to write an algebraic expression, or formula, for this correction because, as stated in Section (11.1), the strength of the gravitational force, and hence acceleration, depends on the mass, size and shape of the objects concerned, and natural topographic features, such as hills, do not have simple geometric shapes.

The best that can be done is to approximate the actual shape with the closest simple geometric figure for which a formula for the gravitational acceleration can be found. Of course, the sphere is the simplest shape to deal with, but few mountains, say, have even an approximately spherical shape. A frequently adopted approach is to picture the local topography as more closely resembling a plateau; i.e., a relatively flat region of low to moderate elevation, with the horizontal dimensions of the plateau much greater than the height of the plateau's top surface. Such a feature often can be approximated, in shape, with a "pillbox": a short, fat, cylinder with vertical height much less than horizontal width.

Even for a pillbox of uniform density, there is no simple formula for the gravitational acceleration at any external point due to the gravitational attraction of the constituent material, even if the pillbox (cylinder) cross section has a simple shape, such as a circle. However, there is a simple formula for the axial component of the gravitational acceleration at a point situated on the flat surface of such a cylinder, as long as the point is not near an edge. That is, in Fig.(11.6), the vertical component of the acceleration due to gravity g_{cyl} at point P, caused by the material of density ρ in the cylinder, is given by

$$g_{\text{cyl}} = 2\pi GH\rho, \tag{11.15}$$

where H is the height of the cylinder. This formula can be established using only second-semester calculus; see Problem (11-23).

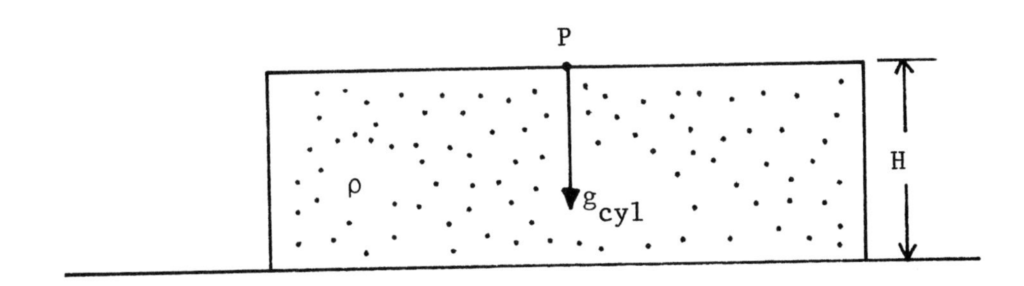

Fig.(11.6) Gravity Due to a Cylinder

If, then, a pill box shaped cylinder is a good approximation to the actual topographic feature encountered, then the second correction to be applied to the reading at point b, Fig.(11.4), to yield the expected reading at the geoid is given by Eq.(11.15). [But Eq.(11.15) does not apply to all features; for example, it would not apply to Eight Mountain, shown on Fig.(11.3).] Evaluate the factor $2\pi G$ numerically and, by multiplying by 1000, adjust the units so that ρ in g/cm^3 (not kg/m^3) and H in m yields g_{cyl} in mGal. This correction Δg_{mat} due to the hill rock is called the *Bouguer correction* and is given by

$$\Delta g_{\mathrm{mat}} = 0.0419\rho H. \tag{11.16}$$

This Bouguer correction is subtracted from the reading at the top of the cylinder (representing the topographic feature).

Putting the free-air and Bouguer corrections together yields

$$g_{\mathrm{B}} = g_{\mathrm{b}} + 0.309H - 0.0419\rho H, \tag{11.17}$$

where ρ in g/cm^3, H in m, and g_{b} in mGal gives g_{B} in mGal also.

It should be emphasized that g_{B} is not the gravity reading that would be obtained at B if a tunnel, say, is dug under the hill to point B and a reading taken. Rather, g_{B} is the reading that would be found at B if the hill is removed. If this corrected reading matches a reading on the geoid at a point well removed from the hill, then the conclusion is that the subsurface rocks (rocks below the geoid) do not vary significantly in density between those two points. In terms of Fig.(11.4), if $g_{\mathrm{B}} = g_{\mathrm{A}}$, then the subsurface rocks do not vary in density, in a significant manner, in the region containing point A and the hill. That is, if there is no variation in density of the subsurface (shallow depth, or surficial) rocks, then $g_{\mathrm{B}} = g_{\mathrm{A}}$, so that

$$g_{\mathrm{A}} = g_{\mathrm{b}} + 0.309H - 0.0419\rho H. \tag{11.18}$$

The conditions as to the units to apply to Eq.(11.18) are the same as those that apply to Eq.(11.17), given above.

****

EXAMPLE 2
The gravity reading at point Q on the 330 m high plateau is 981362.57 mGal, while the reading at point P far away is 981430.14 mGal. The density of the subsurface rocks is the same throughout the region shown on Fig.(11.7). Find the density of the rock making up the plateau.

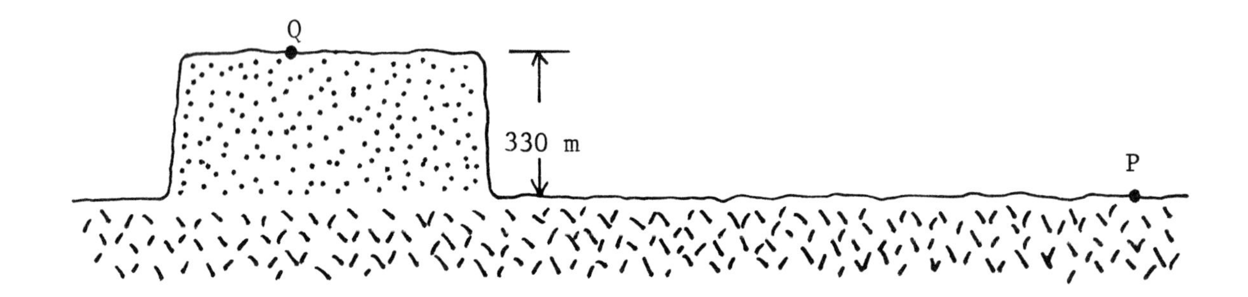

Fig.(11.7) Example 2

Point Q corresponds to point b in Fig.(11.4), and point P to point A. Therefore, Eq.(11.18) yields

$$g_{\mathrm{A}} = g_{\mathrm{b}} + 0.309H - 0.0419\rho H,$$

$$981430.14 = 981362.57 + 0.309(330) - 0.0419\rho(330),$$

$$\rho = 2.49 \text{ g/cm}^3.$$

Note that since the units of the numerical coefficients in Eq.(11.18) are not displayed in that equation, the units of the data when substituted are likewise suppressed. The proper units to use are shown below Eq.(11.17).

****

It is natural to wonder what correction must be applied if the observation point is below the geoid, at a point depressed below mean sea level. For instance, consider the point P in Fig.(11.8) at the bottom of a vertical mine shaft of depth D. By a happy circumstance, there is a simple formula for the gravitational acceleration due to the Earth at point P, a formula valid to the extent that the density distribution in the Earth can be considered to be spherically symmetric. Specifically, at point P,

$$g = \frac{GM_{\mathrm{r}}}{r^2}, \tag{11.19}$$

where M_{r} is the mass of that part of the Earth interior to a sphere of radius r, where $r = R - D$, R the radius of the Earth, the sphere centered at the center of the Earth. It is as if the shell of matter between the surface of the Earth and the sphere whose surface is at depth D is simply stripped away, point P being at the surface of the rump Earth that remains and Eq.(11.4) applying thereto.

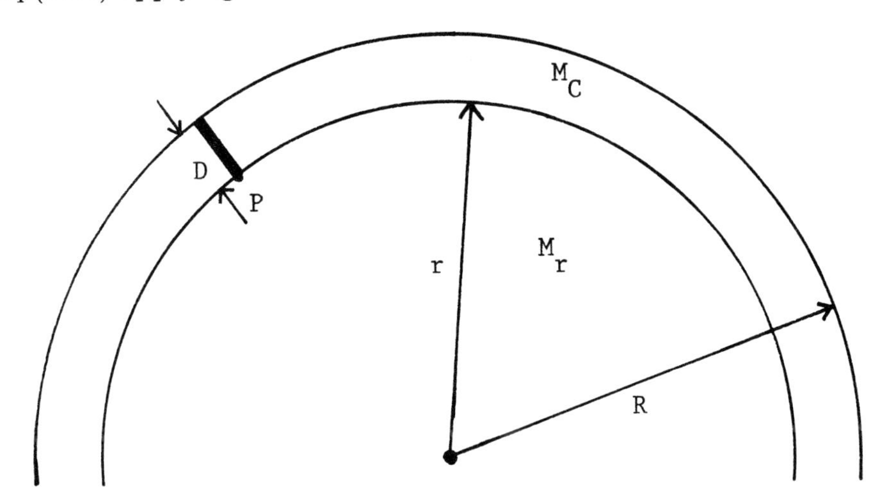

Fig.(11.8) Depression Correction

The correction Δg to be applied to the measured value of gravity at P to yield the value at the surface directly above is given from

$$g + \Delta g = g_S,$$

assuming that the point at the top of the mine shaft is representative of the average surface value of gravity. This correction includes both the free-air and terrain contributions.

As of today (as far as one can tell in advance!), no mine shaft or borehole has penetrated beyond the crust. Hence $D \ll R$ and the approximation of "Eq."(11.9) can be used. In this spirit, at the bottom of the mine at P, the gravity reading can be written, starting from Eq.(11.19), as follows:

$$g = \frac{G(M - M_C)}{(R - D)^2},$$

$$g = \frac{GM}{R^2} \frac{1 - M_C/M}{(1 - D/R)^2},$$

where M_C is the mass of the crust in the shell extending from the surface of the Earth to the concentric spherical surface of radius $R - D$. If ρ_C is the average density of the crust, and ρ_{av} the average density of the whole Earth, then

$$\frac{M_C}{M} = \frac{4\pi R^2 D \rho_C}{\frac{4\pi}{3} R^3 \rho_{av}},$$

$$\frac{M_C}{M} = 3 \frac{\rho_C}{\rho_{av}} \frac{D}{R}.$$

Therefore,

$$g = g_S \left(1 - 3\frac{\rho_C}{\rho_{av}}\frac{D}{R}\right)\left(1 + \frac{2D}{R}\right),$$

and, ignoring the small term in $(D/R)^2$,

$$g = g_S \left(1 + \frac{2D}{R} - 3\frac{\rho_C}{\rho_{av}}\frac{D}{R}\right).$$

Since $\Delta g = g_S - g$,

$$\Delta g = g_S \left(3\frac{\rho_C}{\rho_{av}} - 2\right)\frac{D}{R}. \tag{11.20}$$

For the Earth $\rho_C \approx \frac{1}{2}\rho_{av}$, so that $\Delta g < 0$, indicating that gravity readings in the crust increase with depth. Observed deviations from the result indicated in Eq.(11.20) provides information on local density variations in the vicinity of the borehole or mine.

11.3 Spherical Deposits

Now suppose that the free-air and Bouguer corrections have been applied to the gravity readings made over the region whose subsurface rock distribution is being examined. Many gravity readings will be made over the region. Suppose, now, that these corrected readings made at various locations do not agree. With topographic factors corrected-away, only variations in the density of the rock beneath the surface (geoid) can account for any remaining differences (anomalies) in the readings made at different locations.

The task now is to turn the variations in the corrected gravity readings into specific information on the density distribution of the subsurface rocks. The description given here is similar in approach to that used in illustrating the methods of seismic surveying; i.e., assume a specific density geometry and calculate the gravity readings expected therefrom. If these agree with the actual gravity readings, then specific numerical information on the subsurface density can be extracted. If the readings do not agree, other geometries must be examined until satisfactory agreement is attained.

The specific model to be used to demonstrate gravity surveying methods is shown in Fig.(11.9). Beneath the geoid, the rocks are of density ρ_0, except in a spherical region where the rocks have a different density ρ_d. This spherical region is called a "deposit". The effect of the deposit on the corrected gravity readings over this region must now be determined.

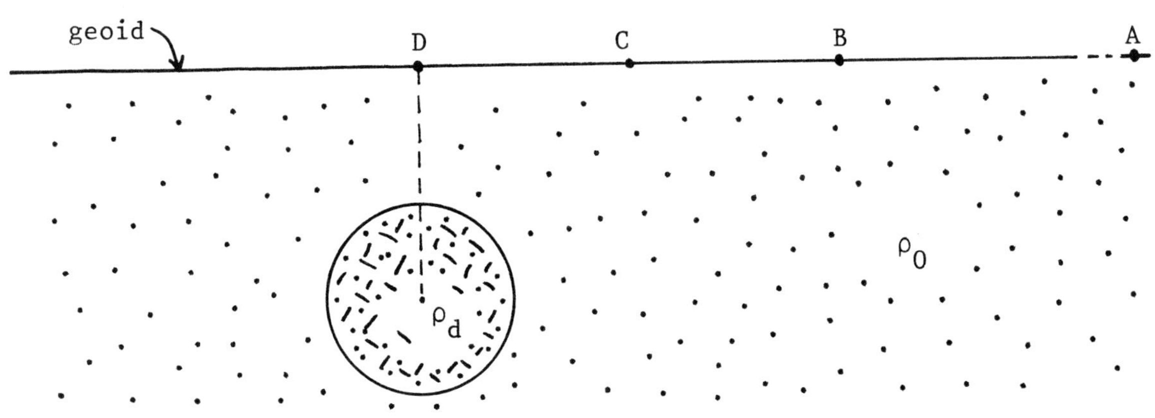

Fig.(11.9) Spherical Deposit

Consider the points A, B, C, D shown on Fig.(11.9). They are aligned on a straight line on the geoid with point D directly above the center of the deposit. Point A is very far away, so that it is actually "off" the figure somewhere to the right. The actual above-surface topography need not be shown, since its effects have been allowed for with the free-air and Bouguer corrections. Gravity readings are made at many points, of which these points are representative. By gravity readings is meant measurements of the vertical component of the acceleration due to gravity recorded by gravimeters placed at the various locations.

At, or near, point A the gravity reading g_{A} is hardly affected at all by the presence of the deposit, since point A is chosen to be "far away" from the deposit. At point B, noticeably closer to the deposit than point A, the deposit has an influence on the value g_{B} of gravity. Of course, the precise location of point B, where the deposit begins to affect the value of gravity, depends on the precision of the gravimeter being used.

If it is assumed that $\rho_{\mathrm{d}} > \rho_0$, then it is expected that $g_{\mathrm{B}} > g_{\mathrm{A}}$; increased mass concentration brings increased gravity. On the other hand, if $\rho_{\mathrm{d}} < \rho_0$, then $g_{\mathrm{B}} < g_{\mathrm{A}}$. The remaining possibility, $\rho_{\mathrm{d}} = \rho_0$ means that $g_{\mathrm{B}} = g_{\mathrm{A}}$; this implies that gravity methods will not reveal a deposit if the deposit rocks have the same density as the surrounding rocks, even if the rocks types are completely different. The greater the absolute value of $\Delta\rho = \rho_{\mathrm{d}} - \rho_0$, then the greater the absolute value of $\Delta g = g - g_{\mathrm{A}}$, where g is the corrected gravity region at any point closer to the deposit than point A. The quantity Δg is called the *gravity anomaly.* If $\Delta g < 0$, then $\rho_{\mathrm{d}} < \rho_0$; $\Delta g > 0$ indicates that $\rho_{\mathrm{d}} > \rho_0$.

Assume that $\rho_{\mathrm{d}} > \rho_0$ in this illustration. Then, at B, $\Delta g > 0$. Proceeding to C, closer to the deposit than B, it is anticipated that $\Delta g_{\mathrm{C}} > \Delta g_{\mathrm{B}}$. That is, the gravity anomaly increases as points closer to the deposit are examined. At point D, directly over the deposit, Δg reaches its maximum value. If readings are taken past point D along the same straight line, Δg will decrease until eventually $\Delta g = 0$ is reached very far from the deposit. Readings would have to be taken along several straight lines, and their greatest readings of Δg compared, to isolate a line that passes directly over the deposit.

Thus, a graph of Δg versus the distance x from the deposit, with $x = 0$ directly over the center of the deposit, should look something like Fig.(11.10). The point D directly over the deposit is identified as the point where Δg has the greatest value over the region. (If $\rho_{\rm d} < \rho_0$, then the values of Δg are negative.)

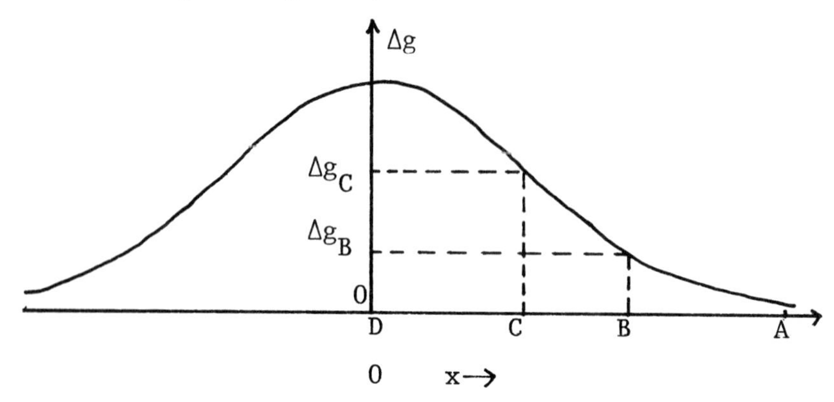

Fig.(11.10) Gravity Anomaly vs. Distance

It is now necessary to obtain a specific equation for the gravity anomaly. The gravity survey, in this case, consists in taking measurements of the gravity anomaly at various positions along straight lines, thereby finding a line that passes over the center of the deposit. After completing the survey, values of Δg with the corresponding values of x are known along a line that passes over the center of the deposit. The density ρ_0 of the surrounding rock is presumed to be known since this rock is accessible at the surface of the Earth and its density can be measured. The quantities sought are $\rho_{\rm d}$, the density of the deposit rocks, the radius R of the deposit (its size), and the depth d to the center of the deposit.

The density distribution shown in Fig.(11.9) can be constructed from a region of uniform density ρ_0 everywhere superimposed on a spherical region of density $\Delta\rho = \rho_{\rm d} - \rho_0$ located at the position of the actual deposit. The corrected gravity reading at any location can then be written as

$$g = g_0 + g_{\rm d} + g_{\rm E}, \tag{11.21}$$

where g_0 is the contribution from the uniform density region, $g_{\rm d}$ the contribution due to the deposit assuming it has density $\Delta\rho$, and $g_{\rm E}$ is the contribution from the rest of the Earth not pictured in Fig.(11.9).

The gravity reading far away from the deposit [point A in Fig.(11.9)] is given by

$$g_{\rm A} = g_0 + g_{\rm E}, \tag{11.22}$$

since, by definition, at points "far away" from the deposit, the gravitational effects of the deposit are taken to be zero. The value of gravity due to the uniform region of density ρ_0 is the same at any location on the geoid. By definition, the gravity anomaly at any location is

$$\Delta g = g - g_{\rm A}. \tag{11.23}$$

By Eqs.(11.21) and (11.22), the anomaly is

$$\Delta g = g_{\mathrm{d}}, \tag{11.24}$$

that is, the gravity anomaly is the vertical component of the gravity due to a deposit visualized as having density $\Delta\rho$ rather than its actual density ρ_{d}.

In Fig.(11.11), the deposit is visualized as having density $\Delta\rho$, with the density of the surrounding rocks thereby represented as being zero. The values of Δg obtained from this distribution of density is precisely the same as that obtained from the actual density distribution shown on Fig.(11.9). The construction of Fig.(11.11) is a mathematical artifice, so there is no need to ask how one is able to stand at a point on a surface beneath which the density is zero.

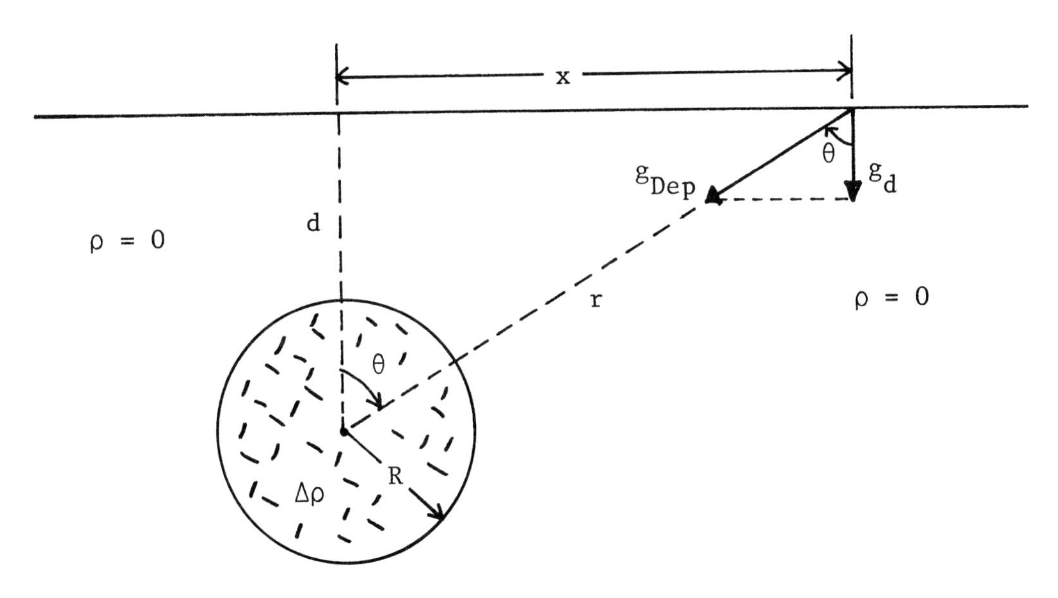

Fig.(11.11) Equivalent Density Distribution

Gravity meters measure the vertical component of the acceleration due to gravity. Recall that gravity due to a sphere of matter is directed toward the center of the sphere, whereas measured and subsequently corrected values of gravity refer to the vertical component only. If g_{d} is the actual measured acceleration due to gravity at the location x in Fig.(11.11) in question, then this figure indicates that

$$g_{\mathrm{d}} = g_{\mathrm{Dep}} \cos\theta, \tag{11.25}$$

where θ is the angle made by g_{Dep} with the vertical, and g_{Dep} is the gravitational acceleration due to the deposit assuming that it has density $\Delta\rho$. From Newton's law of gravitation applied to the spherical distribution of matter shown for the deposit in Fig.(11.11),

$$g_{\mathrm{Dep}} = \frac{GM^*}{r^2}. \tag{11.26}$$

In this expression, M^* is the mass of the deposit imagining it to be composed of material of density $\Delta\rho$; that is,

$$M^* = (\Delta\rho)\frac{4\pi}{3}R^3. \tag{11.27}$$

The actual mass of the deposit is

$$M_\mathrm{d} = \rho_\mathrm{d}(\frac{4\pi}{3}R^3). \tag{11.28}$$

Substituting Eqs.(11.25), (11.26), (11.27) into Eq.(11.24) gives

$$\Delta g = \frac{4\pi G(\Delta\rho)R^3\cos\theta}{3r^2}. \tag{11.29}$$

This expression contains two variables, θ and r, which cannot be immediately measured. But, from Fig.(11.11), it can be seen that

$$\cos\theta = \frac{d}{r}, \tag{11.30}$$

and, by the Pythagorean theorem,

$$r^2 = d^2 + x^2. \tag{11.31}$$

Combining Eqs.(11.29), (11.30), (11.31) leads to

$$\Delta g = \frac{4\pi G(\Delta\rho)R^3 d}{3(d^2 + x^2)^{3/2}}. \tag{11.32}$$

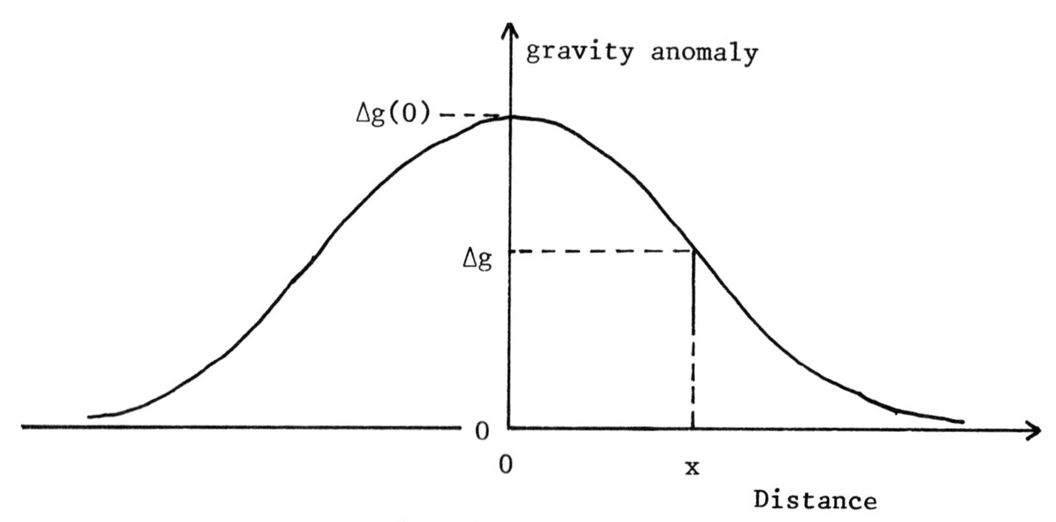

Fig.(11.12) Gravity Profile

This is the final expression for the gravity anomaly since it contains as variables only quantities that either can be measured (Δg, x) or are the object of inquiry ($\Delta\rho$, d, R).

Fig.(11.12) is the graph of Δg vs. x, based on Eq.(11.32). This *gravity profile* has a maximum value at $x = 0$; call this value $\Delta g(0)$. The graph is symmetric about the Δg axis, and decreases to zero as $x \to \pm\infty$. If corrected gravity readings are taken in the field along a line passing directly above a spherical deposit, the curve drawn averaging the data points will look like Fig.(11.12), except that there will be numbers along the x and Δg axes. The value of $\Delta g(0)$ can be read directly from this graph.

Putting $x = 0$ into Eq.(11.32) gives

$$\Delta g(0) = \frac{4\pi G(\Delta\rho)R^3}{3d^2}. \tag{11.33}$$

With this, Eq.(11.32) can be written in the form

$$\Delta g = \Delta g(0)\frac{d^3}{(d^2 + x^2)^{3/2}}. \tag{11.34}$$

Now suppose that from Fig.(11.12) a value of Δg and the associated value of x, other than $x = 0$, are read. The value of $\Delta g(0)$ has already been obtained from the graph. Only d in Eq.(11.34) remains unknown. Solving Eq.(11.34) for d (by raising both sides to the power $\frac{2}{3}$) yields

$$d = \frac{x}{\sqrt{[\frac{\Delta g(0)}{\Delta g}]^{2/3} - 1}}. \tag{11.35}$$

All of the quantities on the right-hand side of Eq.(11.35) can be read from the graph of Fig.(11.12), so that the depth d to the center of the deposit can be obtained.

The values of the deposit radius R and density ρ_d remain elusive; i.e., R and $\Delta\rho$ remain as unknowns. Unfortunately, Eq.(11.33) is one equation and there are still two unknowns. Hence, both R and $\Delta\rho$ cannot be separately calculated. Only if one of these quantities becomes known somehow, Eq.(11.33) can then be used to find the other.

****

EXAMPLE 3

The *gravity profile*, gravity anomaly vs. distance, across a lead-zinc ore deposit is shown in Fig.(11.13). The density of the Pb-Zn ore is 3.65 g/cm^3 and the surrounding rocks have density 2.60 g/cm^3. Modelling the actual deposit geometry as a sphere, find (a) the depth to the center of the deposit, (b) its radius, and (c) the mass of Pb-Zn ore present, in metric tons. (*d*) Show that the gravity profile is consistent with Eq.(11.32).

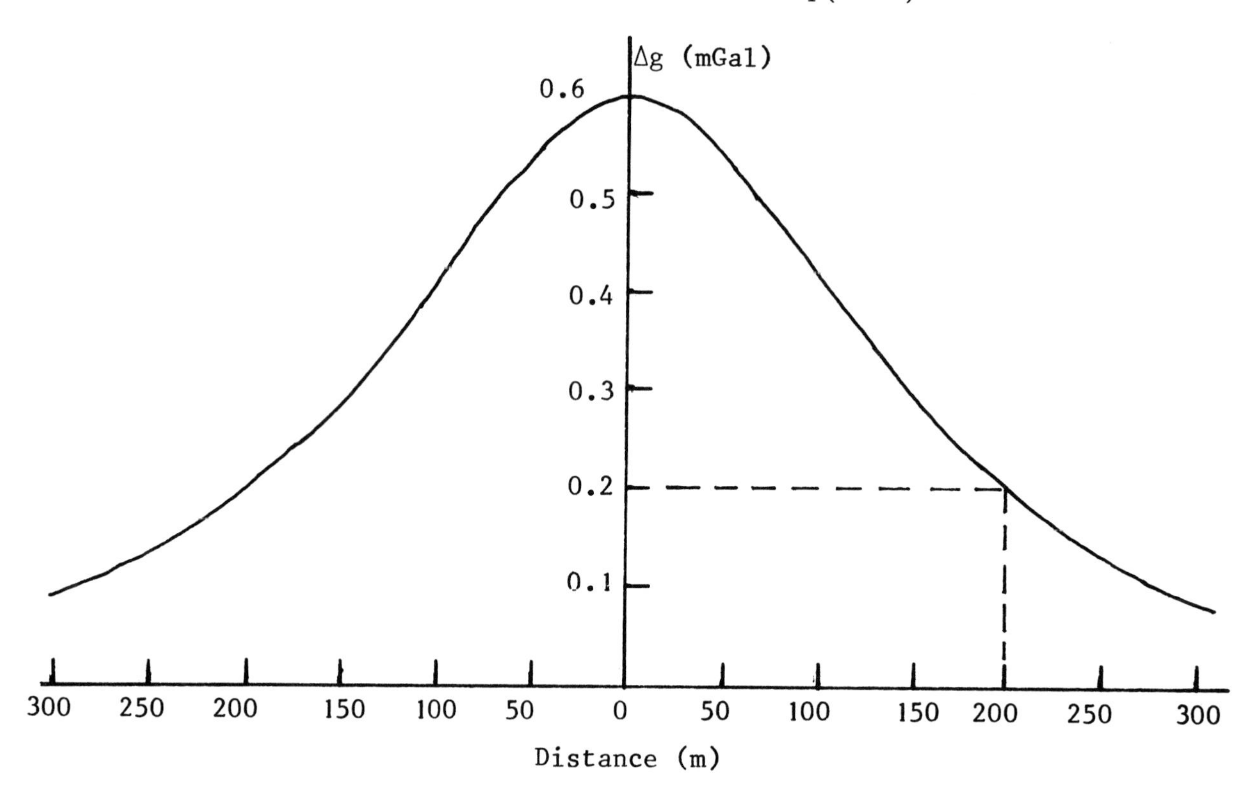

Fig.(11.13)[1] Example 3

(*a*) Use Eq.(11.35). From Fig.(11.13), it is seen that $\Delta g(0) = 0.6$ mGal, the value at $x = 0$. For another data point, choose one so that, if possible, both Δg and x are easy to read from the graph; for example, at $x = 200$ m, $\Delta g = 0.2$ mGal. Using these numbers, and pretending that the graph can be read to three significant figures,

$$d = \frac{x}{\sqrt{[\frac{\Delta g(0)}{\Delta g}]^{2/3} - 1}},$$

$$d = \frac{200 \text{ m}}{\sqrt{[\frac{0.6}{0.2}]^{2/3} - 1}},$$

$$d = 192 \text{ m}.$$

Note that the gravity anomalies enter as a ratio; as long as they are entered in the same units, those units cancel and therefore need not be displayed in the equation.
(*b*) Turn to Eq.(11.33). Unlike Eq.(11.35), only one gravity reading, $\Delta g(0)$, appears; there is no ratio, so the units of gravity do not cancel. Since G is present in Eq.(11.33), $\Delta g(0)$ must

[1]Adapted from: Turcotte, D.L./Schubert, G., 1982, p.215.

be expressed in the same units as G; i.e., SI base units. This means that the gravity anomaly must be expressed as $\Delta g(0) = 0.6 \text{ X } 10^{-5}$ m/s². Rearranging Eq.(11.33) and substituting the data gives

$$\frac{4\pi}{3}R^3 = \frac{d^2[\Delta g(0)]}{G(\Delta\rho)},$$

$$\frac{4\pi}{3}R^3 = \frac{(192 \text{ m})^2[0.6 \text{ X } 10^{-5} \text{ m/s}^2]}{(6.67 \text{ X } 10^{-11} \text{ N}\cdot\text{m}^2/\text{kg}^2)(1050 \text{ kg/m}^3)},$$

$$\frac{4\pi}{3}R^3 = 3.158 \text{ X } 10^6 \text{ m}^3,$$

$$R = 91.0 \text{ m}.$$

(c) The mass M_d of the deposit is

$$M_\text{d} = \rho_\text{d}\frac{4\pi}{3}R^3,$$

$$M_\text{d} = (3650 \text{ kg/m}^3)(3.158 \text{ X } 10^6 \text{ m}^3),$$

$$M_\text{d} = 1.15 \text{ X } 10^{10} \text{ kg},$$

$$M_\text{d} = 1.15 \text{ X } 10^7 \text{ t}.$$

(d) There are many mathematical functions besides Eq.(11.32) whose graph looks very similar to Fig.(11.12). To verify that the data of Fig.(11.13) does match the graph of Fig.(11.12), substitute the values of d and R found in (a) and (b) above into Eq.(11.32) and then plot the resulting function, perhaps using a graphing calculator. It should be found that the numerical values of Δg vs. x match those on Fig.(11.13). For example, calculate the value of Δg for $x = 50$ m expected with $R = 91.0$ m and $d = 192$ m. Put these values into Eq.(11.32), using SI base units, to get

$$\Delta g = (\frac{4\pi}{3}R^3)\frac{G(\Delta\rho)d}{(d^2 + x^2)^{3/2}},$$

$$\Delta g = (3.158 \text{ X } 10^6)\frac{(6.67 \text{ X } 10^{-11})(1050)(192)}{(192^2 + 50^2)^{3/2}},$$

$$\Delta g = 5.4 \text{ X } 10^{-6} \text{ m/s}^2,$$

$$\Delta g = 0.54 \text{ mGal}.$$

Looking at Fig.(11.13), it can be seen that at $x = 50$ m, $\Delta g = 0.54$ mGal, as predicted above. The value of Δg at other values of x should be verified. But how many such values should be confirmed before it is accepted that the deposit is well matched by a spherical configuration? Technically, this must be done for virtually all values of x but, in practice, the successfull checking of three or four values may be sufficiently convincing. Agreement will be limited by the extent to which a spherical idealization of the actual deposit geometry is satisfactory.

****

11.4 Cylindrical Deposit

If a sphere is not a good approximation to the shape of a deposit under examination, then another shape must be used to model the actual deposit. Unfortunately, other than for a sphere, there are no other configurations of matter of finite extent for which a closed, algebraic, formula exists for the resulting gravitational acceleration at any arbitrary point exterior to the configuration.

However, formulas for the gravitational attraction do exist in two (non-sphere) categories: (i) for some shapes of finite extent, at a very specific set of points relative to the gravitating mass; and (ii) for certain shapes of infinite extent that may serve as an approximation to the gravity of necessarily finite-sized objects in the real world.

An example of a formula in category (i) is that given in Problem (11.23) and Eq.(11.15), for the gravity due to a circular cylinder of matter at the center of either of its circular faces. [Actually, a formula exists for gravity at any point on the axis of the cylinder, of which the point at the center of the flat face is a special case.] There is no general formula for gravity valid at every point on the face.

A uniform, circular, cylinder of matter can also provide a formula of category (ii) type. Using calculus, it can be shown that the gravitational acceleration due to a uniform cylinder of matter of mass M and length L, at points outside the cylinder near its midplane (i.e., away from the ends), is given approximately by

$$g = \frac{2GM}{Lr}, \tag{11.36}$$

where r is the distance of the point from the axis of the cylinder; the point must be at a distance $r \geq R$, where R is the radius of the cylinder. The acceleration of gravity is directed toward, and normal to, the axis of the cylinder; see Fig.(11.14).

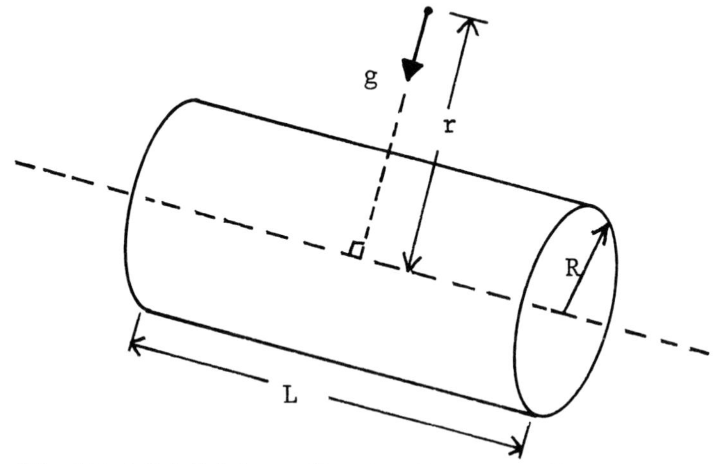

Fig.(11.14) Midplane Gravity Due to a Cylinder

Equation (11.36) is obtained from an exact formula for gravity due to an infinitely long cylinder of matter. Infinitely long objects do not exist in the real world, but they often provide excellent approximations to actual configurations. For Eq.(11.36) to be a good approximation, it is necessary that $r \ll L$.

Suppose, now, that a deposit of rock resembles a long cylinder, rather than a sphere (e.g., an elongated ore vein). Then, Eq.(11.36) can be used to find an expression for the expected gravity anomaly. Assume that the axis of the cylinder is parallel to the geoid (i.e., is horizontal) and that gravity readings are made at points along a straight line on the surface that is perpendicular to the axis of the cylinder. Then Fig.(11.11) can serve as a cross section of this deposit as well as for the spherical deposit, only now the circle is a cross section of the cylinder, not of a sphere. Fig.(11.12) applies also.

The algebraic steps leading to Eq.(11.32) for the sphere can be followed for the cylinder, except that Eq.(11.26) for the sphere be replaced with

$$g_{\text{Dep}} = \frac{2GM^*}{Lr}, \tag{11.37}$$

where

$$M^* = \pi R^2 L(\Delta\rho), \tag{11.38}$$

the term $\pi R^2 L$ being the volume of the deposit. Following the procedure for finding the resulting corrected anomaly for a sphere, but using Eqs.(11.37) and (11.38) instead of Eqs.(11.26) and (11.27) leads to

$$\Delta g = \frac{2\pi G(\Delta\rho)R^2 d}{x^2 + d^2}. \tag{11.39}$$

The depth d to the center of the deposit (cylinder) can be found in a manner analogous to that for the spherical deposit. That is, write the expression for the anomaly above the midplane of the cylinder (i.e., at $x = 0$):

$$\Delta g(0) = \frac{2\pi G(\Delta\rho)R^2}{d}. \tag{11.40}$$

Divide Eq.(11.40) by Eq.(11.39) and solve for d; the result is

$$d = \frac{x}{\sqrt{\frac{\Delta g(0)}{\Delta g} - 1}}. \tag{11.41}$$

Note the similarity between Eq.(11.41) for the cylinder and Eq.(11.35) for the sphere.

****

EXAMPLE 4

Suppose that the deposit of Example 3 is mistakenly assumed to be cylindrical, rather than spherical. (*a*) Using the same data points of $x = 0$ and $x = 200$ m as used in parts (*a*) and (*b*) of Example 3, find the depth and radius of the deposit considering it to be a cylinder.

(*b*) Using the values of d and R found in (*a*), calculate the expected value of Δg at $x = 300$ m. Compare the result with Fig.(11.13) and draw any obvious conclusions.

(*a*) By Eq.(11.41),

$$d = \frac{x}{\sqrt{\frac{\Delta g(0)}{\Delta g} - 1}},$$

$$d = \frac{200 \text{ m}}{\sqrt{\frac{0.6}{0.2} - 1}},$$

$$d = 141 \text{ m}.$$

Using this result in Eq.(11.40) gives, in SI base units,

$$\Delta g(0) = \frac{2\pi G(\Delta\rho)R^2}{d},$$

$$0.6 \text{ X } 10^{-5} = \frac{2\pi(6.67 \text{ X } 10^{-11})(1050)R^2}{141},$$

$$R = 43.8 \text{ m}.$$

These results are very different from those of Example 3.
(*b*) Now use Eq.(11.39) together with $R = 43.8$ m, $d = 141$ m to find Δg at $x = 300$ m; using SI base units,

$$\Delta g = \frac{2\pi G(\Delta\rho)R^2 d}{x^2 + d^2},$$

$$\Delta g = \frac{2\pi(6.67 \text{ X } 10^{-11})(1050)(43.8)^2(141)}{300^2 + 141^2},$$

$$\Delta g = 0.108 \text{ mGal}.$$

But Fig.(11.13) shows $\Delta g \approx 0.085$ mGal. The difference between these two values of Δg at $x = 300$ m is within the precision of the gravity meter to resolve. In view of the substantially different results for R and d using the spherical and cylindrical deposit formulas, it becomes clear that it is necessary to be sure of the shape that best represents the deposit before drawing conclusions about its size and depth. As this example shows, at least three data points are required to make this distinction.

****

11.5 Problems

1. Very accurate values of the local acceleration due to gravity can be obtained by timing the fall of an object from rest through a known vertical distance. Find the value of g, in mGal, at a location where the free fall time from rest through a distance of 1.25000 m is found to be 504.93 ms.

2. Gravity surveyors assert that at a particular location gravity has the value 979062.50 mGal. If this value is correct, how long should a simple pendulum 120.000 cm long take to complete 175.000 oscillations at this location? The period T of a simple pendulum of length L is given by $T = 2\pi\sqrt{L/g}$.

3. Calculate the free-air correction for the planet Mars, which has a mass of 6.40 X 10^{23} kg and a radius equal to 3395 km.

4. Far away from Slab Mesa, Fig.(11.15), on the surface of the Earth, the gravity reading is 982986.358 mGal. But on top of the mesa, near its middle, the gravity reading is 982885.337 mGal. Slab Mesa is 570 m high. Find the density of the mesa rocks. (The density of the subsurface rocks does not vary over the region.)

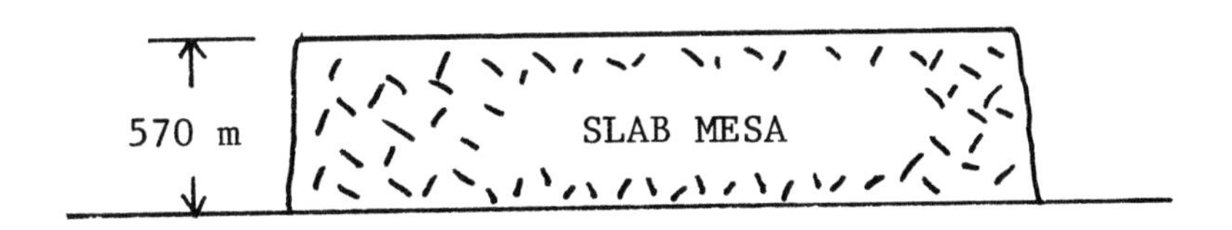

Fig.(11.15) Problem 4

5. The measured value of the acceleration due to gravity at the top of Block Hill, Fig.(11.16), is 980911.63 mGal; far away, on the Earth' surface, the value is 980987.68 mGal. The density of Block Hill rock is 2200 kg/m^3. Find the height H of Block Hill. (Assume no variation in the density of subsurface rocks.)

Fig.(11.16) Problem 5

6. The gravity reading is 981637.4 mGal at a central location on the top of a plateau 221 m high and composed of rock with density 2420 kg/m^3. Engineers drill a narrow, vertical, borehole from this location through the plateau down to the geoid. What gravity reading can be expected at the bottom of this borehole?

7. Mount Spherical, a spherical mountain shown in Fig.(11.17), has a height of $H = 340$ m. The density of the mountain rock is 2.82 g/cm^3. Far away, on the surface of the Earth, the gravitational acceleration is 981987.86 mGal. Find the value of the gravitational acceleration at the summit of Mount Spherical. (The surficial rocks have uniform density over the region.)

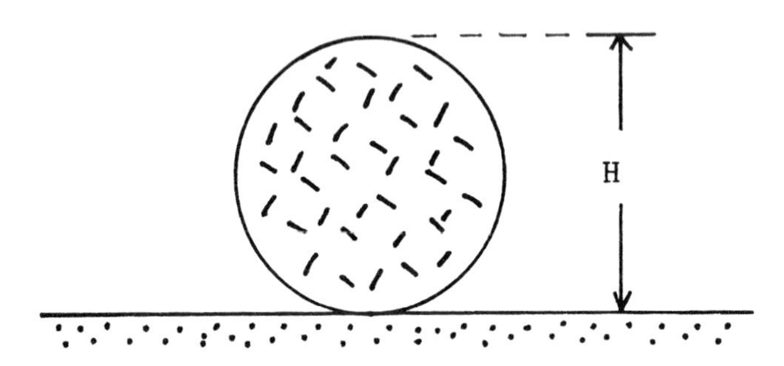

Fig.(11.17) Problems 7 and 8

8. At the top of Mount Sphere, with $H = 440$ m but otherwise apparently identical to Mount Spherical in Problem (11.7), the gravity reading is 982873.943 mGal. Far away on the surface of the Earth, gravity is 117.705 mGal greater, even though the surficial rocks have the same density throughout the region. Find the density of the rocks composing Mount Sphere. See Fig.(11.17).

9. Block Plateau, 190 m high, consists of rock with density 2.32 g/cm^3, except for a spherical inclusion of density 2.87 g/cm^3 that just fits in the plateau, as shown in Fig.(11.18). By how much does the gravity reading at point P differ from the reading on the geoid a long distance away? The density of the underground rocks is the same throughout the region.

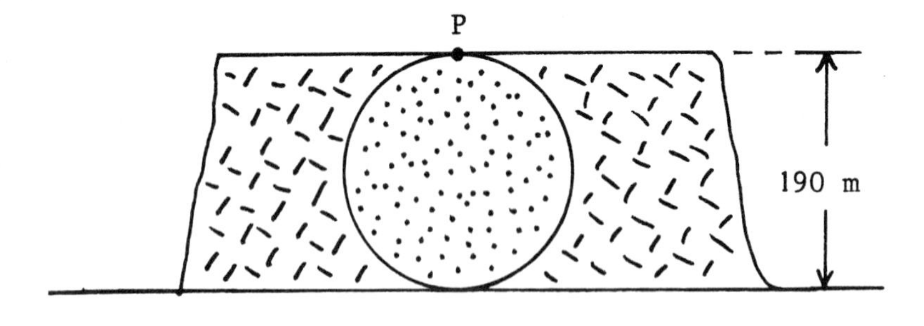

Fig.(11.18) Problem 9

10. Show that if the Earth was of uniform density throughout its entire interior, the depression correction at depth D would be given by

$$\Delta g = 0.154D,$$

where D in m gives Δg in mGal.

11. Show that the gravity anomaly due to a spherical deposit can be written as

$$\Delta g = (0.02794)\frac{R^3 d(\Delta\rho)}{(d^2 + x^2)^{3/2}},$$

where R, d, and x, in m and $\Delta\rho$ in g/cm^3 gives Δg in mGal.

12. The top of a spherical mineral deposit is just at the surface of the Earth, as shown in Fig.(11.19). The mineral rocks have density 2.30 g/cm^3 and the surrounding rocks have density 2.70 g/cm^3. At the top of the deposit gravity equals 982986.070 mGal, whereas far away it is 982987.858 mGal. Find the radius of the deposit.

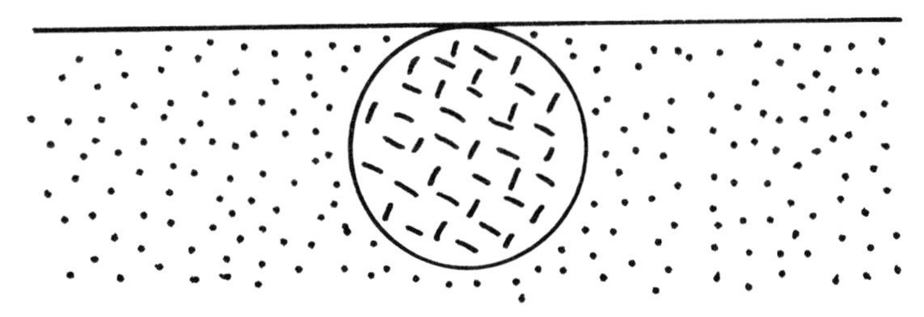

Fig.(11.19) Problem 12

13. A mineral deposit with a block-like shape 77.9 m thick is found in a region where the surrounding rocks have density 2.40 g/cm^3. The top of the deposit is at the surface of the Earth, as shown on Fig.(11.20), and occupies a land area of 4.71 ha. The measured value of gravity at the top of the deposit, near its middle, is 979992.60 mGal; far away gravity equals 979988.97 mGal. Find the mass of the deposit.

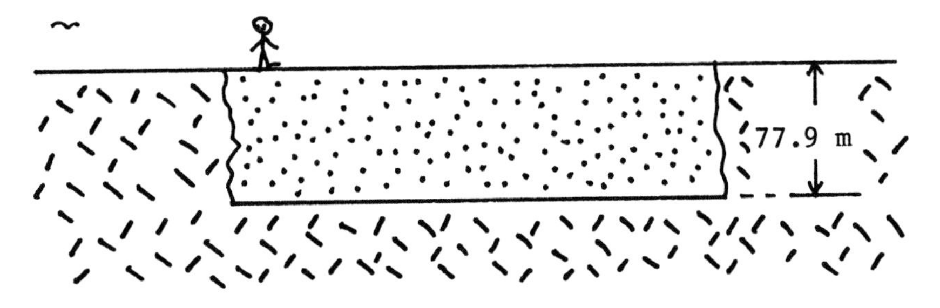

Fig.(11.20) Problem 13

14. A gravity reading is taken on the Earth's surface at a point directly above a flooded spherical cavern known from old mining records to have a radius of 54.0 m. Surrounding rocks have density 3.10 g/cm^3. The value of gravity recorded is 982986.484 mGal; far away from the cavern gravity equals 982987.858 mGal. Find the minimum depth to the roof of the cavern.

15. By how much does a gravity reading halfway up Eight Hill (Fig.11.3) differ from the average reading taken over the Earth's surface?

16. Directly above a spherical deposit the gravity anomaly is +13.0 mGal, and at a point 248 m away is +8.30 mGal. The deposit rocks have density 4.70 g/cm^3 and the surrounding country rock has a density of 2.80 g/cm^3. Calculate (*a*) the minimum depth to the top of the deposit, and (*b*) the mass of the deposit.

17. Gravity surveyors find corrected gravity anomalies varying from −0.17 mGal to −1.24 mGal over a distance of 70.0 m. Assume that the more negative anomaly is recorded directly over a spherical cavern known to be in the region. The surrounding rock has density 2260 kg/m^3. (*a*) Find the depth to the center of the cavern and its radius assuming it to be empty. (*b*) Suppose that the cavern is completely flooded with water. What do the gravity readings now indicate for its depth and radius?

18. You are trapped at the bottom of an empty spherical cavern. The top of the cavern is at the surface of the Earth, where your friend (who was smart enough not to go into the cavern) has created a small opening. Far away from the cavern, gravity equals 982987.858 mGal. Just before making the hole, your friend measured gravity at the top of the cavern and found it to be 982983.790 mGal. Surrounding rocks have density 2.60 g/cm^3. Find the length of rope needed to reach you.

19. You find yourself alone at the bottom of an empty spherical cavern, as shown in Fig.(11.21). A narrow shaft will be drilled from point A, directly above the center of the cavern, and a rope lowered to you. At A, the gravity anomaly is −6.00 mGal. At point B, which is 130 m from point A, the gravity anomaly is −1.11 mGal. The surrounding rocks

have unit weight 28.8 kN/m^3. How long should the rope be?

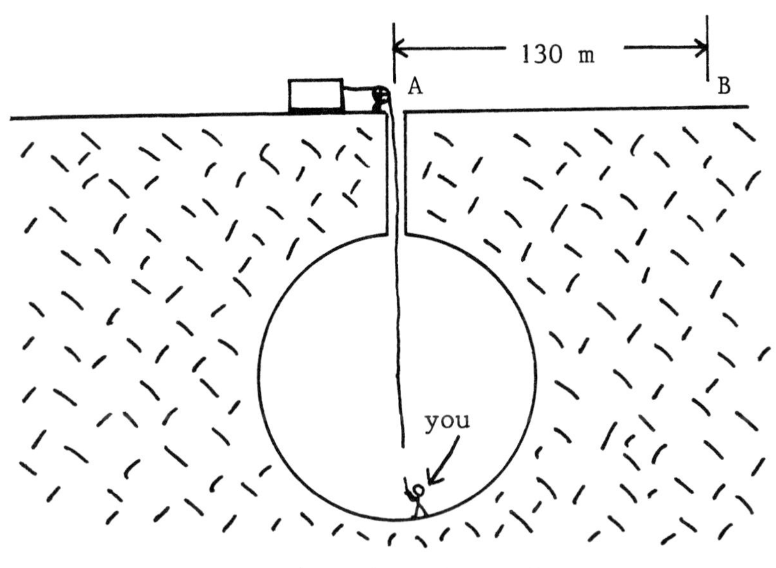

Fig.(11.21) Problem 19

20. Prospectors are doing a gravity survey over an abandoned mining region. They obtain the following data transverse to an old, horizontal, circular mine tunnel:

x (m)	Δg (mGal)
0	−0.0250
40.0	−0.0203

The tunnel is 3.50 km long and has a radius of 4.00 m; its depth is unknown. The surrounding rocks have a density of 3.10 g/cm^3. (*a*) Find the depth to the center of the tunnel. (*b*) Is the tunnel empty, or flooded with water?

21. Gravity surveyors record the following data:

x (m)	Δg (mGal)
0	1.536
50.0	0.585
100	0.205

The density of the deposit is 4.43 g/cm^3 and the density of the surrounding rocks is 3.00 g/cm^3. (*a*) Is the deposit a sphere or a horizontal cylinder? (*b*) Find the mass of the deposit. (If it is cylindrical, assume that the cylinder is 2300 m long.)

22. A horizontal, cylindrical deposit of rock of density 3.06 g/cm^3 has a radius of 3.82 m and a length of 1.73 km. It is embedded in rock of density 1.93 g/cm^3 at a depth to its axis

of 12.6 m. Find the gravity anomaly duc to thc dcposit at a distance of 18.0 m from a point directly above the deposit axis near the midplane of the deposit.

23. A uniform, solid, circular cylinder of matter has height H, radius R, and density ρ. (*a*) Using the methods of calculus, show that the acceleration due to gravity of the cylinder at a point at the center of one of the circular faces is given by

$$g = 2\pi G\rho[(H + R) - \sqrt{H^2 + R^2}].$$

(*b*) Show also that if $H \ll R$, then the formula above reduces to

$$g = 2\pi G\rho H,$$

as quoted in Eq.(11.15).

24. Without realizing it, you are standing on top of a loosely-covered abandoned vertical mine shaft. The gravity anomaly you measure there is -0.303 mGal. The surrounding rocks have density 2.80 g/cm^3. Upon stepping back and clearing the ground, you find the diameter of the shaft to be 5.20 m. Find the depth of the shaft. See Problem (11.23).

25. Answer Problem (11.16) assuming that the gravity measurements were taken by the first human explorers on the planet Mars. See Problem (11.3) for data on Mars.

Chapter 12

Radioactivity in the Earth

12.1 Radioactivity

All of the naturally occurring isotopes of all of the elements with atomic number Z greater than 83 are radioactive, as are some of the isotopes of most of the elements with Z less than, or equal to, 83. The element with $Z = 83$ is bismuth (chemical symbol Bi) . Put another way, then, none of the elements beyond bismuth (i.e., with $Z > 83$) has a stable isotope.

Isotopes of the same element have different numbers of neutrons in their nucleus; being of the same element, they necessarily have the same number of protons.

The number of protons in the nucleus of an atom is called the *atomic number* Z of the element of which the atom is an isotope. Atoms of different elements have different values of Z. The number of neutrons in a nucleus is written as N. Atoms of different elements can have the same value of N, so specifying N does not specify the element, the way specifying Z does. The *atomic mass number* A of a nucleus is defined by

$$A = Z + N. \tag{12.1}$$

Isotopes of the same element have the same Z, but different N and therefore, by virtue of Eq.(12.1), different A. The numbers Z, N, A have no units.

An atom of a radioactive isotope undergoes the process underlying radioactivity by spontaneously (i.e., without outside prompting) emitting a particle or photon (a "particle" of radiation). The "type" of radioactivity is specified from the identity of the particle that is emitted. Only two types of radioactivity will be of concern here. The first is called *alpha decay* . In this case, the nucleus emits an alpha particle; "alpha particle" is the name given to an assembly of two neutrons and two protons bonded together as a single entity. An alpha particle is identical to the nucleus of an atom of the isotope of helium (He, $Z = 2$) that contains two neutrons; i.e., with $A = 4$. Writing α for the alpha particle, $\alpha = {}^{4}_{2}\mathrm{He}$, using the notation that the superscript to a chemical symbol used to identify an element is A and the subscript is Z. (Sometimes the subscript indicating Z is omitted.)

The other type of radioactive decay of immediate concern is called *beta decay*. As can probably be guessed, in beta decay the nucleus emits a beta particle, symbol β. A beta particle is an electron. Nuclei do not contain electrons. Hence, the nucleus must create the electron that is then immediately ejected from the nucleus. The notation for an electron is ${}_{-1}^{0}\mathrm{e}$; the subscript is -1 since the charge on an electron is negative in exactly the same amount that the charge of a proton is positive, and a proton has $Z = +1$; the electron's mass number is zero, since nuclei do not contain electrons, so they cannot contribute to the mass of a nucleus.

It is a feature of radioactivity that in any individual decay, the A number of the nucleus before the decay equals the total of the A numbers of the product particles after the decay, the product particles meaning the particle emitted from the nucleus and the nucleus remaining after the decay. The same rule applies to the Z number. In both alpha and beta decay, the particle emitted has a Z number that is not zero. Therefore, the Z number of the nucleus that remains after the decay will be different from that of the nucleus beforehand. The nuclei before and after the decay are of different elements. Specifically, in alpha decay,

$$ {}_{Z}^{A}\mathrm{P} \longrightarrow {}_{Z-2}^{A-4}\mathrm{D} + {}_{2}^{4}\mathrm{He}, \tag{12.2} $$

and in beta decay

$$ {}_{Z}^{A}\mathrm{P} \longrightarrow {}_{Z+1}^{A}\mathrm{D} + {}_{-1}^{0}\mathrm{e}. \tag{12.3} $$

In these reaction equations, P stands for "parent" nucleus (not for Phosphorus), the nucleus before the decay took place; D stands for "daughter" nucleus, the nucleus remaining after the decay. This, almost sexist, terminology is a relic from the vocabulary introduced by the scientists who originally investigated radioactivity; the terms alpha particle and beta particle are other examples, terms introduced before the identity of these particles became known.

In beta decay, a *neutrino*, a particle similar in some ways to a photon, is also emitted; many alpha decays are followed soon after by the emission of a *gamma ray*, a very high energy photon. It is not necssary to consider these particles in the present discussion.

The term "decay" to denote a nucleus emitting a particle is rather an unfortunate term for, ordinarily, a decay is thought of as a slow, insidious process (like tooth decay!), whereas radioactive emission is almost an instantaneous process. Sometimes the term "disintegration" is used, but this is not much better: disintegration implies breaking up into many parts, whereas the parent nucleus in radioactivity emits only a single particle. But these terms of historical origin are still in vogue, and so are used in the present work.

An example of Eq.(12.2) is the alpha decay of the $A = 238$ isotope of uranium (U):

$$ {}_{92}^{238}\mathrm{U} \longrightarrow {}_{90}^{234}\mathrm{Th} + {}_{2}^{4}\mathrm{He}, \tag{12.4} $$

where Th is thorium. The $A = 40$ isotope of potassium (K) decays by beta decay, Eq.(12.3), so its reaction equation reads

$$ {}_{19}^{40}\mathrm{K} \longrightarrow {}_{20}^{40}\mathrm{Ca} + {}_{-1}^{0}\mathrm{e}. \tag{12.5} $$

In both of these reactions, the identity of the daughter isotope is by its value of Z. Identification of elements by Z number can be found in a Periodic Table, published in virtually all college physics and chemistry textbooks.

As is well known, radioactivity is associated with danger to health. One reason is that the alpha particle emitted in alpha decay, and the electron emitted in beta decay, are ejected from the nucleus at high speed, and thereby carry a biologically-significant amount of energy. This means that if one of these particles subsequently strikes another atom that is a component of a molecule, the particle can knock the atom out of the molecule, thereby changing the nature of the molecule. Living tissue contains many complex molecules that are susceptible to this kind of damage. If the radiation exposure is severe enough, too many molecules are altered in this manner for the human body to repair the damage.

12.2 Law of Radioactive Decay

It is not possible to tell in advance just at what moment in time a particular atom of a radioactive isotope will undergo the decay process associated with that isotope. However, given an initially pure sample of a radioactive isotope with a very large number of atoms, it is possible to calculate, by means of an equation deduced from observation, just how many of the nuclei will undergo the decay process in any subsequent interval of time.

These two statements are not contradictory. As an analogous situation, consider a large number N of Susan B. Anthony dollar coins, laying on a flat surface all with "heads up". Now suppose that the surface is shaken vigorously enough to toss all the coins. When they all have landed again, it is fairly certain that the number of "heads" will be very close to $\frac{1}{2}N$. But it is not possible to say before the toss which particular coins will show a head after the toss, only that each coin has a probability of $\frac{1}{2}$ of doing so.

Suppose, then, that at a certain time, a pure sample of a particular radioactive isotope contains N_0 atoms of this isotope. The law of radioactive decay states that at a time t later there will only be N atoms of this isotope present, where

$$N = N_0 \left(\frac{1}{2}\right)^{t/T} . \tag{12.6}$$

In Eq.(12.6), the quantity T is called the *half-life* of the particular radioactive isotope. The half-life has units of time, since the exponent must be dimensionless. At time $t = T$, Eq.(12.6) shows that $N = \frac{1}{2}N_0$; that is, the half-life is the time required for one-half of the atoms initially present to undergo decay, leaving one-half left still as parent atoms. (Of course, it is the nucleus of each atom that undergoes the decay.)

Each radioactive isotope has a specific value for the half-life. For the ^{238}U isotope, which decays by Eq.(12.4), the half-life is T = 4.47 X 10^9 y, or T = 4.47 Gy. The half-life of ^{40}K is 1.28 Gy. Clearly, these isotopes have very long half-lives. Fortunately, in order to measure the half-life of a particular radioactive isotope, it is not necessary to actually wait one half-life. Some isotopes have very short half-lives; for example, ^{211}Po (polonium) has a half-life of only 0.52 s. Even shorter yet are the half-lives of the boron isotope ^{13}B of 17.4 ms, or, for what must be close to the shortest, the half-life of the beryllium isotope ^{8}Be of 2 X 10^{-16} s.

****

EXAMPLE 1
The nickel isotope $^{65}_{28}$Ni decays by beta decay to the non-radioactive isotope of copper $^{65}_{29}$Cu. The half-life of the nickel isotope is 2.52 h. An initially pure sample of ^{65}Ni contains 5.27 X 10^{20} atoms. How much time will pass until 8.37 X 10^{19} atoms of ^{65}Cu have been created?

When 8.37 X 10^{19} atoms of ^{65}Cu have been created, there must be 5.27 X 10^{20} – 8.37 X 10^{19} = 4.433 X 10^{20} atoms of ^{65}Ni remaining undecayed. Therefore N = 4.433 X 10^{20} and N_0 = 5.27 X 10^{20}. The half-life is T = 2.52 h and the elapsed time t is sought. It is convenient to solve Eq.(12.6) algebraically for the time t, since t is often the unknown quantity. Taking the natural logarithm of both sides of Eq.(12.6) after dividing by N_0 gives

$$\frac{N}{N_0} = \left(\frac{1}{2}\right)^{t/T},$$

$$\ln\left(\frac{N}{N_0}\right) = \frac{t}{T}\ln\left(\frac{1}{2}\right),$$

$$\ln\left(\frac{N_0}{N}\right) = \frac{t}{T}(\ln 2),$$

$$t = \frac{T\ln(N_0/N)}{\ln 2}.$$

With the values of N and N_0 found above, $N_0/N = 1.189$. Hence,

$$t = \frac{(2.52\text{ h})\ln(1.189)}{\ln 2},$$

$$t = 0.629\text{ h}.$$

****

A quantity used to describe the properties of a certain sample of a specific radioactive isotope is the sample's *activity* R. The activity is the sample's instantaneous rate of decay. The decay of one atom in the sample is registered by detecting the particle emitted in the decay process: a helium nucleus in the case of alpha decay, an electron in beta decay. If Δn of these particles are detected from a sample in the time Δt, then the activity is

$$R = \frac{\Delta n}{\Delta t}. \tag{12.7}$$

Each emitted particle signifies one atom decaying, reducing the number of parent atoms remaining by one. Hence $\Delta n = -\Delta N$, where ΔN is the change in the number of parent atoms present in the sample. Therefore

$$R = -\frac{\Delta N}{\Delta t}.$$

For this to be an instantaneous rate of decay, Δt must be much less than the half-life: $\Delta t \ll T$. In the limit $\Delta t \to 0$, the activity becomes

$$R = -\frac{dN}{dt}. \tag{12.8}$$

Applying Eq.(12.8) to Eq.(12.6) leads to

$$R = \frac{N \ln 2}{T}. \tag{12.9}$$

Equation (12.9) can be used to calculate the very long half-lives of some isotopes after the current activity of a sample has been measured, and the number of parent atoms in the sample has been determined from the mass of the sample.

Since the number of parent atoms present, N, has no units, the SI base units of R are s^{-1}, sometimes written as decays/s (the word "decay" has no units). A decay/s is also called a becquerel (Bq) , but use of this term is not commonplace (unlike, say, the unit hertz, Hz, for oscillation/s, a term which is in common use). The more widely used, non-SI, unit for activity is the *curie*, Ci, defined by

$$1 \text{ Ci} = 3.7 \text{ X } 10^{10} \text{ s}^{-1}. \tag{12.10}$$

One curie is a very substantial activity. For samples with small activities, the unit curie often is combined with the usual SI prefixes.

The activity is a property of the size of the sample as well as of the identity of the isotope. For any isotope, larger samples (greater N) have greater activities than smaller samples. For samples of different isotopes with the same number N of parent atoms, the samples of isotopes with the shorter half-lives have the greater activities, since one-half of the atoms must decay in a shorter time. The value of the half-life T itself is independent of the size of the sample.

The radioactive decay law, Eq.(12.6), and the activity, Eq.(12.9), both are written in terms of the number N of atoms present. This number must be found by calculation, since laboratory balances yield the mass m of a sample, not the number of atoms present. It is usually of sufficient accuracy to use for the mass m_{atom} of an atom the relation

$$m_{\text{atom}} = A \text{ u}, \tag{12.11}$$

where A is the atomic mass number of the isotope, and u is the atomic mass unit :

$$1 \text{ u} = 1.661 \text{ X } 10^{-27} \text{ kg}. \tag{12.12}$$

This means that the number N of atoms present in a pure sample of mass m is given by

$$N = \frac{m}{A \text{ u}}, \tag{12.13}$$

since the number of atoms must equal the mass of the sample divided by the mass per atom.

Note that both the activity R and mass m of an initially pure sample of a certain radioactive isotope are proportional to the number N of atoms present. Therefore, the activity and total mass of parent atoms present obey the same decay law followed by the number of atoms itself. That is, by Eq.(12.6),

$$R = R_0 \left(\frac{1}{2}\right)^{t/T}, \tag{12.14}$$

and

$$m = m_0 \left(\frac{1}{2}\right)^{t/T}, \tag{12.15}$$

where R_0 and m_0 are the initial activity and initial parent mass of the sample, and R and m are the activity and parent mass after a time t has elapsed.

****

EXAMPLE 2

The radioactive isotope of polonium ${}^{210}_{84}\text{Po}$ decays with a half-life of 138.4 d. What mass of this isotope has an activity of 9.00 mCi?

The curie is defined in terms of decays/s, so the half-life must be expressed in seconds. Since 1 d = 86,400 s, the half-life of the isotope is $T = 1.196 \text{ X } 10^7$ s. The activity of the sample is

$$R = (9.00 \text{ X } 10^{-3} \text{ Ci})(3.7 \text{ X } 10^{10} \text{ s}^{-1}/\text{Ci}),$$

$$R = 3.33 \text{ X } 10^8 \text{ s}^{-1}.$$

The number of polonium atoms in the sample can be found from Eq.(12.9):

$$R = \frac{N \ln 2}{T},$$

$$3.33 \text{ X } 10^8 \text{ s}^{-1} = \frac{N \ln 2}{1.196 \text{ X } 10^7 \text{ s}},$$

$$N = 5.746 \text{ X } 10^{15}.$$

Finally, Eq.(12.13) yields the mass of the sample:

$$m = NA \text{ u},$$

$$m = (5.746 \text{ X } 10^{15})(210)(1.661 \text{ X } 10^{-27} \text{ kg}),$$

$$m = 2.00 \ \mu\text{g}.$$

For such a small sample, it is more convenient to express the mass in grams, as above, rather than in kilograms.

****

12.3 Environmental Radioactivity

Radioactivity in the environment arises both from natural and human-generated sources. All isotopes of all elements with $Z > 83$ are naturally radioactive, as are some isotopes of elements with $Z \leq 83$. The isotopes that have very long half-lives are present today because not enough time has elapsed since their creation for them to decay away to immeasurably small abundances. Those of much shorter half-lives that are still present in noticeable amounts must have been created recently (compared to their half-lives) as the daughter products of some other radioactive parent; the parent itself may, or may not, still be present depending on the value of its half-life.

Some of the "heavier" (large A value) radioactive isotopes of long half-life that are found in rocks are the uranium isotopes $^{235}_{92}\mathrm{U}$ ($T = 704$ My), $^{238}_{92}\mathrm{U}$ ($T = 4.47$ Gy), and the thorium isotope $^{232}_{90}\mathrm{Th}$ ($T = 14.1$ Gy). These three isotopes decay by alpha decay. They are most commonly found in igneous rocks, especially granite. The concentration of each isotope is generally specified as so-many parts per million, by mass (ppm).

****

EXAMPLE 3
A 185-kg block of granite contains $^{232}\mathrm{Th}$ at a concentration of 21.5 ppm. Find the activity of the block due to this isotope.

The mass m of thorium in the block, mass M, is related to the ppm concentration by

$$m = (\text{ppm X } 10^{-6})M.$$

Hence, the mass of thorium in the block is

$$m = (21.5 \text{ X } 10^{-6})(185 \text{ kg}),$$

$$m = 0.00398 \text{ kg}.$$

The number N of atoms of the thorium isotope present in the block is given from Eq.(12.13):

$$N = \frac{m}{A\,\mathrm{u}},$$

$$N = \frac{0.00398 \text{ kg}}{(232)(1.661 \text{ X } 10^{-27} \text{ kg})},$$

$$N = 1.033 \text{ X } 10^{22}.$$

The activity R can now be calculated from Eq.(12.9):

$$R = \frac{N \ln 2}{T},$$

$$R = \frac{(1.033 \text{ X } 10^{22})(\ln 2)}{(14.1 \text{ X } 10^9 \text{ y})(3.16 \text{ X } 10^7 \text{ s/y})},$$

$$R = 1.61 \text{ X } 10^4 \text{ s}^{-1}.$$

This can be written more compactly as 16.1 kBq. Each second, 16,100 alpha particles are emitted within the granite block due to the decay of the same number of the thorium atoms. In curie units, this activity is

$$R = \frac{1.61 \text{ X } 10^4 \text{ Bq}}{3.7 \text{ X } 10^{10} \text{ Bq/Ci}},$$

$$R = 0.435\ \mu\text{Ci}.$$

****

These uranium and thorium isotopes as found in rocks are not usually dangerous to the general public, since the concentrations are low (typically between 4 and 16 ppm in granite, but up to 50 ppm in some sedimentary rocks), and the isotopes are bound in the minerals making up the rock. People do not eat rock, so consumption of these isotopes is not an issue. Also, the intact rock itself absorbs most of the emitted alpha particles. However, if the rock is crushed to form a dust that can be inhaled, then the isotopes can pose a health risk. This can be a danger in uranium mining, for example.

There is one radioactive isotope connected with ^{238}U that can be a significant health hazard to the general public. This isotope of concern is the isotope of radon $^{222}_{86}$Rn. It arises in the following manner. The daughter nucleus to ^{238}U is ^{234}Th. Now the ^{234}Th is itself radioactive and decays to the protactinium isotope $^{234}_{91}$Pa. But ^{234}Pa is also radioactive and decays to yet another radioactive isotope. Hence, the decay of ^{238}U produces a "chain" of radioactive nuclei. The chain ends at an isotope of lead, $^{206}_{82}$Pb, which is stable (i.e., not radioactive). The segment of the chain involving radon is the following:

$$^{238}_{92}\text{U} \longrightarrow \cdots \longrightarrow {}^{226}_{88}\text{Ra} \longrightarrow {}^{222}_{86}\text{Rn} \longrightarrow {}^{218}_{84}\text{Po} \longrightarrow \cdots \longrightarrow {}^{206}_{82}\text{Pb}$$

The half-lives of the three isotopes in the "central" part of the chain are

$$^{226}\text{Ra} : T = 1622 \text{ y},$$

$$^{222}\text{Rn} : T = 3.823 \text{ d},$$

$$^{218}\text{Po} : T = 3.110 \text{ min}.$$

The elements involved are radium (Ra), radon (Rn), polonium (Po).

It is the radon that is the origin of the potential health hazard because radon is a gas. It is therefore potentially free to percolate through the subsurface rock via the connected pores and cracks in the rock. The gas can enter homes and buildings though their foundations. If the building is not well ventilated, the gas can accumulate. (Chemically, radon gas is inert; unlike "natural" gas, it will not explode.)

Now this radon isotope has a half-life of a little less than four days. Therefore, if the radon remains in a house for even only a few hours, a significant fraction will undergo decay there. The daughter nucleus is the polonium isotope listed above. These polonium atoms adhere to the dust in the air. Some of the dust drawn into the human respiratory system with each breath will stick to those parts of the lungs that are designed to trap dust. The polonium has a half-life of only a few minutes and therefore if trapped will decay in the body. Some of the polonium nuclei decay by alpha decay, others by beta decay, so the internal organs of the body are thereby exposed to both alpha and beta particles.

Currently, the Environmental Protection Agency considers an indoor radon activity per liter of air that is greater than $R = 4.0$ pCi to be hazardous. Per cubic meter of air, this threshold activity is 148 Bq.

Another naturally produced radioactive isotope present in the environment is tritium, which is an isotope of hydrogen. Hydrogen has three isotopes: ${}^1_1$H (hydrogen), ${}^2_1$H (deuterium), and ${}^3_1$H (tritium). (Hydrogen is the only element whose isotopes have their own names.) The first isotope listed is by far the most abundant; deuterium follows and tritium is the most rare. Only tritium is radioactive; it decays by beta decay to the helium nucleus ${}^3_2$He; the half-life of tritium is 12.26 y.

Since tritium has such a short half-life, its presence in the environment means that it must be continually being produced. The natural source of tritium is as the product of a nuclear reaction between hydrogen in the atmosphere and high-energy cosmic ray particles, which continually bombard the Earth from outer space. Chemically, tritium is identical to "ordinary" hydrogen and therefore can combine with hydrogen and oxygen to form water. The molecule THO (T for tritium) is one form of *heavy water*; DHO (D for deuterium) is another. Tritium is found in rainwater and thereby in water obtained from rainwater.

Tritium concentration often is expressed in a special *tritium unit* TU:

$$1 \text{ TU} = 1 \text{ tritium atom per } 1 \text{ X } 10^{18} \text{ ordinary H atoms.}$$

If tritium is in a particular sample of water, then the water is thereby rendered radioactive and its activity can be calculated. Consider one liter of water. Since the density of water is 1000 kg/m^3, and 1 m^3 = 1000 L, then one liter of water has a mass of 1 kg. The number of molecules of water in 1 kg of water can be found from Eq.(12.13) using $A = 18$. [Since most of the water is H_2O, use $A = 2(1) + 16$, $A = 18$ ($A = 16$ for oxygen).] The number of molecules turns out to be 3.345 X 10^{25}. There are two atoms of hydrogen in each molecule of water, so there are 6.690 X 10^{25} atoms of hydrogen present. With a tritium concentration of 1 TU, this means that there are 6.69 X 10^7 atoms of tritium. By Eq.(12.9), the associated activity is

$$R = \frac{N \ln 2}{T},$$

$$R = \frac{(6.69 \text{ X } 10^7)(\ln 2)}{(12.26 \text{ y})(3.16 \text{ X } 10^7 \text{ s/y})},$$

$$R = 0.120 \text{ Bq.}$$

Since 1 pCi = (1 X 10^{-12})(3.7 X 10^{10} Bq), this is $R = 3.24$ pCi. Thus, 1 TU in one liter of water implies a radioactivity due to the tritium of 3.24 pCi.

Tritium was part of the reaction cycle in the first generation of nuclear fusion weapons (H bombs). Consequently, atmospheric testing of these devices beginning in the early 1950's produced elevated levels of tritium around the world. Before nuclear testing, the measured tritium level was about 10 TU in rainwater, so this is apparently the natural concentration to be expected without human intervention.

Nuclear power plants produce vast quantities of nuclear waste at varied levels of activity that must be safely stored. This is no easy task. Nuclear waste products may be very active chemically, and have long half-lives. Radioactive material buried underground may leak through ruptured containers, and enter and be carried along by ground water. The detection of levels of radioactivity significantly above that expected from natural sources at any site will trigger an investigation of contaminant penetration into adjacent regions.

****

EXAMPLE 4

Radioactive waste from a disposal site is leaking into a nearby groundwater aquifer. A well very close to the disposal site records a tritium activity of 260 pCi per liter of water. At a well 450 m from the site the activity is 13.0 pCi per liter. Assume that no rainwater entered the aquifer between the two wells, and find the speed at which the groundwater is moving.

Since the distance between the wells is given, it is only necessary to find out how long it took for the water to move the 450 m. Use Eq.(12.14), with $T = 12.26$ y. Since no rainwater entered, the decrease in activity between the two wells is due only to the decay of the tritium from the disposal site and not to any dilution from added water. The pCi units cancel; omitting the time unit for clarity in calculation gives

$$R = R_0 \left(\frac{1}{2}\right)^{t/T},$$

$$13.0 = 260 \left(\frac{1}{2}\right)^{t/12.26},$$

$$\ln 13.0 = \ln 260 + \left(\frac{t}{12.26}\right) \ln \left(\frac{1}{2}\right),$$

$$\ln 13.0 = \ln 260 - \left(\frac{t}{12.26}\right) \ln 2,$$

$$t = 53.0 \text{ y}.$$

Hence, the speed of the water is $v = (450 \text{ m})/(53.0 \text{ y})$, $v = 8.49$ m/y, or about 2.32 cm/d.

****

12.4 Age of Rocks

At first thought, it may be considered that investigating the age of rocks is an esoteric endeavor of no practical value. But there are correlations between the ages of certain types of rocks and their possibility of containing industrially important metals and minerals in sufficient abundance to justify mining. Coal and oil deposits are most prevalent in sedimentary rocks of certain ages. The ability to determine the age of rocks can assist in locating significant deposit of these materials.

Now rocks that contain radioactive isotopes thereby contain an internal clock, so to speak, since the decays takes place at a rate determined by the concentration of the isotope in the rock and by the half-life of the isotope. The half-lives of radioactive isotopes have been measured in the laboratory, and can be considered to be known quantities. There is no indication that half-lives change with time.

Consider a rock that has just formed by solidification from a molten magma. Sealed in the solid rock are, most probably, several radioactive isotopes. Focus on one of these isotopes, the parent isotope denoted as P. Assume, for the moment, that this parent decays directly, in one decay, to a stable daughter isotope D; the half-life of the decay is T. The solidification may have taken place hundreds of millions, or even billions, of years ago.

By examining the rock today (now), the mass $m_{\rm P}$ of parent isotope and the mass $m_{\rm D}$ of daughter isotope present now can be determined. From the mass and the atomic mass number A, the number $N_{\rm P}$ of parent atoms and the number $N_{\rm D}$ of daughter atoms present now can be calculated from Eq.(12.13).

The number of parent atoms present now $N_{\rm P}$ is related to the age of the rock by the law of radioactive decay, Eq.(12.6), applied to the parent atoms; that is,

$$N_{\rm P} = N_{\rm P0}\left(\frac{1}{2}\right)^{t/T}. \tag{12.16}$$

In Eq.(12.16), t is the age of the rock and $N_{\rm P0}$ is the number of parent atoms present at the moment of solidification (then). To solve Eq.(12.16) for the age t, with $N_{\rm P}$ and T known, the value of $N_{\rm P0}$ must be independently determined.

Since there is no way to go back in time and measure $N_{\rm P0}$, an assumption must be made. The simplest to make is to assume that no daughter atoms were present when the rock solidified, and that all of the daughter atoms present now originated, one for one, by the decay of the parent atoms. In this event,

$$N_{\rm P0} = N_{\rm P} + N_{\rm D}. \tag{12.17}$$

For Eq.(12.17) to be valid, the rock must not have been contaminated with daughter atoms from its environment, nor must it have given up, or lost, any of the created daughter atoms. For example, groundwater should not have percolated through the rock for any significant length of time, for groundwater can carry daughter atoms into, and/or out of, the rock.

But suppose that the rock has remained intact and uncontaminated since its solidification, so that Eq.(12.17) applies. Then substitute Eq.(12.17) into Eq.(12.16) and rearrange, to get

$$2^{t/T} = 1 + \frac{N_D}{N_P},$$

$$t = \left(\frac{T}{\ln 2}\right) \ln\left(1 + \frac{N_D}{N_P}\right), \tag{12.18}$$

for the age of the rock. Apply Eq.(12.13) to N_P and N_D separately to find that, in terms of the masses of the parent and daughter isotopes present,

$$\frac{N_D}{N_P} = \frac{m_D}{m_P}\frac{A_P}{A_D}, \tag{12.19}$$

where A_P and A_D are the mass numbers of the parent and daughter isotopes. (The atomic mass unit u cancelled out in forming the ratio.) In terms of masses, the age of the rock is

$$t = \left(\frac{T}{\ln 2}\right) \ln\left(1 + \frac{m_D}{m_P}\frac{A_P}{A_D}\right). \tag{12.20}$$

In developing Eq.(12.20), it is presumed that the parent isotope decays to a stable daughter in one decay. Some radioactive isotopes do just this. But others are similar to ^{238}U, discussed above: that is, their decay starts a chain of subsequent decays that eventually leads to a stable isotope. The use of these "chain-decaying" isotopes cannot be avoided since they have the very long half-lives needed to determine the million and billion year ages of rocks. If the half-life of an isotope used to determine an age is very short compared to the age, i.e., if $T \ll t$, then very little parent isotope will be left, so that an accurate assessment of m_P is difficult. On the other hand, if $T \gg t$, then very little daughter isotope has been created, and m_D cannot be measured to great accuracy. In either case, m_D/m_P will be poorly determined, leading to an inaccurate value of t. The optimum isotopes to use are those for which $T \approx t$. But t is not known in advance, so several isotopes may have to be tried to find the isotope that gives the most accurate determination of the age.

Consider the thorium isotope ^{232}Th as an example of a decay chain as it pertains to age determination. As noted in Example 3, this isotope has the very long half-life of $T = 1.41$ X 10^{10} y,. The complete decay chain of ^{232}Th is given in Table (12.1) below. (The SI prefix n is 1 X 10^{-9}.) Note that there are two step 9's. This is not a typographical error: ^{212}Bi decays by two different modes, either to ^{212}Po or to ^{208}Tl. But each of these isotopes decays to the same stable end product, the lead isotope ^{208}Pb, so the dual-channel decay of ^{212}Bi is of no consequence for age studies.

Step	Element	Isotope	Half-life
0.	Thorium	^{232}Th	14.1 Gy
1.	Radium	^{228}Ra	5.77 y
2.	Actinium	^{228}Ac	6.13 h
3.	Thorium	^{228}Th	1.913 y
4.	Radium	^{224}Ra	3.64 d
5.	Radon	^{220}Rn	55 s
6.	Polonium	^{216}Po	0.15 s
7.	Lead	^{212}Pb	10.64 h
8.	Bismuth	^{212}Bi	60.6 min
9.	Polonium	^{212}Po	304 ns
9.	Thallium	^{208}Tl	3.10 min
10.	Lead	^{208}Pb	STABLE

Table(12.1) The ^{232}Th (Thorium) Decay Chain

Examine the half-lives. The half-life of the first step in the chain, the decay of ^{232}Th to ^{228}Ra, is 1.41 X 10^{10} y. All of the other half-lives are very small, to say the least, compared with this; they range from a few years to nanoseconds. This means that, on the average, an atom of ^{232}Th will remain an atom of ^{232}Th for about 14 Gy before it decays to ^{228}Ra. Once it has decayed to ^{228}Ra, the subsequent decays occur very rapidly indeed, so that the atom quickly becomes one of ^{208}Pb, which is stable.

Put it this way: suppose a rock is formed containing some atoms of ^{232}Th and no atoms of each of the other products 1.→10. of the decay chain. The rock is examined now, millions or perhaps billions of years later. The atoms that were ^{232}Th now either are still ^{232}Th atoms, or are atoms of ^{208}Pb. Virtually no atoms will be found of the products 1.→9. because, on average, atoms of these isotopes exist only for a time definitely very short compared with geologic time. This means that, for all practical purposes,

$$N_{\mathrm{Th},0} = N_{\mathrm{Th}} + N_{\mathrm{Pb}}, \tag{12.21}$$

where the left-hand side of Eq.(12.21) refers to "then" (time of rock solidification) and the right-hand side refers to "now" (the present). In effect, the decay chain acts as if it proceeded in a single step, with the half-life of 14.1 Gy. Equation (12.17) is satisfied, and the age of the rock follows from Eq.(12.20), where P (parent) is ^{232}Th and D (daughter) is ^{208}Pb.

Two other geologically important isotopes (i.e., that are found in rocks) besides ^{232}Th that can be used in age determinations are shown, along with ^{232}Th, in Table (12.2). They also produce chains of decay, with the first decay having a half-life much longer than the half-lives of the other isotopes in the chain. Hence, their decay chain can be viewed as a single step, so that Eqs.(12.17) and (12.20) apply. These chains also produce an isotope of radon, which is listed with its half-life. The rock must remain intact for the analysis of the age to be valid, for the radon gas must decay in the rock, but whichever isotope of radon is produced, its half-life is short, and therefore it only has a short time as a gas to escape.

Parent	Stable Daughter	Steps	Half-life (Gy)	Radon	Radon Half-life
^{232}Th	^{208}Pb	10	14.1	^{220}Rn	55 s
^{238}U	^{206}Pb	14	4.47	^{222}Rn	3.823 d
^{235}U	^{207}Pb	11	0.713	^{219}Rn	4.0 s

Table (12.2) Decay Chain for Rock Ages

There are other isotopes that can be used for age determinations. Some of these really do proceed in a single step. For example, the beta decay of the rubidium isotope $^{87}_{37}$Rb to the strontium isotope $^{87}_{38}$Sr, with a half-life of 47 Gy. Also, the decay of the potassium isotope $^{40}_{19}$K to the argon isotope $^{40}_{18}$Ar, which takes place when the nucleus of a potassium isotope absorbs an atomic electron; the half-life is 1.28 Gy. There are certain complications in using these isotopes (for example, the argon is a gas, which brings problems of retention), but they can be used to obtain ages in certain rocks.

****

EXAMPLE 5

A 1.28 kg rock contains 4.67 ppm of ^{238}U. The rock is thought to be 2.78 Gy old. If this quoted age is correct, what mass of ^{206}Pb should it contain?

The uranium is the parent, so if M is the mass of the rock,

$$m_{\mathrm{P}} = (\text{ppm X } 10^{-6})M,$$

$$m_{\mathrm{P}} = (4.67 \text{ X } 10^{-6})(1.28 \text{ kg}),$$

$$m_{\mathrm{P}} = 5.978 \text{ mg}.$$

The parent's mass number is $A_{\mathrm{P}} = 238$. The daughter isotope is ^{206}Pb, so $A_{\mathrm{D}} = 206$. The mass m_{D} is the quantity sought. For this decay chain, the half-life is $T = 4.47$ Gy (Table 12.2). With $t = 2.78$ Gy, Eq.(12.20) yields

$$t = \left(\frac{T}{\ln 2}\right) \ln\left(1 + \frac{m_{\mathrm{D}}}{m_{\mathrm{P}}} \frac{A_{\mathrm{P}}}{A_{\mathrm{D}}}\right),$$

$$2.78 = \left(\frac{4.47}{\ln 2}\right) \ln\left(1 + \frac{m_{\mathrm{D}}}{5.978} \frac{238}{206}\right).$$

The units Gy of t and T cancel; the units of m_{P} are omitted for clarity in the equation. Doing some arithmetic gives

$$0.4311 = \ln(1 + 0.1933 m_{\mathrm{D}}),$$

$$e^{0.4311} = 1 + 0.1933 m_{\mathrm{D}},$$

$$m_{\mathrm{D}} = 2.79 \text{ mg}.$$

The units of m_{D} must be milligrams (mg) since the value of m_{P} substituted is in milligrams.

****

12.5 Age of the Earth

The Earth must be at least as old as the oldest indigenous rocks found on the Earth. The qualification "indigenous" is necessary because rocks from interplanetary space fall on the Earth as meteorites; also, some rocks from the planet Mars apparently have found their way to the Earth after being expelled in some manner from the surface of that planet. The greatest measured solidification age of indigenous Earth rocks is about 3.7 Gy.

It is possible to push the determination of the Earth's age back even farther in time. By examining the abundances of certain isotopes, it is possible to estimate the age of the magma from which the rock solidified. Applying these techniques to the rock of solidification age 3.7 Gy, the age of the magma from which it formed can be determined to be about 4.5 Gy. This value stands as the best estimate of the age of the Earth; no older rock has been found.

The value of 4.5 Gy correlates with three other age determinations. The oldest rocks brought back from the Moon are about 4.6 Gy old. The oldest meteorites that have been recovered on Earth are also about 4.6 Gy old. Finally, the age of the Sun, as obtained from its mass, composition, and rate of energy generation, is also about 4.6 Gy. Apparently, the Earth, the Sun, and the rest of the solar system were formed at about the same time; the Moon may have formed at that time, or "split off" from the Earth later, perhaps as the result of a collision. Evidently one thing is absolutely certain: the Earth is very old compared with any human timescales.

12.6 Problems

1. The lead isotope $^{212}_{82}$Pb has a half-life of 10.64 h. An initially pure sample of this isotope contains 4.77 X 10^{21} atoms. How many of these atoms are left after 1.00 d?

2. Using Eqs.(12.6), (12.9), and (12.13), verify Eqs.(12.14) and (12.15) for the time-dependence of the activity and mass of an initially pure radioactive sample.

3. By taking the derivative of N in Eq.(12.6), show that the activity R of a sample of radioactive material, defined by $R = -dN/dt$, is also given by Eq.(12.7).

4. The initial activity of a certain pure sample of the polonium isotope $^{210}_{84}$Po is 4.79 μCi, but 75.0 d later the activity has fallen to 3.29 μCi. Calculate the half-life of this polonium isotope.

5. For contamination of drinking water by radium ($^{226}_{88}$Ra, $T = 1622$ y), an upper limit of 5.00 pCi/L of water has been set for the water to be considered "safe". The drinking water in a certain municipal water system is found to exceed the limit by 3.00 pCi/L. What volume of this water contains 1.00 mg of radium?

6. Calculate the activity in curies of 3.62 μg of the thorium isotope $^{227}_{90}$Th, which has a half-life of 18.5 d.

7. An initially pure sample of tritium has a mass of 4.33 mg. How much is left after 5.00 y? The half-life of tritium is 12.26 y.

8. With reference to the EPA radon guidelines, show that 4.0 pCi/L = 148 Bq/m^3.

9. In some localities, real estate transactions require certification that the radon level is less than 4.0 pCi per liter of air. If the radon is all ^{222}Rn, find the mass of radon in a house of dimensions 15 m, 8.4 m, 4.9 m that just meets this requirement.

10. The activity of radon ^{222}Rn in a house is 12.6 pCi/L of air. Assuming that all the windows are sealed and that no new radon seeps into the house, how long will it take for the radon activity to fall to 4.0 pCi/L of air?

11. Consider an unfortunate house that experiences the greatest possible invasion of ^{222}Rn radon gas; that is, all of the air in the house has been replaced with the pure radon gas. Calculate the activity of 1.00 L of the radon. Assume that the pressure is 101.3 kPa (i.e.,

normal atmospheric pressure), and the temperature is 21.0°C.

12. Discarded tailings from a uranium mine must be covered with enough earth so that the escape of radon gas (^{222}Rn) from radioactive elements in the tailings is less than 2.00 pCi/m^2·s. A particular earth-covered tailings lagoon measures 120 m by 75.0 m. If the lagoon just meets the requirment, how many radon atoms escape in a period of 24.0 h?

13. What mass of radon ^{219}Rn has the same activity as 1.48 mg of radon ^{222}Rn?

14. The tritium abundance in water in a particular well is 1.44 TU. The tritium content of recently fallen rainwater is 4.76 TU. How long ago did the water in the well fall as precipitation? (Assume that the tritium content of rainwater has not changed.)

15. Calculate the activity in curies of 1.00 kg of 100% "superheavy" water T_2O; i.e., in all of the water molecules, both hydrogen atoms have been replaced with tritium atoms.

16. Each of the decays in the chain from ^{232}Th to ^{208}Pb is either an alpha decay or a beta decay. Assign the appropriate decay mode to each isotope in the chain.

17. How old is a rock that contains 8.00 X 10^{16} atoms of ^{208}Pb and 4.00 X 10^{17} atoms of ^{232}Th, assuming that the rock contained no lead atoms when formed?

18. A rock contains 6.83 mg of ^{238}U and 2.27 mg of ^{206}Pb. (*a*) How many atoms of ^{238}U did the rock contain at solidification? (*b*) How many atoms of ^{222}Rn has the rock produced until now?

19. How old is a rock that contains 3.86 mg of ^{235}U and 5.34 mg of ^{207}Pb?

20. A rock 2.77 Gy old contains 1.28 mg of ^{238}U. How many atoms of ^{206}Pb does it contain?

21. What is the age of a rock that is 7.38 ppm ^{207}Pb and 2.63 ppm ^{235}U?

22. A rock is found to contain: 2.02 mg of ^{238}U, 1.34 mg of ^{235}U, 1.08 mg of ^{207}Pb, 0.93 mg of ^{208}Pb. Calculate the age of the rock.

23. A rock contains: 2.71 mg of ^{238}U, 1.15 mg of ^{206}Pb, 9.78 mg of ^{232}Th, 3.64 mg of ^{208}Pb. How much ^{208}Pb was in the rock when it solidified?

Appendix A

The International System of Units (SI)

Units

Quantity	SI Base Unit	Other Units	Conversion Factor
Length	meter (m)	centimeter (cm)	1 cm = 1 X 10^{-2} m
Area	m^2	cm^2 hectare (ha)	1 cm^2 = 1 X 10^{-4} m^2 1 ha = 1 X 10^4 m^2
Volume	m^3	cm^3 liter (L)	1 cm^3 = 1 X 10^{-6} m^3 1 L = 1 X 10^{-3} m^3
Mass	kilogram (kg)	gram (g) metric ton (t) atomic mass unit (u)	1 g = 1 X 10^{-3} kg 1 t = 1 X 10^3 kg 1 u = 1.661 X 10^{-27} kg
Time	second (s)	minute (min) hour (h) day (d) year (y)	1 min = 60 s 1 h = 3600 s 1 d = 86,400 s 1 y = 3.16 X 10^7 s
Density	kg/m^3	g/cm^3	1 g/cm^3 = 1 X 10^3 kg/m^3
Force	newton (N)		
Unit Weight	N/m^3		
Stress	pascal (Pa)		1 Pa = 1 N/m^2
Pressure	Pa		
Torque	N·m		
Energy	joule (J)		
Radioactivity	becquerel (Bq)	 curie (Ci)	1 Bq = 1 s^{-1} 1 Ci = 3.7 X 10^{10} Bq

SI Prefixes

Symbol	Prefix	Factor
T	tera-	1×10^{12}
G	giga-	1×10^{9}
M	mega-	1×10^{6}
k	kilo-	1×10^{3}
c	centi-	1×10^{-2}
m	milli-	1×10^{-3}
μ	micro-	1×10^{-6}
n	nano-	1×10^{-9}
p	pico-	1×10^{-12}

Appendix B

Answers To Odd-Numbered Problems

Chapter 1

1. 27.9 kN/m^3. **3.** 1.52 m^3. **5.** 8230 cm^3. **7.** 9.38 m^3. **9.** 14.1 kN. **11.** 0.403. **13.** 1.19 m. **15.** (*a*) 64.4%. (*b*) 531 cm^3. **19.** 0.279. **21.** 5.59 X 10^8 gallons. **23.** 0.660. **25.** 34,100 barrels. **27.** 0.676 m^3. **29.** 35.8%. **31.** (*a*) 0.395. (*b*) 2.03 m. **35.** 0.273. **37.** 119 g.

Chapter 2

1. 126 kPa. **3.** 41.1 MPa. **5.** 26.1 kN. **7.** 143 MPa. **9.** 9.01 MPa. **11.** 68.6 MPa. **13.** 2.78 m^2. **15.** (*a*) 21. (*b*) 16.4 m^2. **17.** (*a*) 4490 m^3. (*b*) 15.7 MPa. **19.** 590 m **21.** 235 m. **23.** 11. **25.** (*a*) 8.23 MPa. (*b*) 14.8 MPa. (*c*) 3.43 g/cm^3. (*d*) 36.1 MPa. **29.** 168 m. **33.** 787 m. **35.** (*a*) 623 kN. (*b*) 4.66 cm. **37.** 3.43 kPa.

Chapter 3

1. −0.55 mm. **3.** 77.53 cm. **5.** 513 kN. **7.** (*a*) −6.99 mm. (*b*) 0.152 mm. **9.** −48.6 MPa. **11.** (*a*) 0.4845. (*b*) 67.3 GPa. **13.** −0.386 cm^3. **15.** 4.50 GPa. **19.** 183 MPa. **21.** 255 kPa. **23.** 2400 cm^3. **27.** $E = 95.7$ GPa, $\nu = 0.118$.

Chapter 4

1. 35.1 kPa. **3.** 1.79. **5.** 4.78 MN. **7.** 297 kN. **9.** 48. **11.** (*a*) 1.89. (*b*) 9.73 MN. **13.** 10. **15.** (*a*) 113. (*b*) 783 MPa. **17.** (*a*) 16. (*b*) 213. **19.** (*a*) 65. (*b*) 1.45. **21.** 22.1 m^2. **23.** 4.51 MPa. **25.** (*a*) 47.8°. (*b*) 41.0°. **27.** 53.1 kPa. **29.** (*a*) 2.04. (*b*) 3.37. **31.** I and II.

Chapter 5

1. 2.22 g/cm^3. **3.** (*a*) 0.667. (*b*) 2.60 g/cm^3. **5.** 334 L. **7.** (*a*) 392 cm^3. (*b*) 0.361. **9.** 23. **11.** (*b*) 0.877. **13.** $0.40 \text{ mm} \leq D \leq 2.0 \text{ mm}$. **15.** 8.28. **17.** 32.9%. **21.** 36.7% silt, 25.3% sand, 38.0% clay. **25.** 22.6 m **27.** 1.16 MPa. **29.** 19.6%. **31.** 15.0°. **33.** 20.2° **37.** (*a*) 18.6 MPa. (*b*) 154 MPa. **39.** 27.0°.

Chapter 6

1. 28.5 m. **3.** 4.43 m. **5.** (*a*) 201 m^3/s. (*b*) 1550 m^3/s. **7.** 3.53 m. **9.** 9.7 X 10^6 m^3. **11.** 4010 km^2. **13.** 65 cm. **15.** 88.3 cm. **17.** (*a*) 18.8 m. (*b*) 2.30 m. **19.** (*a*) 25.1 m. (*b*) 7.60 m/s. **21.** 1.89 X 10^5 t. **23.** 1240 s. **25.** $a = 0.163$ SI base units, $b = 0.440$. **27.** 4.1 g/cm^3. **29.** (*a*) 3.14 m. (*b*) 1.57 m. **31.** 6.30 m. **33.** 7.11 m/s.

Chapter 7

1. (*a*) 3.37 GN·m. (*b*) 37.1 m. **3.** 41.0 m. **5.** (*a*) 50.1 m. (*b*) 9.01. **7.** (*a*) 0.788. (*b*) 15.1 GN·m. **9.** (*a*) 826 MN. (*b*) 0.529. **13.** $FS = 3\rho_d/\rho$. **17.** 1.49 kN. **19.** (*a*) 26.0%. (*b*) 1.41 X 10^5 m^3. **21.** 6.40 d.

Chapter 8

1. 28.3%. **3.** (*a*) 112 m/d. (*b*) 5.54 X 10^5 m^3/d. **7.** (*a*) 0.139. (*b*) 2.50 X 10^5 m^3. (*c*) 1.85 X 10^7 m^3. **9.** (*a*) 812 d. (*b*) 3.16 X 10^5 m^3. (*c*) 1.45 X 10^7 m^3. **13.** 15.0 m. **15.** 109 m/d. **17.** 5.22 y.

Chapter 9

1. (*a*) 2.00 km/s. (*b*) 3.70 km/s. **3.** 2.48 km/s. **5.** 20.5 GPa. **7.** 0.251. **9.** (*a*) 7.27 km/s. (*b*) $E = 84.9$ GPa, $k = 101$ GPa, $G = 31.2$ GPa. **11.** $(\cos i)/v_1 = (\cos r)/v_2$. **13.** 26.6°. **15.** (*a*) Not possible. (*b*) 55.3°. (*c*) 2.08 km/s. (*d*) 2.24 km/s. (*e*) 39.7°. **17.** 93.0 m. **19.** 196 ms. **21.** (*a*) 1400 m/s, 2500 m/s. (*b*) 70.0 m. (*c*) 94.6 m. **25.** 120 m. **27.** (*a*) 301 m. (*b*) 7490 m/s. **29.** (*a*) 345 m. (*b*) 398 m. **35.** (*a*) 1500 m/s. (*b*) 4000 m/s. (*c*) 142 m. (*d*) 189 ms. (*e*) 115 m.

Chapter 10

1. 1:25 pm. **3.** (*a*) 4500 km. (*b*) 2:04 am. (*c*) 2:25 am. **7.** $0.371A$. **9.** 57 min. **11.** 0.28. **13.** 5.93 g/cm^3. **15.** 4.00 X 10^{24} kg. **17.** 6.35. **19.** (*a*) 0.267 Mton. (*b*) 7.68. (*c*) 8.43 Mton. **21.** 27 cm. **23.** 1.97 MPa.

Chapter 11

1. 980570 mGal. **3.** (0.218 mGal/m)H. **5.** 351 m. **7.** 981896.19 mGal. **9.** Less by 38.8 mGal. **13.** 1.29 X 10^{10} kg. **15.** Less by 49.4 mGal. **17.** (*a*) 42.1 m, 32.7 m. (*b*) 42.1 m, 39.7 m. **19.** 174 m. **21.** (*a*) Cylinder. (*b*) 3.22 X 10^{10} kg. **25.** 69.1 m, 8.50 X 10^{11} kg.

Chapter 12

1. 9.99 X 10^{20}. **5.** 1.22 X 10^5 m^3. **7.** 3.26 mg. **9.** 1.6 X 10^{-14} kg. **11.** 1.42 MCi. **13.** 17.7 ng. **15.** 2.65 MCi. **17.** 3.79 Gy. **19.** 971 My. **21.** 1.47 Gy. **23.** 2.46 mg.

Bibliography

Bell, F. G., 1993: ENGINEERING GEOLOGY. Blackwell Scientific Publications.

Costa, J. E., Baker, V. R., 1981: SURFICIAL GEOLOGY, Building with the Earth. John Wiley.

Dennen, W. H., Moore, R. R., 1986: GEOLOGY AND ENGINEERING. Wm. C. Brown,

Fetter, C. W., 1994: APPLIED HYDROGEOLOGY, 3rd ed. Prentice Hall.

Goodman, R. E., 1989: INTRODUCTION TO ROCK MECHANICS, 2nd ed. John Wiley.

Halliday, D., Resnick. R., Krane, K. S., 1992: PHYSICS, 4th ed. John Wiley.

Hartmann, W. K., 1983: MOON AND PLANETS, 2nd ed. Wadsworth Publishing.

Johnson, R. B., DeGraff, J. V., 1988: PRINCIPLES OF ENGINEERING GEOLOGY. John Wiley.

Kehew, A. E., 1995: GEOLOGY FOR ENGINEERS AND ENVIRONMENTAL SCIENTISTS, 2nd ed. Prentice Hall.

Keller, E. A., 1996: ENVIRONMENTAL GEOLOGY, 7th ed. Prentice Hall.

Krane, K. S., 1988: INTRODUCTORY NUCLEAR PHYSICS. John Wiley.

Langmuir, D., 1997: AQUEOUS ENVIRONMENTAL GEOCHEMISTRY. Prentice Hall

Plummer, C. C., McGeary, D., 1991: PHYSICAL GEOLOGY, 5th ed. Wm. C. Brown.

Rahn, P. H., 1996: ENGINEERING GEOLOGY, An Environmental Approach, 2nd ed. Prentice Hall.

Robinson, E. S., 1982: BASIC PHYSICAL GEOLOGY. John Wiley.

Robinson, E. S., Çoruth, C., 1988: BASIC EXPLORATION GEOPHYSICS. John Wiley.

Sheriff, R. E., 1989: GEOPHYSICAL METHODS. Prentice Hall.

Tarbuck, E. J., Lutgens, F. K., 1996: EARTH, An Introduction to Physical Geology, 5th ed. Prentice Hall.

Wehr, M. R., Richards Jr, J. A., Adair III, J. W., 1985: PHYSICS OF THE ATOM, 4th ed. Addison Wesley.

West, T. R., 1995: GEOLOGY Applied to Engineering. Prentice Hall.

Credits

The entries in Table (1.1) are adapted from Goodman (1989), p. 31, and Kehew (1995), p. 40.

The carbon phase diagram in Problem (2.24) is a slight revision of a diagram in Robinson (1982), p. 215.

Problem (6.4) suggested by text and figure in Robinson (1982), p. 571-572.

The wave speeds in Example (9.6) are from Rahn (1996), p.497.

Data in Problem (9.24) from Kapitsa., et. al., NATURE, Vol. 381, p. 684-686.

Table (12.1) from data in Wehr, Richards, Adair (1985).

Index

The Author

Edward Derringh was born in Oxford UK in 1943. He came to the United States in 1955. The author received a BS in Astronomy from Case Institute of Technology (now Case-Western Reserve University) in 1965, and a Ph.D. in Astronomy from Rensselear Polytechnic Institute in 1974. He has taught at Rensselear Polytechnic Institute, College of Saint Rose, and Siena College.

Dr. Derringh is the author of instructor solution supplements for the books:
Eisberg, R., Resnick, R., 1985: QUANTUM PHYSICS of ATOMS, MOLECULES, SOLIDS, NUCLEI and PARTICLES, 2nd ed., John Wiley;
Resnick, R., Halliday, D., 1985: BASIC CONCEPTS in RELATIVITY AND EARLY QUANTUM THEORY, 2nd ed., John Wiley.
He also wrote the student solution supplement to
Halliday, D., Resnick, R., Krane, K. S., 1992: PHYSICS, 4th ed., John Wiley.

Since 1979, Dr. Derringh has been Associate Professor of Physics at Wentworth Institute of Technology, Boston, Massachusetts.

The author typeset the text of this work using LaTeX 2_ε, Personal TeX, Inc.